A FIRST COURSE IN MATHEMATICAL MODELING

A FIRST COURSE
IN MATHEMATICAL MODELING

FRANK R. GIORDANO

U.S. Military Academy, West Point, New York

MAURICE D. WEIR

Naval Postgraduate School, Monterey, California

Brooks/Cole Publishing Company
Monterey, California

To my mother, Margaret McCarthy Giordano
and in loving memory of my father, Samuel Rudolf Giordano

Brooks/Cole Publishing Company
A Division of Wadsworth, Inc.
© 1985 by Wadsworth, Inc., Belmont, California 94002. All rights reserved. No part of this book may
be reproduced, stored in a retrieval system, or transcribed, in any form or by any means—electronic,
mechanical, photocopying, recording, or otherwise—without the prior written permission of the publisher,
Brooks/Cole Publishing Company, Monterey, California 93940, a division of Wadsworth, Inc.

Printed in the United States of America
10 9 8 7 6 5 4 3 2 1

Library of Congress Cataloging in Publication Data

Giordano, Frank R., [date]
 An introductory course in mathematical modeling.

 Includes bibliographical references and index.
 1. Mathematical models.   I. Weir, Maurice D.
II. Title.
QA401.G55   1983   511′.8   84-7703
ISBN 0-53-403367-9

Sponsoring Editor: *Craig Barth*   Production Editor: *Gay L. Orr*   Manuscript Editor: *Jonas Weisel*
Permissions Editor: *Carline Haga*   Photo Editor/Researcher: *Judy Blamer*   Interior and Cover
Design: *Katherine Minerva*   Cover Photo: *NASA Photo, Hansen Planetarium, Salt Lake City, Utah*
Art Coordinator: *Judy Macdonald*   Interior Illustration: *Lori Heckelman*   Typesetting: *Syntax
International*   Printing and Binding: *R. R. Donnelley & Sons, Inc.*

Photo Credits p. 4, Cornell Capa, © Magnum Photos; p. 16, Mark Godfrey, © Archive Pictures;
p. 57, Emilio A. Mercado, © 1977 Jeroboam, Inc.; p. 64, Joseph Schuyler, © Stock Boston; p. 70,
Robert Rubic, DPI; p. 74, Jonathan Rawle, © Stock Boston; p. 79, Sybil Shelton, Monkmeyer;
p. 136, David Strickler, Monkmeyer; p. 146, UPI; p. 151, Bruce Forrester, © Jeroboam, Inc.; p. 165,
Ian Berry, © Magnum Photos; p. 237, Lloyd Ulberg, California Academy of Sciences; p. 244, Dresser
Industries; p. 249, Mark Antman, © Stock Boston; p. 255, NASA; p. 260, Hazel Carew, Monkmeyer;
p. 280, Robert Eckert, EKM-Nepenthe; p. 290, Hugh Rogers, Monkmeyer; p. 305, John Maher,
© 1980 EKM-Nepenthe; p. 316, Jeff Dunn, © Stock Boston; p. 348, Roy Pinney, Monkmeyer; p. 355,
Leonard Lee Rue, © Monkmeyer; p. 369, Christian Poveda, © Stock Boston.

PREFACE: TO THE INSTRUCTOR

The Mathematical Association of America's Committee on the Undergraduate Program in Mathematics (CUPM) recommended in 1981 that "Students should have an opportunity to undertake 'real world' mathematical modeling projects, either as term projects in an operations research course, as independent study, or as an internship in industry."* That report goes on to add that a *modeling experience* should be included within the *common core* of all mathematical sciences majors. Further, this experience in modeling should begin early: ". . . to begin the modeling experience as early as possible in the student's career and reinforce modeling over the entire period of study."† We would like to describe how our book is designed for a *first course* in modeling.

Goals and Orientation

While this text can be used in a variety of ways, its main purpose is to provide the textual material supporting a modeling course that can be taught as soon as possible after the introductory engineering or business calculus sequence. The course is a bridge between calculus and the applications of mathematics to various fields, and it is a transition to the significant modeling experiences recommended by CUPM. The course affords the student an early opportunity to see how the pieces of an applied problem fit together. By using fundamental calculus concepts in a modeling framework, the student investigates meaningful and practical problems chosen from common

---

* Mathematical Association of America, Committee on the Undergraduate Program in Mathematics, *Recommendations for a General Mathematical Sciences Program* (Mathematical Association of America: Washington, D.C., 1981), p. 13.
† Ibid., p. 77.

v

experiences encompassing many academic disciplines, including the mathematical sciences, operations research, engineering, and the management and life sciences.

This text provides an introduction to the entire modeling process. The student will have occasions to practice the following facets of modeling:

1. *Creative and Empirical Model Construction*: Given a real-world scenario, the student learns to identify a problem, make assumptions and collect data, propose a model, test the assumptions, refine the model as necessary, fit the model to data if appropriate, and analyze the underlying mathematical structure of the model in order to appraise the sensitivity of the conclusions.

2. *Model Analysis*: Given a model, the student learns to work backward to uncover the implicit underlying assumptions, assess critically how well those assumptions fit the scenario at hand, and estimate the sensitivity of the conclusions when the assumptions are not precisely met.

3. *Model Research*: The student investigates a specific area to gain a deeper understanding of some behavior and learns to use what already has been created or discovered.

It is our perception that many mathematics students lack real problem solving capability, and we have designed our modeling course to help rectify that deficiency. For purposes of discussion we identify the following steps of the problem solving process:

1. Problem identification
2. Model construction or selection
3. Identification and collection of data
4. Model validation
5. Calculation of solutions to the model
6. Model implementation and maintenance

In many instances the undergraduate mathematical experience consists almost entirely of doing step 5: calculating solutions to models that are given. There is relatively little experience with "word problems," and that is spent with problems that are short (in order to accommodate a full syllabus) and often contrived. Such problems require the student to apply the mathematical technique currently being studied, from which a *unique* solution to the model is calculated with great *precision*. For lack of experience, consequently, students often feel anxious when given a scenario for which the model is *not* given or for which there is no unique solution, and then are told to identify a problem and construct a model addressing the problem "reasonably well."

With this in mind, we feel that in an introductory modeling course students should spend a significant amount of time on the first several steps of the process described above—learning how to identify problems, construct or select models, and figure out what data needs to be collected—progressing from relatively easy scenarios to more difficult ones. It is probably unreasonable to expect an average student to excel in a semester-long project on the first attempt. It takes time and experience to develop skill and confidence in the modeling process. We have found that involving students in the mathematical modeling process as early as possible, beginning with short projects, facilitates their progressive development and confidence in mathematics and modeling.

Many modeling texts present "type models" such as various inventory models for determining optimal inventory strategies. Students then learn to select an appropriate model for a particular situation. This approach has merit, and model selection is a valid step in the problem-solving process. However, undergraduate students often do not comprehend the assumptions behind a model, and only rarely do they take into consideration the appropriateness and sensitivity of those assumptions. Therefore, we emphasize *model construction* in order to promote student creativity and to demonstrate the artistic nature of model building, including the ideas of experimentation and simulation. Although we do discuss fitting data to chosen model types, our concentration is still on the entire model-building process, leaving the study of type models for more advanced courses.

Student Background and Course Content

Since our desire is to initiate the modeling experience as early as possible in a student's program, the only prerequisite for this text is a basic understanding of single-variable differential and integral calculus. (Occasionally we use partial derivatives, but other than a quick explanation by the instructor demonstrating how to compute such derivatives, the student does not need multivariable calculus.) While some unfamiliar mathematical ideas are taught as part of the modeling process, the emphasis is on using mathematics already known by the student after completing calculus. The modeling course will then motivate students to study the more advanced courses such as linear algebra, differential equations, optimization and linear programming, numerical analysis, probability, and statistics. The power and utility of these subjects are intimated throughout the text.

Although there are strong arguments to include such courses as advanced calculus, linear algebra, differential equations, linear programming, probability, and statistics as prerequisites to an introductory modeling course, this requirement necessitates postponing the course until the junior or senior undergraduate year, delaying the student's exposure to real-world applications. It also cuts off a number of student beneficiaries (namely, those nonmathematics majors who cannot satisfy all the prerequisites). Though our philosophy differs somewhat, this text still serves the more advanced student who has taken more mathematics courses. Certain sections of the text can be covered more rapidly by the advanced student, allowing more time for deeper extensions of the material as suggested by the projects for each chapter. The advanced student might also solve some of the models—such as a linear programming model—that would be beyond the capabilities of a student knowing only calculus.

While the text is designed so that it can be studied early in a student's program, readers at all levels should find the scenarios and problems both interesting and challenging. They are not designed for the application of a particular mathematical technique. Instead, they demand thoughtful ingenuity in using fundamental concepts to find reasonable solutions to "open-ended" problems. Certain mathematical techniques (such as dimensional analysis, curve fitting, and Monte Carlo simulation) are presented because they often are not formally covered at the undergraduate level. As an instructor, you should find great flexibility in adapting the text to meet the particular needs of your students through the problem assignments and student projects. We have used this material to teach courses to both undergraduate and graduate students, and even as a basis for faculty seminars.

Organization of the Text

The organization of the text is best understood with the aid of Figure 1. Part One consists of the first three chapters and is directed toward creative model construction and to an overview of the entire modeling process. We begin with the construction of graphical models, which provides us with some concrete models to support our discussion of the modeling process in Chapter 2. This approach also naturally extends the student's calculus experience, providing a transition into model construction by first involving the student in *model analysis*. Next we classify models and analyze the modeling process. At this point students can really begin to analyze scenarios, identify problems, and determine the underlying assumptions and principal variables of interest in a problem. This work is preliminary to the models they will create later in the course. (The order of Chapters 1 and 2 may be reversed, although we have found the current order best for capturing student interest and reducing student anxiety.) In their first modeling experience, students are quite anxious about their "creative" abilities and how they are going to be evaluated. For these reasons we have found it advantageous to start them out on familiar ground by appealing to their understanding of graphs of functions and having them learn model analysis. The book blends mathematical modeling techniques with the more creative aspects of modeling for variety and confidence building, and gradually the

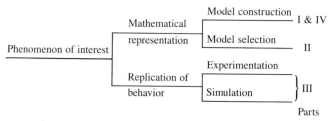

FIGURE 1. The organization of the text follows the above classification of the various models.

transition is made to the more difficult creative aspects. Students will find Parts Three and Four more challenging than the first parts of the text.

In Chapter 3 we present the concepts of proportionality and geometric similarity and use them to construct mathematical models for some of the previously identified scenarios. The student formulates tentative models or submodels and begins to learn how to test the appropriateness of the assumptions.

After Chapter 3 the book provides a number of models that can be related to curve fitting and optimization. These models motivate the study in Part Two. In Chapter 4 model fitting is discussed, and in the process several optimization models are developed. These optimization models are then analyzed in Chapter 5. The area of optimization is so rich in practical applications for modeling that it is tempting to teach optimization solution techniques (such as linear programming) as part of the course. However, such an approach detracts from the time allowable for model formulation and construction. Thus we have chosen to give students the opportunity to practice model construction while addressing a wide variety of scenarios. In a subsequent section the students are asked to solve those optimization problems requiring only calculus. Although students may not be able to solve some of the optimization models they will formulate here, nevertheless they do obtain needed additional practice in model construction, through which they gain confidence in their modeling skills. Moreover, the motivation is provided for studying linear optimization later in a full course. For those instructors who do wish to cover linear programming or other optimization topics as part of their course, we have suggested a sequence of UMAP (Undergraduate Mathematics and Its Applications Project) modules that provide excellent introductory material.* The modules include both graphical and analytical treatments of the Simplex method, for instance. While Part Two can be viewed more as model "solving," nevertheless, once completed, it enables a student to fit constructed or selected models to a set of data.

Part Three of the book consists of Chapters 6–8 and is dedicated to empirical model construction. It begins with fitting simple one-term models to collected sets of data and progresses to more sophisticated interpolating models, including polynomial smoothing models and cubic splines. The next topic is dimensional analysis and it presents a means of significantly reducing the experimental effort required when constructing models based on data collection. We also include a brief introduction to similitude. Finally, simulation models are discussed. An empirical model is fit to some collected data, and then Monte Carlo simulation is used to duplicate the behavior being investigated. The presentation motivates the eventual study of probability and statistics.

---

* UMAP modules are developed and distributed through COMAP, Inc./UMAP, 271 Lincoln St., Lexington, MA 02173.

In Part Four dynamic (time varying) scenarios are treated. Modeling based on differential equations in lieu of difference equations is motivated by our desire to emphasize the use of mathematics that students already know, namely the calculus. We begin by modeling initial value problems in Chapter 9 and progress to interactive systems in Chapter 10, with the student performing a graphical stability analysis. Students with a good background in differential equations can pursue analytical and numerical stability analyses as well, or they can investigate the use of difference equations or numerical solutions to differential equations by completing the projects.

The text is arranged in the order we prefer in teaching our modeling course. However, once the material in Chapters 1–4 is covered, the order of presentation may be varied to fit the needs of a particular instructor or group of students. Figure 2 shows how the various chapters are interdependent or independent, allowing progression through the chapters without loss of continuity.

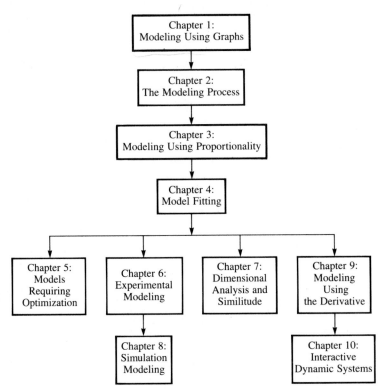

FIGURE 2. Chapter organization and progression.

Student Projects

Student projects are an essential part of any modeling course. This text includes projects in creative and empirical model construction, model analysis, and model research. Thus we recommend a course consisting of a mixture of projects in all these three facets of modeling. These projects are most instructive if they address scenarios that have no unique solution. Some projects should include *real* data that the student is either given or can *readily* collect. A combination of individual and group projects can also be valuable. Individual projects are appropriate in those parts of the course where the instructor wishes to emphasize the development of individual modeling skills. However, the inclusion of a group project early in the course gives students the exhilaration of a "brainstorming" session. A variety of projects is suggested in the text, such as constructing models for various scenarios, completing UMAP modules, or researching a model presented as an example in the text or class. It is valuable for each student to receive a mixture of projects requiring either model construction, model analysis, or model research for variety and confidence building throughout the course. Students might also choose to develop a model in a scenario of particular interest, or analyze a model presented in another course. We recommend six to eight short projects in a typical modeling course. Detailed suggestions on how the student projects can be assigned and used are included in the Instructor's Manual written for this text.

In terms of the number of scenarios covered throughout the course, as well as the number of homework problems and projects assigned, we have found it better to pursue a few that are developed very carefully and completely. Two or three good problems are about the maximum that an average student can handle in one week. We have provided many more problems and projects than can reasonably be assigned in order to allow for a wide selection covering many different application areas.

The Role of Computation

Although computing capability is *not* a requirement in using this book, computation does play a role of increasing importance after Part One. We have found a combination of programmable calculators, microcomputers, and mainframe computers to be advantageous throughout the course. Students who have programming experience can write computer code as part of a project, or software can be provided by the instructor as needed. We include some programs in the Instructor's Manual for this text. Typical applications for which students will find computers useful are in graphical displays of data, transforming data, least-squares curve fitting (and possibly the Simplex method if the Chebyshev criterion is pursued with more advanced students), divided difference tables and cubic splines, programming simulation models, and numerical solutions to differential equations. The use of computers has the added advantage of getting the student to think early about numerical methods and strategies, and it provides insight into how "real-world" problems are actually attacked in business and industry. Students appreciate being provided with or developing software that can be taken with them after completion of the course.

Resource Materials

We have found material provided by the Undergraduate Mathematics and Its Applications Project (UMAP) to be outstanding and particularly well suited to the course we propose. UMAP started under a grant from the National Science Foundation and has as its goal the production of instructional materials to introduce applications of mathematics into the undergraduate curriculum.

The individual modules may be used in a variety of ways. First, they may be used as instructional material to support several lessons. (We have incorporated several modules in the text in precisely this manner.) In this mode a student completes the self-study module by working through its exercises (the detailed solutions provided with the module can be removed conveniently before it is issued). Another option is to put together a block of instruction covering, for example, linear programming or difference equations, using as instructional material one or more UMAP modules suggested in the projects sections of the text. The modules also provide excellent sources for "model research" since they cover a wide variety of applications of mathematics in many fields. In this mode, a student is given an appropriate module to be researched and is asked to complete and report on the module. Finally, the modules are excellent resources for scenarios for which students can practice model construction. In this mode the teacher writes a scenario for a student project based on an application addressed in a particular module and uses the module as background material, perhaps having the student complete the module at a later date. Information about Project UMAP (now COMAP) may be obtained by writing the director, Ross L. Finney.

Several other excellent sources can be used as background material for student projects. The comprehensive four-volume series *Modules in Applied Mathematics,* edited by William F. Lucas and published by Springer-Verlag, provides important and realistic applications of mathematics appropriate for undergraduates. The volumes treat differential equations models, political and related models, discrete and system models, and life science models. Another book, *Case Studies in Mathematical Modelling,* edited by D. J. James and J. J. McDonald and published by Halsted Press, provides case studies explicitly designed to facilitate the development of mathematical models. Finally, scenarios with an industrial flavor are contained in the modular series edited by J. L. Agnew and M. S. Keener of Oklahoma State University. While we feel that this text provides abundant material for an introductory modeling course, any of the preceding references can be used selectively to complement the text and suit the tastes of the individual instructor.

Acknowledgments

It is always a pleasure to acknowledge individuals who have played a role in the development of a book. Several colleagues were especially helpful to us. We are particularly grateful to Colonel Jack M. Pollin, United States Military Academy, and Professor Carroll Wilde, Naval Postgraduate School, for stimulating our interest in teaching modeling and for support and guidance in our careers. We also thank Jack Gafford, Naval Postgraduate School, for his many suggestions for student projects and problems, and for his insights in how to teach modeling. We are indebted to many colleagues for reading the manuscript and suggesting modifications and problems: Rickey Kolb, John Kenelly, Robert Schmidt, Stan Leja, Bard Mansager, and especially Steve Maddox and Jim McNulty. We thank the following individuals who reviewed preliminary versions of the manuscript: Gilbert G. Walter, University of Wisconsin; Peter Salamon, San Diego State University; Peter A. Morris, Applied Decision Analysis, Inc.; Christopher Hee, Eastern Michigan University; Eugene Spiegel, University of Connecticut; David Ellis, San Francisco State University; David Sandell, U.S. Coast Guard Academy; Jeffrey Arthur, Oregon State University; Courtney Coleman, Harvey Mudd College; Don Snow, Brigham Young University.

We are indebted to a number of individuals who authored or co-authored UMAP materials that appear in the text: David Cameron, Brindell Horelick, Michael Jaye, Bruce King, Sinan Koont, Stan Leja, Michael Wells, and Carroll Wilde. In addition we thank Ross Finney, Solomon Garfunkel, and the entire UMAP staff for their cooperation on this project. We acknowledge William F. Lucas for editing *Modules In Applied Mathematics*, and the National Science Foundation, the Mathematical Association of America, and CUPM for their support of modeling courses.

The production of any mathematics text is a complex process and we have been especially fortunate in having a superb and creative production staff at Brooks/Cole. In particular we express our thanks to Craig Barth, our editor and friend; to Katherine Minerva for an attractive design and stunning cover; to Gay Orr, our production editor; to Judy Macdonald, our art editor; to Judy Blamer, our photo editor; and to Carline Haga, our permissions editor.

Finally, we are grateful of our wives who provided not only support but also assisted in producing this book: to Judi Giordano for processing a large portion of the manuscript and suggesting several scenarios to be modeled, and to Gale Weir for preliminary editing. And we thank our children Amy, Meg, Debbie, and Sammy Giordano and Maia and Renee Weir for their loving support and seemingly inexhaustible supply of cookies.

Frank R. Giordano
Maurice D. Weir

CONTENTS

* Optional Section

* Optional Section

PART ONE

CREATIVE MODEL CONSTRUCTION
AND THE MODELING PROCESS

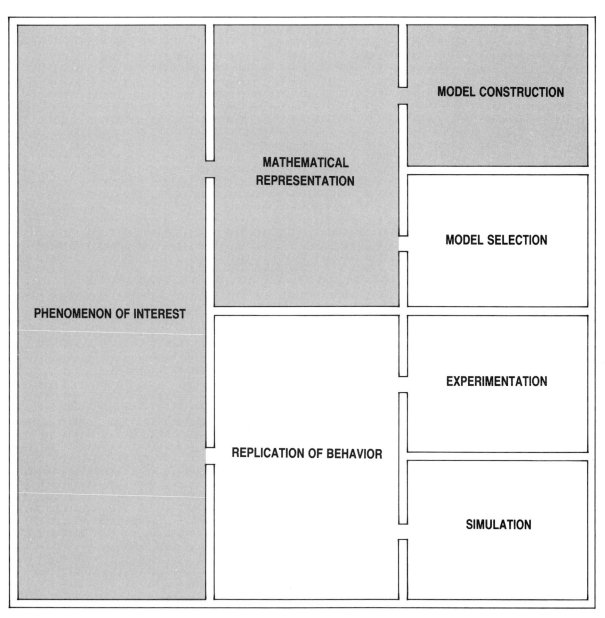

GRAPHS OF FUNCTIONS AS MODELS

INTRODUCTION

Quite often we are interested in analyzing complex situations in order to predict qualitatively the effect of some course of action. For example, in a two-country nuclear arms race that is in a state of relative equilibrium, what will be the effect on the number of nuclear missiles possessed by each side if one of the countries introduces mobile launching pads? In view of the many factors affecting the behavior of the parties involved, the assumptions of our model will necessarily be rather crude. Initially, we will be satisfied with a model that simply captures the general trend of the nuclear arms situation. Another example of a complex situation exists in the economics of the oil industry. What is the effect of a surcharge tax on gasoline at the pump? Will the tax actually reduce consumer demand? Will it contribute to inflation? In cases like those just described, graphical models are very useful in helping us understand the situation being studied.

In mathematical modeling we often attempt to construct a mathematical function relating variables to serve as a model. In subsequent chapters we attempt to determine a precise description of this function, but in this chapter the graph of the function is used simply to gain a qualitative understanding of the behavior under investigation. A graphical model has the important advantage of appealing to our visual intuition. It gives us a picture and a "feel" for what is happening that often eludes us in more symbolic analyses. Graphs are very good for gaining qualitative information. On the other hand, graphical analysis does limit the number of variables that can be studied effectively and the precision that can be attained. Nevertheless, these limitations are no disadvantage in cases where precision is restricted by the crudeness of the data, or where the complexity of the situation confines our expectations of the model. Graphical analysis is also useful as a prelude to a more detailed analytical model, often providing clues as to which factors should be considered more thoroughly in a subsequent analysis. We will use graphical models throughout the text to support other types of analyses.

In a graphical model we are primarily interested in the general shapes of the curves representing functional relationships that are being studied and the curve's intercepts with the coordinate axes. Recall from calculus that a **function** consists of a domain and a rule. The domain may be represented by one or several independent variables, and the rule assigns to each member of the domain one and only one value. For instance, the rule that assigns to every real number x the value $3x^2 + 1$ is a functional relationship of one independent variable whose graph is a parabola in the Cartesian plane. Likewise, the rule that assigns to every ordered pair of real numbers (x, y) the value $4 - x^2 - y^2$ is a functional relationship of two independent variables whose graph is a parabolic surface in Cartesian space. As the number of independent variables increases, a graphical analysis can become quite complex. In order to obtain visualizations in 2-dimensional or 3-dimensional space, it becomes necessary to fix the values of all but one or two of the independent variables, like taking a "snapshot" frozen for those particular values. Then new values can be assigned and another "snapshot" taken. We will discuss some of the ways that the number of independent variables can be reduced in Chapters 3 and 7 in the text.

Let's review briefly some of the fundamental geometric ideas associated with the graph of a function of a single independent variable, say $y = f(x)$. Several concepts from calculus are useful in determining the shape of the graph of a function. The first derivative gives the slope of the line tangent to the graph at a point, so the sign of the first derivative indicates if the function is **increasing** (positive sign) or **decreasing** (negative sign). The magnitude of the first derivative gives the instantaneous rate of change of the function at the point. For example, the function depicted in Figure 1-1 has a first derivative that is positive and increasing at every point.

Since the first derivative itself is a function, the sign of its derivative (that is, the sign of the second derivative) indicates whether the first derivative is increasing or decreasing. To interpret the second derivative geometrically, consider the function depicted in Figure 1-2. It has a second derivative that is everywhere positive because the first derivative is steadily increasing from negative to positive values. Notice that the tangent line turns continuously in the counterclockwise direction as x advances, and that the graph lies everywhere above the tangent line. In this situation we say that the graph is **concave upward**. A graph is **concave downward** if the second derivative is negative. A graph that is shaped concave upward is cupped to "hold water"; when it is shaped concave downward, it would "spill water." Note also in Figure 1-2 that the first derivative at the point x_1 is zero. Points where **relative maxima** and **minima** occur must have a first derivative equal

function

increasing
decreasing

concave upward
concave downward

relative maxima
relative minima

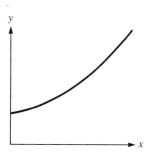

FIGURE 1-1 A graph showing y as an increasing function of x.

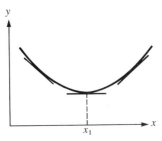

FIGURE 1-2 A steadily increasing first derivative corresponds to a positive second derivative.

to zero, provided the derivative exists. We will use these basic concepts from calculus to determine the general shapes of the curves in our graphical models.

MODEL

1.1 A NUCLEAR ARMS RACE

You might ask, "Why study the arms race?" One reason is that almost all modern wars are preceded by unstable arms races. Strong evidence suggests that an unstable arms race between great powers, characterized by a sharp acceleration in military capability, is an "early warning indicator" of war. In a 1979 article, Michael Wallace of the University of British Columbia studied 99 international disputes during the 1816–1965 period.* He found that disputes preceded by an unstable arms race escalated to war 23 out of 28 times, whereas disputes *not* preceded by an arms race resulted in war only 3 out of 71 times. Wallace calculated an "arms race index" for the two nations involved in each dispute that correctly predicts war or no war in 91 out of the 99 cases studied. His findings do not mean that an arms race between the powers necessarily results in war nor that there is a causal link between arms races and conflict escalation. They do establish, however, that rapid competitive military growth is strongly associated with the propensity to

* Michael Wallace, "Arms Races and Escalation: Some New Evidence," in *Explaining War*, ed. J. David Singer (Beverly Hills, Calif.: Sage, 1979), pp. 240–252.

war. Thus, by studying the arms race, we have the potential for predicting war, and if we can predict war, then there is hope that we can learn to avoid it.

There is another reason for studying the arms race. If the arms race can be approximated by a mathematical model, then it can be understood more concretely. You will see that the answers to such questions as, "Will civil defense dampen the arms race?" or "Will the introduction of mobile missile launching pads help to reduce the arms requirements?" are not simply matters of political opinion. There is an objective reality to the arms race that the mathematical model intends to capture.

The Soviet Union and the United States are engaged in a nuclear arms race. How should the United States react to changes in numbers and sophistication of the Soviet nuclear arsenal? To answer the difficult question of "How many weapons are enough?" several factors must be considered, including American objectives, Soviet objectives, and weapon technology. Former chairman of the Joint Chiefs of Staff, General Maxwell D. Taylor, has suggested the following objectives for American strategic forces:

> The strategic forces, having the single capability of inflicting massive destruction, should have the single task of deterring the Soviet Union from resorting to any form of strategic warfare. To maximize their deterrent effectiveness they must be able to survive a massive first strike and still be able to destroy sufficient enemy targets to eliminate the Soviet Union as a viable government, society and economy, responsive to the national leaders who determine peace or war.[*]

Note especially that Taylor's strategy assumes the worst possible case: the Soviet's launching a preemptive first attack to destroy our nuclear force.

How many weapons would be necessary to accomplish the objectives General Taylor suggests? After describing an appropriate system of Soviet targets (generally population and industrial centers), he goes on to say:

> The number of weapons we shall need will be those required to destroy the specific targets within this system of which few will be hardened silos calling for the accuracy and short flight time of ICBMs. As a safety factor, we should add extra weapons to compensate for losses that may be suffered in a first strike and for uncertainties in weapon performance. The total weapons requirement should be substantially less than the numbers available to us in our present arsenal.[†]

Thus a minimum number of missiles would be required to destroy specific enemy targets (generally population and industrial centers) chosen to inflict unacceptable damage upon the enemy. Additional missiles would be required to compensate for losses incurred in the Soviet's presumed first strike. Implicitly, the number of such additional missiles depends upon the size and effectiveness of the Soviet missile forces. Taylor concludes that meeting these objectives allows for a reduction in our present nuclear arsenal.

---

[*] M. D. Taylor, "How To Avoid a New Arms Race," *The Monterey Peninsula Herald*, January 24, 1982, p. 3C.
[†] Ibid.

In response to a question on expenditures for national defense, Admiral Hyman G. Rickover testified before a congressional committee as follows:

> For example, take the number of nuclear submarines, I'll hit right close to home. I see no reason why we have to have just as many as the Russians do. At a certain point you get where it's sufficient. What's the difference whether we have 100 nuclear submarines or 200? I don't see what difference it makes. You can sink everything on the ocean several times over with the number we have and so can they. That's the point I'm making.*

Again Admiral Rickover concludes that a reduction in arms is possible. We are going to develop a graphical model of the nuclear arms race based on the preceding remarks.

Developing the Graphical Model

Suppose two countries, Country X and Country Y, are engaged in a nuclear arms race and that each country adopts the following strategies.

Friendly Strategy: To survive a massive first strike and inflict unacceptable damage upon the enemy.

Enemy Strategy: To conduct a massive first strike to destroy the friendly missile force.

That is, each country follows the friendly strategy when determining its missile force and presumes the enemy strategy for the opposing country. Note especially that the friendly strategy implies targeting population and industrial centers while the enemy strategy implies targeting missile sites.

Now let's define the following variables:

$$x = \text{number of missiles possessed by Country X}$$
$$y = \text{number of missiles possessed by Country Y}$$

Next, let $y = f(x)$ denote the function representing the minimum number of missiles required by Country Y to accomplish its strategies when Country X has x missiles. Similarly, let $x = g(y)$ represent the minimum number of missiles required by Country X to accomplish its objectives.

We begin by investigating the nature of the curve $y = f(x)$. Since a certain number of missiles y_0 is required by Country Y to destroy the selected population and industrial centers of Country X, y_0 is the intercept when $x = 0$. That is, Country Y considers that it needs y_0 missiles even if Country X has none (basically, a psychological defense). As Country X increases its missile force, Country Y must add additional missiles since it assumes Country X is following the enemy strategy and targeting its missile force. Let's assume that the weapons technology is such that Country X can destroy no more than one of Country Y's missiles with each missile fired. Then

* Hyman G. Rickover, testimony before Joint Economic Committee; *The New York Review*, March 18, 1982, p. 13.

the number of additional missiles Country Y needs for each missile added by Country X depends upon the effectiveness of Country X's missiles. Convince yourself that the curve $y = f(x)$ must lie between the limiting lines shown in Figure 1-3. Line A, having slope 0, represents a state of absolute invulnerability of Country Y's missiles to any attack. At the other extreme Line B, having slope 1, says that Country Y must add one new missile for each missile that is added by Country X. To determine more precisely where $y = f(x)$ does lie, we will analyze what happens for various cases relating the relative sizes of the two missile forces.

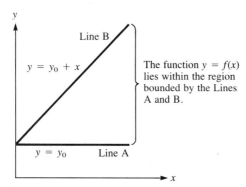

FIGURE 1-3 Bounding the function $y = f(x)$.

Case 1 $x < y$ If Country X attacks in this situation, it fires x missiles at the same number of Country Y's missiles. Since the number $(y - x)$ of Country Y's missiles could not be attacked, at least that many would survive. Of the number x of Country Y's missiles that were fired upon, a percentage s would survive, where $0 < s < 1$. Thus the total number of missiles surviving the attack is $y - x + sx$. Now Country Y must have y_0 missiles survive in order to inflict unacceptable damage on Country X. Hence,

$$y_0 = y - x + sx \qquad \text{for } 0 < s < 1$$

or solving for y,

$$y = y_0 + (1 - s)x \tag{1.1}$$

Equation (1.1) gives the minimum number of missiles Country Y must have in order to be confident that y_0 missiles will survive an attack by Country X.

Case 2 $y = x$ In this scenario Country X fires exactly one of its missiles at each of Country Y's missiles. Assuming the percentage s to survive the attack, the number $sx = sy$ survive, in which case Country Y needs

$$y = \frac{y_0}{s} \tag{1.2}$$

missiles in order to inflict unacceptable damage on Country X.

Case 3 $y < x < 2y$ Here Country X targets each of Country Y's missiles once and a portion of them twice, as illustrated in Figure 1-4.

Country X:

Country Y:

FIGURE 1-4 An example of $y < x < 2y$.

Convince yourself that $x - y$ of Country Y's missiles would be targeted twice and $y - (x - y) = 2y - x$ would be targeted once. Of those targeted once, a percentage $s(2y - x)$ will survive as before. Of those targeted twice, the percentage $s(x - y)$ will survive the first round. Of those that survive the first round, the percentage $s[s(x - y)] = s^2(x - y)$ will survive the second round. Hence Country Y must have

$$y_0 = s^2(x - y) + s(2y - x)$$

missiles survive, or, solving for y,

$$y = \frac{y_0 + x(s - s^2)}{2s - s^2} \tag{1.3}$$

is the minimum number of missiles required by Country Y.

Case 4 $x = 2y$ Now Country X will fire exactly two missiles at each of Country Y's missiles. If we reason as in Case 2, the number s^2y will survive so that

$$y = \frac{y_0}{s^2} \tag{1.4}$$

is the minimum number of missiles required by Country Y.

Now let's combine all of the preceding scenarios into a single graph. For convenience we are going to assume that the discrete situation of the minimum number of missiles can be represented by a continuous model (giving rise to fractions of missiles). First observe that Equations (1.1) and (1.3) both represent straight-line segments: one segment for $x < y$ and the second segment for $y < x < 2y$. In Case 1, when $x < y$, we obtained the equation

$$y_0 = y - x + sx$$

As x approaches y, this last equation becomes (in the limit) $y_0 = sy$. In Case 3, when $y < x < 2y$, we obtained the equation

$$y_0 = s^2(x - y) + s(2y - x) \tag{1.5}$$

Again, as x approaches y, the equation becomes $y_0 = sy$. Thus the two line segments meet at $x = y$ with the common value $y = y_0/s$.

Finally, as x approaches $2y$, Equation (1.5) becomes $y_0 = s^2y$. These observations mean that the two line segments defined by (1.1) and (1.3) form a continuous curve meeting the lines $y = x$ and $2y = x$. Moreover, since the slope for the line segment represented by (1.3) is less than the slope of the line segment represented by (1.1), we say that the curve is *piecewise linear with decreasing slopes*. The graphical model is depicted in Figure 1-5. Notice how the graph does lie within the cone-shaped region between Lines A and B as discussed previously.

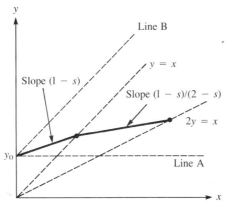

FIGURE 1-5 A graphical model relating the number of missiles for Country Y to the number of missiles for Country X when $0 \leqslant x \leqslant 2y$.

We could continue to analyze additional cases, such as what happens when $2y < x < 3y$. However, since we are interested only in qualitative information, let's see if we can determine the general shape of the curves more simply.

Generalizing from our analysis in Cases 2 and 4, let's consider the following model:

$$y = \frac{y_0}{s^{x/y}} \qquad \text{for } 0 < s < 1 \qquad \textbf{(1.6)}$$

An inspection of Equation (1.6) reveals that for every ratio x/y we can find y. That is, the curve $y = f(x)$ crosses each line $x = y$, $x = 2y$, ..., $x = ny$, as illustrated in Figure 1-6.

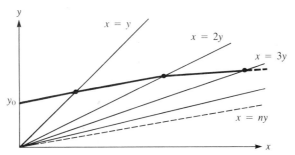

FIGURE 1-6 The curve $y = f(x)$ must cross every line $x = ny$.

The situation for Country X is entirely symmetrical. Thus its minimal number of missiles is represented by a continuous curve $x = g(y)$ that crosses every line $y = x$, $y = 2x, \ldots, y = nx$. Thus the two curves must intersect.

The preceding discussion leads us to consider two idealized continuously differentiable curves like those drawn in Figure 1-7. Since the curve $y = f(x)$ represents the minimum number of missiles required by Country Y, the region above the curve represents missile levels satisfactory to Country Y. Likewise, the region to the right of the curve $x = g(y)$ represents missile levels satisfactory to Country X. Thus the darkest region in Figure 1-7 represents missile levels satisfactory to both countries.

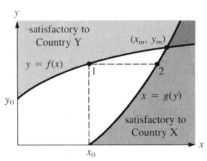

FIGURE 1-7 Regions of satisfaction to Country X and Country Y.

The intersection point of the curves $y = f(x)$ and $x = g(y)$ represents the minimum level at which both sides are satisfied. To see that this is so, assume Country Y has y_0 missiles and observes that Country X has x_0 missiles. In order to meet its objectives, Country Y will have to add sufficient missiles to reach point 1 in Figure 1-7. In turn, Country X will have to add sufficient missiles to reach point 2 in Figure 1-7. This process will continue until both sides are satisfied simultaneously. Notice that any point in the darkest region will suffice to satisfy both countries, and there are many points in the darkest region that are quite likely to occur. The intersection point (x_m, y_m) in Figure 1-7 represents the minimum force levels required of both countries to meet their objectives.

We would like to know if the intersection point is unique. Note from Equation (1.6) that as the ratio x/y increases, y must increase; likewise, $x = g(y)$ increases. Since both curves are increasing, it is tempting to conclude that the intersection point is unique. However, consider Figure 1-8. In the figure both curves are steadily increasing; the curve $y = f(x)$ crosses every line $x = ny$, and $x = g(y)$ crosses every line $y = nx$. Yet the curves have multiple intersection points. How can we ensure a unique intersection point? Notice that the slope of the curve $y = f(x)$ in Figure 1-8 is steadily decreasing until the point $x = x_1$ when it begins to increase. Thus the first derivative changes from a decreasing to an increasing function at $x = x_1$. That is, the

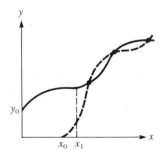

FIGURE 1-8 Steadily increasing curves with multiple intersection points.

second derivative changes sign. If we can show such a sign change is impossible, then we can conclude that the intersection point is unique. In fact, we will show that the second derivative of $y = f(x)$ is always negative.

Taking the logarithm of Equation (1.6) yields

$$\ln y = \ln y_0 - \frac{x}{y} \ln s$$

Multiplying both sides of the previous equation by y and simplifying gives

$$y \ln y - y \ln y_0 = -x \ln s$$

Differentiating implicitly with respect to x and simplifying yields

$$y'(1 + \ln y - \ln y_0) = -\ln s$$

or

$$y' = \frac{-\ln s}{1 + \ln y - \ln y_0}$$

Differentiating this last equation for the second derivative gives

$$y'' = \frac{-(-\ln s)\dfrac{1}{y} y'}{(1 + \ln y - \ln y_0)^2}$$

Next we determine the sign of y'. Rewrite y' as

$$y' = \frac{\ln s}{-1 + \ln \dfrac{y_0}{y}}$$

Since $0 < s < 1$, $\ln s$ is negative. Now, for the cases we are considering, $y > y_0$, which implies that $\ln (y_0/y) < 0$. Thus $y' > 0$ everywhere, in which case $y'' < 0$ everywhere. Therefore, we can conclude that a unique intersection point does in fact exist. The model has the general shape shown in Figure 1-9.

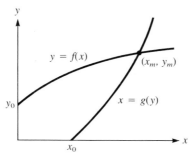

FIGURE 1-9 A graphical model of the nuclear arms race.

Model Interpretation

Example 1   Let's use the graphical model to analyze the effects on the intersection point of strategies likely to be considered by Countries X and Y. For instance, suppose Country X doubles its annual budget for civil defense. Presumably, Country Y will then need more missiles to inflict an unacceptable level of damage on Country X's population centers. Thus y_0 increases. Since the effectiveness of Country X's weapons has not changed, the shape of the curve $y = f(x)$ is the same. The curve $x = g(y)$ does not change either because there is no change in Country Y's population centers nor in its weapons effectiveness. The net effect of the increased civil defense is shown in Figure 1-10. The dashed curve is the new position of the function $y = f(x)$ resulting from the civil defense of Country X. The point (x_m', y_m') is the new intersection point. Note that although the course of action seemed fairly passive, the net effect is to increase the minimum number of missiles required by both sides because $x_m' > x_m$ and $y_m' > y_m$.

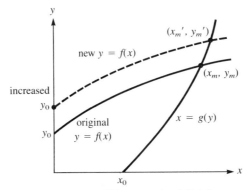

FIGURE 1-10 Country X increases its civil defense posture.

Example 2   Next assume that Country X puts its missiles on mobile launching pads, which can be relocated during times of international crisis. There is no change in x_0 since Country X still requires the same number of missiles to inflict unacceptable damage on Country Y's population and industrial centers. However, since Country X's missiles are less vulnerable than before, the curve $x = g(y)$ would rise more steeply, as shown by the dashed curve in Figure 1-11. The curve $y = f(x)$ does not change since (1) Country Y requires the same number of missiles to inflict unacceptable damage on Country X as before and (2) there is no change in the effectiveness of Country X's weapons. The net effect on the arms race of the mobile launching pads is depicted in Figure 1-11 and shows the new position of the equilibrium point (x_m', y_m'). Note that although this strategy is far less passive than the civil defense strategy, this alternative does lead to a reduction in the minimum number of missiles required by both countries because $x_m' < x_m$ and $y_m' < y_m$.

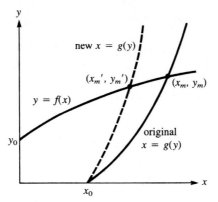

FIGURE 1-11 Country X employs mobile launching pads.

Example 3 In the next scenario suppose that Country X and Country Y both employ multiple warheads that can be targeted independently. Assume that each missile is armed with 16 smaller missiles each possessing its own warhead. Since it still takes the same number of warheads to destroy the opponent's population and industrial centers, it is reasonable to expect the number of larger missiles, x_0 and y_0, to be reduced by the factor 16. Now consider the slope of the curve $y = f(x)$. Since the warheads can be targeted independently, an increase in x by 1 gives the capability of destroying 16 of Country Y's missiles. Thus $y = f(x)$ must rise more sharply than before to compensate for the increased destruction if it is to be able to meet its friendly strategy objective. A similar analysis applies to Country X and the curve $x = g(y)$. The new curves are represented graphically in Figure 1-12. Since the reduction in values of the intercepts x_0 and y_0 coupled with the changes in the slopes of the curves give different effects on the new location of the intersection point (see Figure 1-12), it is difficult to tell from a graphical analysis whether the minimum number of missiles actually increases or decreases. This analysis demonstrates a limitation of graphical models. To determine the location of the equilibrium point (x_m', y_m') would require a more detailed analysis.

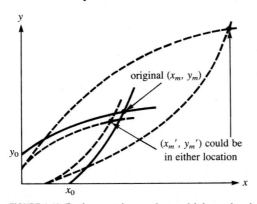

FIGURE 1-12 Both countries employ multiple warheads targeted independently.

Example 4   While we were unable to predict the effect of multiple warheads on the minimum number of missiles required by each side in the preceding example, we can analyze the total number of warheads under this strategy. Let x and y now represent *the number of warheads* possessed by Country X and Country Y, respectively. The number of warheads needed by each country to inflict unacceptable damage on the opponent remain at the levels x_0 and y_0, as before. However, an increase in x by 1 warhead enables Country X to destroy an additional 16 of Country Y's warheads since they are all located on a single missile. Thus Country Y will need more warheads, and $y = f(x)$ rises more steeply than before, as illustrated in Figure 1-13. The same argument applies to Country X and the curve $x = g(y)$. Note that both countries require *more warheads* if multiple warheads are introduced because $x_m' > x_m$ and $y_m' > y_m$ in Figure 1-13.

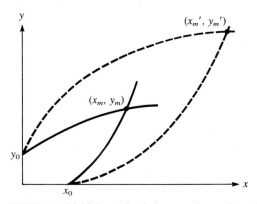

FIGURE 1-13 Multiple warheads increase the total number of warheads required by each side.

1.1 PROBLEMS

1. Analyze the effect on the arms race of each of the following strategies:
   a. Country X increases the accuracy of its missiles by using a better guidance system.
   b. Country X increases the payload (destructive power) of its missiles without sacrificing accuracy.
   c. Country X is able to retarget its missiles in flight so that it can aim for missiles that previous warheads have failed to destroy.
   d. Country Y employs sea-launched ballistic missiles.
   e. Country Y adds long-range intercontinental bombers to its arsenal.
   f. Country X develops sophisticated jamming devices that dramatically increase the probability of neutralizing the guidance systems of Country Y's missiles.
2. Discuss the appropriateness of the assumptions used in developing the nuclear arms race model. What is the effect on the number of missiles if

each country believes the other country is also following the friendly strategy? Is disarmament possible?

3. Develop a graphical model based on the assumptions that each side is following the "enemy strategy." That is, each side desires a "first strike" capability for destroying the missile force of the opposing side. What is the effect on the arms race if Country X now introduces antiballistic missiles?

4. Discuss how you might go about validating the nuclear arms race model. What data would you collect? Is it possible to obtain the data?

1.1 PROJECTS

For Projects 1 through 4, complete the requirements in the referenced UMAP module, and prepare a short summary for classroom discussion.

1. "The Distribution of Resources," by Harry M. Schey, UMAP 60, 61, 62 (one module). The author investigates a graphical model that can be used to measure the distribution of resources. Provided here is an excellent review of the geometric interpretation of the derivative as applied to the economics of the distribution of a resource. Numerical calculation of the derivative and definite integral is also discussed.

2. "Nuclear Deterrence," by Harvey A. Smith, UMAP 327. The stability of the arms race is analyzed assuming objectives similar to those suggested by General Taylor. Analytic models are developed using probabilistic arguments. An understanding of elementary probability is required.

3. "The Geometry of the Arms Race," by Steven J. Brams, Morton D. Davis, and Philip D. Straffin, Jr., UMAP 311. In this module the possibilities of both parties disarming is analyzed introducing elementary game theory. Interesting conclusions are based upon Country X's ability to detect Country Y's intentions and vice versa.

4. "The Richardson Arms Race Model," by Dina A. Zinnes, John V. Gillespie, and G. S. Tahim, UMAP 308. A model is constructed based upon the classical assumptions of Lewis Fry Richardson. Difference equations are introduced.

1.1 FURTHER READING

Saaty, Thomas L. *Mathematical Models of Arms Control and Disarmament*. New York: Wiley, 1968.

Schrodt, Philip A. "Predicting Wars With the Richardson Arms-Race Model." *BYTE* 7, no. 7 (July 1982): 108–134.

Wallace, Michael D. "Arms Races and Escalation: Some New Evidence." In *Explaining War*, edited by J. David Singer, pp. 240–252. Beverly Hills, Calif.: Sage, 1979.

MODEL

1.2 MANAGING NONRENEWABLE RESOURCES: THE ENERGY CRISIS

Over the last century the United States has shifted into nearly complete dependence upon nonrenewable energy sources. Petroleum and natural gas now constitute about three-fourths of the nation's fuel, and nearly half of our crude oil comes from foreign sources. The rise of the Organization of Petroleum-Exporting Countries cartel (OPEC) has caused some analysts to fear for supply security, especially during periods of political unrest when we are threatened by constraints on the supply of foreign oil, such as oil embargoes. Thus there are significant attempts to conserve energy so as to reduce our long-run oil consumption. There is also interest in more drastic short-term reductions in order to survive a crisis situation.

Various solutions have been proposed to address these long- and short-term needs. One solution is gas rationing. Another is to place a surcharge tax on each gallon of gasoline sold at the local pump. Basically, the idea behind this solution is that gasoline companies will pass the tax on to the consumer by increasing the price per gallon by the amount of the tax. Accordingly, it is supposed the consumer will reduce consumption because of the higher price. Let's study this proposal by constructing a graphical model and qualitatively addressing the following questions:

1. What is the effect of the surcharge tax on short- and long-term consumer demand?
2. Who actually pays the tax, the consumer or the oil companies?
3. Does the tax contribute to inflation?

Stated more succinctly, the problem is to determine the effect of a surcharge tax on the market price and consumer demand of gasoline. In the following analysis we are concerned with gaining a qualitative understanding of the principal factors involved with the problem. A graphical analysis is appropriate to gain this understanding, especially since precise data would be difficult to obtain. We begin by graphically analyzing some pertinent general economic principles. In the ensuing sections we interpret the conclusions of the graphical model as they apply to the oil situation.

Constructing a Graphical Model

Suppose a firm within a large competitive industry produces a single product. A question facing the firm is how many units to produce in order to maximize profits. Assume that the industry in question is so large that any particular firm's production has no appreciable effect on the market price. Hence the firm may assume the price of the product as constant, and need consider only the difference between the price and the firm's costs in producing the product. Individual firms encounter *fixed costs*, which are independent of the amount produced over a wide range of production levels. These costs include rent and utilities, equipment capitalization costs, management costs, and the like. The *variable costs* depend upon the quantity produced. Variable costs include the cost of raw materials, taxes, labor, and so forth. When the fixed costs are divided by the quantity produced, the share apportioned to each unit is obtained. This per unit share is relatively high when production levels are low. However, as production levels increase, not only does the per unit share of the fixed costs diminish, but economies of scale (such as buying raw materials in large quantities at reduced rates) often reduce some of the variable cost rates as well. Eventually, the firm reaches production levels that strain the capabilities of the firm. At this point the firm is faced with hiring additional employees, paying overtime, or capitalizing additional machinery or similar costs. Because the per unit costs tend to be relatively high when production levels are either very low or very high, one intuitively expects the existence of a production level q^* that yields a maximum profit over the range of production levels being considered. This idea is illustrated in Figure 1-14. Next, consider the characteristics of q^* mathematically.

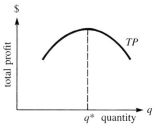

FIGURE 1-14 Profit is maximized at q^*.

At a given level of production q, total profit $TP(q)$ is the difference between total revenue $TR(q)$ and total cost $TC(q)$. That is,

$$TP(q) = TR(q) - TC(q)$$

A necessary condition for a relative maximum to exist is that the derivative of TP with respect to q must be 0:

$$TP' = TR' - TC' = 0$$

or, at the level q^* of maximum profit,

$$TR'(q^*) = TC'(q^*) \tag{1.7}$$

Thus, at q^* it is necessary that the slope of the total revenue curve equal the slope of the total cost curve. This condition is depicted in Figure 1-15.

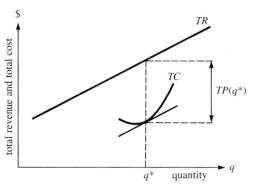

FIGURE 1-15 At q^* the slopes of the total cost and total revenue curves are equal.

Let's interpret economically the meaning of the derivatives TR' and TC'. From the definition of the derivative,

$$TR'(q) \approx \frac{TR(q + \Delta q) - TR(q)}{\Delta q}$$

for Δq small (where the symbol $\approx$ means "is approximately equal to"). Thus, if $\Delta q = 1$, you can see that $TR'(q)$ approximates $TR(q + 1) - TR(q)$, which is the revenue generated by the next unit sold, or the *marginal revenue MR* of the $q + $ 1st unit. Since total revenue is the price per unit times the number of units, it follows that the marginal revenue of the $q + $ 1st unit is the price of that unit less the revenue lost on previous units resulting from price reductions (see Problem 4). Similarly, $TC'(q)$ represents the *marginal cost MC* of the $q + $ 1st unit—that is, the *extra* cost in changing output to include one additional unit. If Equation (1.7) is interpreted in these new terms, a necessary condition for maximum profit to occur at q^* is that marginal revenue equal marginal cost:

$$MR(q^*) = MC(q^*) \tag{1.8}$$

For the critical point defined by Equation (1.8) to be a relative maximum, it is sufficient that the second derivative TP'' be negative. Since $TP' = MR - MC$, we have

$$TP''(q^*) = MR'(q^*) - MC'(q^*) < 0$$

or

$$MR'(q^*) < MC'(q^*) \tag{1.9}$$

In words, at the level q^* of maximum profit the slope of the marginal revenue curve is less than the slope of the marginal cost curve. The results (1.8) and

(1.9) together imply that the marginal revenue and marginal cost curves intersect at q^* with the marginal cost curve rising more rapidly. These results are illustrated in Figure 1-16.

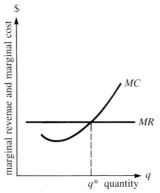

FIGURE 1-16 At q^*, $MR = MC$ and $MR' < MC'$.

Interpreting the Graphical Model

Now, let's interpret the graphical model represented by Figure 1-16. The MR curve represents the revenue generated by the next unit sold. The curve is drawn horizontally since in a large competitive industry, the amount one particular firm produces seldom influences the market price so there is no loss in revenue on previous units resulting from price reduction. Thus the MR curve represents the (constant) price of the product. Given a market price determined by the entire industry and aggregate consumer demand, a firm attempting to maximize profits will continue to produce units until the cost of the next unit produced exceeds its market price. Verify that this situation is suggested by the graphical model in Figure 1-16.

1.2 PROBLEMS

1. Justify mathematically, and interpret economically, the graphical model for the theory of the firm given in Figure 1-17. What are the major assumptions upon which the model is based?
2. Show that for total profit to reach a relative minimum, $MR = MC$ and $MC' < MR'$.
3. Suppose the large competitive industry is the oil industry, and the firm within that industry is a gasoline station. How well does the model depicted in Figure 1-16 reflect the reality of that situation? How would you adjust the graphical model in order to make improvements?
4. Verify the result that the marginal revenue of the $q + 1$st unit equals the price of that unit minus the loss in revenue on previous units resulting from price reduction.

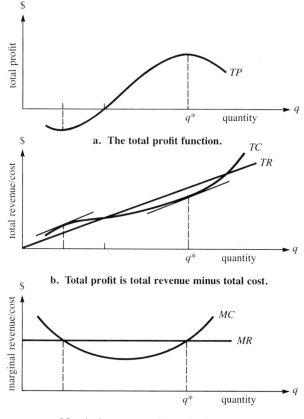

a. The total profit function.

b. Total profit is total revenue minus total cost.

c. Marginal revenue and marginal cost curves.

FIGURE 1-17 A graphical model for the theory of the firm.

1.3 EFFECTS OF TAXATION ON THE ENERGY CRISIS

Let's suppose a firm currently is maximizing its profits; that is, given a market price MR, it is producing q^* units as suggested by Figure 1-16. Assume further that a tax is added to each unit sold. Since the firm must pay the government the amount of the tax for each unit sold, the marginal cost to the firm of each unit increases by the amount of the tax. Geometrically, that means the marginal cost curve shifts upward by the amount of the tax. Assume for the moment that the entire industry is able to increase the market price by simply adding on the amount of the tax to the price of each unit. Under this condition the marginal revenue curve also shifts upward by the amount of the tax. This situation is depicted in Figure 1-18. Note from the figure that the optimal production quantity is still q^*. Hence the model predicts no change in production as a result of the tax. Rather, the firm will produce the same amount but charge a higher price, thereby contributing to inflation. Note too that it's the consumer who pays the full amount of the tax in the form of a price increase.

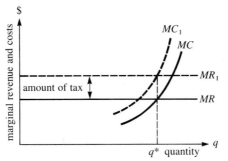

FIGURE 1-18 Both the marginal revenue and marginal cost curves shift upward by the amount of the tax leaving the optimum production at the same level q^*.

The shortcoming with the model in Figure 1-18 is that it does not reveal whether the entire industry can in fact continue to sell the same quantity at the higher price. In order to find out we need to construct a model for the industry. Thus, for each firm in the industry consider the intersection of the firm's various marginal revenue curves with each marginal cost curve. (Remember that each horizontal MR curve corresponds to the price of a unit.) This situation is depicted for one firm in Figure 1-19a.

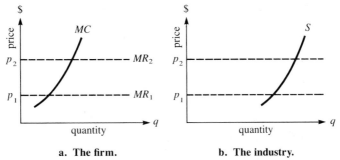

a. The firm.   **b. The industry.**

FIGURE 1-19 The industry's supply curve is obtained by summing together the amounts the firms would produce at each price level.

For each price, sum over all firms in the industry the amount each firm would optimally produce. This summing procedure yields a curve for the entire industry. Since this curve represents the amount the industry would supply at various price levels, it is called a *supply curve*, an example of which is depicted in Figure 1-19b. Qualitatively, as the market price increases, the industry is willing to produce greater quantities.

Next, consider aggregate consumer demand for the product at various market price levels. From a consumer's point of view, the quantity demanded is a function of the market price. However, it is traditional to plot market price as a function of quantity (see Figure 1-20a). Conceptually, for each price level, individual consumer demands could be summed as in the procedure for obtaining the industry's supply curve. This summation is depicted graphically in Figure 1-20. Qualitatively, as the price increases, we expect the aggregate demand for the product to decrease as consumers begin to use less or substitute cheaper alternative products (see Figure 1-20b).

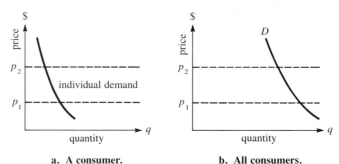

a. A consumer.   b. All consumers.

FIGURE 1-20 The industry's demand curve D represents the aggregate demand for the product at various price levels and is obtained by summing individual consumer demands at those levels. Notice that we plot price versus quantity for demand curves rather than vice versa.

Finally, consider the industry's supply and demand curves together. Suppose the two curves intersect at a unique point (q^*, p^*) as depicted in Figure 1-21. If the industry supplies q^* and charges p^* (supply curve), then consumers are willing to buy the amount q^* at the price p^* (demand curve). Thus there is equilibrium in the sense that no excess supply exists at that price, and both consumers and suppliers are satisfied.

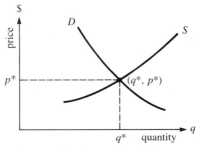

FIGURE 1-21 The intersection of the supply and demand curves gives a market price and a market quantity that satisfy both consumers and suppliers alike.

Obviously an industry does not know the precise demand curve for its product. Therefore, it is important to determine what occurs if the industry supplies an amount other than q^*. For example, suppose the industry supplies an amount q_1 greater than q^* (see Figure 1-22a). Then the consumers are willing to buy an amount as large as q_1 only if the price is as low as

p_1, forcing a reduction in price. However, if the market price drops to p_1, the industry is willing to supply only q_2 units and would cut production back to that level. Then at q_2 the unsatisfied consumers would drive the price up to p_2. Convince yourself that the process converges to (q^*, p^*) in Figure 1-22a, where the supply curve is steeper than the demand curve. Unlike the arms race model, there are forces that actually drive supply and demand to the equilibrium point. On the other hand, consider Figure 1-22b in which the supply curve is more horizontal than the demand curve. In this case the equilibrium point (q^*, p^*) will not be achieved by the iterative process just described. Instead, there is likely to be wild fluctuation in the amount supplied and the market price as the industry and consumers search for the equilibrium point. (Convince yourself from Figure 1-22b that the equilibrium point is difficult to achieve when the supply curve is not as steep as the demand curve.)

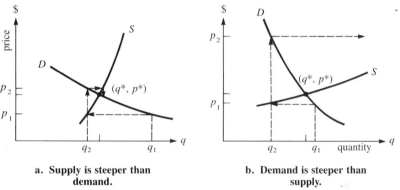

a. Supply is steeper than demand.   **b. Demand is steeper than supply.**

FIGURE 1-22 The ease with which the equilibrium point of supply and demand is achieved depends upon the relative slopes of the supply and demand curves.

Now consider the effect of a tax on the supply and demand curves. Suppose that a particular industry is in an equilibrium market position (q^*, p^*) when a tax is added to each unit sold. Since each firm has to pay the tax to the government, each marginal cost curve shifts upward by the amount of the tax (see Figure 1-18). These individual shifts cause the aggregate supply curve for the industry to shift upward by the amount of the tax as well. This phenomenon is depicted in Figure 1-23. If there is no reason for a shift in the demand curve, the intersection of the demand curve with the new supply curve shifts upward toward the left to a new equilibrium point (q_1, p_1), indicating an increase in the equilibrium market price with a corresponding decrease in market quantity. Furthermore, notice from Figure 1-23 that the increase in price, $p_1 - p^*$, is less than the tax. Thus the model predicts that the consumer and the industry share the tax. Study Figure 1-23 carefully and convince yourself that the proportion of the tax that the consumer pays and the relative reduction in the quantity supplied at equilibrium depend on the slopes of the supply and demand curves at the time the tax is imposed.

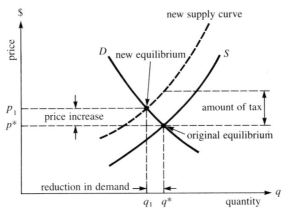

FIGURE 1-23 A tax added to each item sold causes a decrease in the quantity produced and an increase in the price.

1.3 PROBLEMS

1. Show that when the demand curve is very steep, a tax added to each item sold will fall primarily on consumers. If the demand curve is more nearly horizontal, show that the tax is paid mostly by the industry. What if the supply curve is very steep? What if the supply curve is nearly horizontal?

2. Consider the oil industry. Discuss the conditions for which the demand curve will be steep near the equilibrium. What are the situations for which the demand curve will be more horizontal (or flat)?

3. Criticize the following quotation:

    The effect of a tax on a commodity might seem at first sight to be an advance in price to the consumer. But an advance in price will diminish the demand. And a reduced demand will send the price down again. It is not certain, therefore, after all, that the tax will really raise the price.*

4. Suppose the government pays producers a subsidy for each unit produced instead of levying a tax. Discuss the effect on the equilibrium point of the supply and demand curves. What happens to the new price and the new quantity? Discuss how the proportion of the benefits to the consumer and to the industry depends on the slopes of the supply and demand curves at the time the subsidy is given (see Problem 1).

1.4 AN OIL EMBARGO

Now, let's consider the energy crisis. Suppose an oil embargo is imposed at a time when the vast majority of the population depends upon the auto-

* H. D. Henderson, *Supply and Demand* (Chicago: University of Chicago Press, 1958), p. 22.

mobile to get to work and, further, that no alternative mass transportation system is immediately available. Assume also that, in the short run, most people cannot switch to more fuel-efficient cars because they are not readily available or easily affordable. These assumptions suggest qualitatively a demand curve that is steep over a wide range of values because, in order to get to work, consumers will suffer a high increase in price before significantly cutting back on demand. Of course, eventually consumers reach price levels where it no longer pays to go to work. The demand curve is portrayed in Figure 1-24. Notice as q increases, consumers enter regions where additional gasoline being used is for leisure. In such flat regions the consumer is most sensitive to price changes.

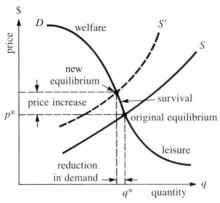

FIGURE 1-24 In the short run the demand curve for gasoline is likely to be quite steep.

Next consider the oil industry's supply curve. If the initial embargo catches the industry by surprise, most likely it will seek to find and develop alternative sources of oil. Possibly the industry will be forced to turn to more expensive sources in order to provide the same quantities as before the embargo. Furthermore, in the short run, the oil industry will be more sensitive to price because it will be very difficult for the industry to provide immediate increases in supply. These arguments suggest an upwards shift of the supply curve, which may become more vertical as well. Study Figure 1-24 and convince yourself that if the demand curve is steep, a significant price increase may result, but that it will take an appreciable shift in the supply curve in order to reduce demand significantly. Decide if the new equilibrium point can easily be attained in Figure 1-24.

Now consider the supply–demand curves depicted in Figure 1-25. Suppose the government is dissatisfied with the reduction in demand for oil resulting from the shift in the supply curve so it imposes a tax on each gallon of gasoline to reduce demand further. As discussed in the previous section, the tax causes the supply curve to shift upward (see Figure 1-25). If the consumers are less sensitive to price than the industry, the new equilibrium point will be difficult to achieve. Furthermore, the new equilibrium point suggests that consumers will pay most of the tax in the form of a price increase. In summary, the graphical model suggests that large fluctuations in price are probable, only modest reductions in demand are achievable, and the consumer bears the large portion of the tax burden.

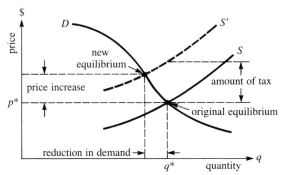

FIGURE 1-25 If the demand curve is relatively steep, the consumer pays the lion's share of the tax while reductions in demand are modest.

What about the long-range effects of the oil embargo? After the embargo is lifted, the oil industry will again have the foreign oil as well as the new sources that were developed during the crisis. These two sources cause the supply curve to shift downward and perhaps become more horizontal than during the embargo. Meanwhile, a change in consumer demand has occurred as well. Car pools have been formed, mass transportation systems are in place, and a larger proportion of the people have switched to fuel-efficient cars. These changes in supply and demand effectively transform the x-axis for the consumers. That is, for the same amount of gasoline, the consumer is operating closer to the leisure range, where the demand curve is flatter (see Figure 1-26). The effect of these shifts in the supply and demand curves promise lower prices, but it is difficult to determine whether or not a significant reduction in demand will occur from the qualitative model depicted in Figure 1-26 (see Problem 1).

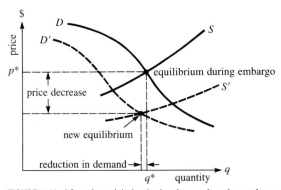

FIGURE 1-26 After the crisis both the demand and supply curves shift.

Finally, suppose the government is still dissatisfied with the level of demand and imposes a tax to reduce demand further. The supply curve shifts
upward by the amount of the tax as before (see Figure 1-27). Notice from
Figure 1-27 that since the demand curve is more horizontal than the supply
curve at the original equilibrium, the increase in price due to the shift in
the supply curve caused by the tax is small compared to the amount of the
tax. In essence, the oil industry suffers the burden of the tax. Moreover,
notice that the reduction in demand is more significant with this flatter
demand curve. Finally, the new equilibrium position is more easily obtained.

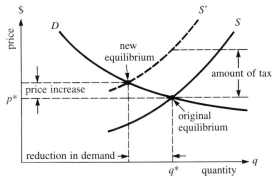

FIGURE 1-27 If the demand curve is relatively flat, the industry pays the larger portion of the
tax, and the reduction in demand is more significant.

1.4 PROBLEMS

1. Consider the graphical model in Figure 1-26. Argue that if the demand
   curve fails to shift significantly to the left, there could actually occur an
   increase in demand after the crisis.
2. Consider the situation in which demand is a fixed curve but there is an
   increase in supply so the supply curve shifts downward. Discuss how the
   slope of the demand curve affects the change in price and the change in
   quantity: How does the price change, and when does it change the most?
   When does it change least? Answer similar questions for the quantity.
3. Criticize the graphical model of the oil industry. Name some major factors
   that have been neglected. Which of the underlying assumptions are not
   satisfied by the crisis situation? Did the graphical model help you identify
   some of the key factors and their interactions? How could you adjust the
   model?

1.4 PROJECTS

For Projects 1 through 4, complete the requirements in the referenced UMAP
module and prepare a short summary for classroom discussion.

1. "Differentiation, Curve Sketching, and Cost Functions," by Christopher H. Nevison, UMAP 376. In this module costs and revenue for a firm are discussed using elementary calculus. The author discusses several of the economic ideas presented in this chapter.

2. "Price Discrimination and Consumer Surplus: An Application of Calculus to Economics," by Christopher H. Nevison, UMAP 294. The topics in the title are analyzed in a competitive market, and two-tier price discrimination is also discussed. The module discusses several of the economic ideas presented in this chapter.

3. "Economic Equilibrium: Simple Linear Models," by Philip M. Tuchinsky, UMAP 208. In this module linear supply and demand functions are constructed, and the equilibrium market position is analyzed for an industry producing one product. The result is then extended to n products. The author concludes by briefly considering nonlinear and discontinuous functions.

4. "I Will If You Will . . . A Critical Mass Model," by Jo Anne S. Growney, UMAP 539. A graphical model is presented to treat the problem of individual behavior in a group when the individual makes a choice dependent upon his or her perception of the behavior of fellow group members. The model can provide insight into paradoxical situations in which members of a group prefer one type of behavior but actually engage in the opposite behavior (like not cheating versus cheating in a class).

1.4 FURTHER READING

Asimakopulos, A. *An Introduction to Economic Theory: Microeconomics.* New York: Oxford University Press, 1978.

Cohen, Kalman J., and Richard M. Cyert. *Theory of the Firm.* Englewood Cliffs, N.J.: Prentice-Hall, 1975.

Henderson, Hubert D. *Supply and Demand.* Chicago: University of Chicago Press, 1958.

Thompson, Arthur A., Jr. *Economics of the Firm: Theory and Practice.* Englewood Cliffs, N.J.: Prentice-Hall, 1973.

THE MODELING PROCESS

INTRODUCTION

In Chapter 1 we presented graphical models representing a two-country arms race and the energy crisis. Let's now examine more closely the process of mathematical modeling.

To gain an understanding of the processes involved in mathematical modeling, consider the two "worlds" depicted in Figure 2-1. Suppose we want to understand some behavior or phenomenon in the real world. We may wish to make predictions about that behavior in the future and analyze the effects various situations have on that behavior. In the energy crisis, for instance, we were interested in predicting the effect of a surcharge tax on short- and long-term consumer demand for gasoline. As another example, when studying the populations of two interacting species, we may wish to know if the species can coexist within their environment or if one species will eventually dominate and drive the other to extinction. Or in the management of a fishery, it may be important to determine the optimal sustainable yield of a harvest and the sensitivity of the species to population fluctuation caused by

Real-World Systems	**Mathematical World**
Observed behavior or phenomenon	Models Mathematical operations and rules Mathematical conclusions

FIGURE 2-1 The real and mathematical worlds.

harvesting. How can we construct and use models in the mathematical world to help us better understand real-world systems? Before discussing how we link the two worlds together, let's consider what we mean by a real-world system and why we would be interested in constructing a mathematical model for a system in the first place.

system   A **system** is an assemblage of objects joined in some regular interaction or interdependence. The modeler is interested in understanding how a particular system works, what causes changes in the system, and the sensitivity of the system to certain changes. He or she is also interested in predicting what changes might occur and when they occur. How might such information be obtained? For instance, suppose the goal is to draw some conclusions about an observed phenomenon in the real world. One procedure would be to conduct some real-world trials or experiments and observe their effect on the real-world behavior. This is depicted in the left side of Figure 2-2. While such a procedure might minimize the loss in fidelity incurred by a less direct approach, there are many situations in which one would not want to follow such a course of action. For instance, there may be prohibitive costs for conducting even a single experiment, such as determining the level of concentration at which a drug proves to be fatal, or studying the radiation effects of a failure in a nuclear power plant near a major population area. Or we may not be willing to accept even a single experimental failure, such as when investigating different designs for a heat shield for a manned spacecraft. Moreover, it may not even be possible to produce a trial, as in the case of investigating some specific change in the composition of the ionosphere and its corresponding effect on the polar ice cap. Further, we may be interested in generalizing the conclusions beyond the specific conditions set by one trial (such as a cloudy day in New York with temperature 82 degrees, wind 15–20 miles per hour, humidity 42%, and so forth). Finally, even though we succeed in predicting the real-world behavior under some very specific conditions, we have not necessarily *explained* why the particular behavior occurred. (While the ability to predict and explain are often closely related, the ability to predict a behavior does not necessarily imply an understanding of it. In Part III of the book, we will study techniques specifically

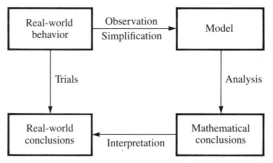

FIGURE 2-2 Reaching conclusions about the behavior of real-world systems.

designed to help us make predictions even though we cannot explain satisfactorily all aspects of the behavior.) The preceding discussion underscores the need to develop indirect methods for studying real-world systems.

An examination of Figure 2-2 suggests an alternative way of reaching conclusions about the real world. First, we make some specific observations about the behavior being studied and identify the factors that seem to be involved. Usually we cannot consider, or even identify, all the factors involved in the behavior, so we make simplifying assumptions that eliminate some factors. For instance, we may choose to neglect the humidity in New York City, at least initially, when studying radioactive effects from the failure of a nuclear power plant. Next, we conjecture tentative relationships among the factors we have selected, thereby creating a rough "model" of the behavior. Having constructed a model, we then apply appropriate mathematical analysis leading to conclusions about the model. Note that these conclusions pertain only to the model, not to the actual real-world system under investigation. Since we made some simplifications in constructing the model, and since the observations upon which the model is based invariably contain errors and limitations, we must carefully account for these anomalies before drawing any inferences about the real-world behavior. In summary we have the following "rough" modeling procedure:

1. Through observation, identify the primary factors involved in the real-world behavior, possibly making simplifications.
2. Conjecture tentative relationships among the factors.
3. Apply mathematical analysis to the resultant "model."
4. Interpret mathematical conclusions in terms of the real-world problem.

Figure 2-3 portrays the entire modeling process as a closed system. Given some real-world system, we gather sufficient data to formulate a model. Next we analyze the model and reach mathematical conclusions about it. Then we interpret the model and make predictions or offer explanations. Finally we test our conclusions about the real-world system against new observations and data. We may then find we need to go back and refine the model in order to improve its predictive or descriptive capabilities. Or perhaps we

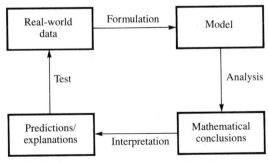

FIGURE 2-3 The modeling process as a closed system.

will discover that the model really doesn't "fit" the real world at all accurately, so we must formulate a new model. You will study the various components of this modeling process in detail throughout the book.

2.1 MATHEMATICAL MODELS

Until now we have been intentionally vague about the nature of the model itself. The models used to represent or approximate a real-world system can be very different in both appearance and purpose. For instance, one kind of model is a miniature replication of a real-world object of interest, like a model spacecraft that might be used to study certain design features under experimental conditions. Another kind of model is a mathematical model. In this book our main concern is with the latter type. For our purposes we define a **mathematical model** as a mathematical construct designed to study a particular real-world system or phenomenon. We include graphical, symbolic, simulation, and experimental constructs. An example of a graphical model is that of the energy crisis presented in Chapter 1. A symbolic model can be a formula, equation, or system of equations, describing how the underlying factors of the model are interrelated. Still another kind of mathematical model is a simulation model. An example of a simulation is a computer program that generates 1000 integers randomly from 1 to 6 as representative of 1000 tosses of a six-sided die. Another example is using a scaled-down model to simulate the drag force on a proposed design for a submarine.

mathematical model

Mathematical models can be differentiated further. There are mathematical models that already exist that can be identified with some particular real-world phenomenon and selected to study it. Then there are those mathematical models that we construct specifically to study a special phenomenon. Figure 2-4 depicts this differentiation between models. Starting from some real-world phenomenon, we can represent it mathematically by constructing a new model or selecting an existing model. On the other hand, we can replicate the phenomenon experimentally or with some kind of simulation.

When it comes to the question of constructing a mathematical model, a variety of conditions can cause us to abandon hope of achieving any success. The mathematics involved may be so complex and intractable that there is little hope of analyzing or solving the model, thereby defeating its utility.

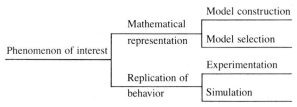

FIGURE 2-4 The nature of the model.

This complexity can occur when attempting to use a system of partial differential equations or a system of nonlinear algebraic equations, for instance. Or the problem may be so large (in terms of the number of factors involved) that it is impossible to capture all the necessary information in a single mathematical model. Predicting the global effects of the interactions of population, use of resources, and pollution is an example of such an impossible situation. In such cases we may attempt to replicate the behavior *directly* in some manner, or we might observe the behavior directly by conducting various experimental trials. Then we collect data from these trials and analyze the data in some way, possibly using statistical techniques or curve-fitting procedures. From the analysis, certain conclusions could be reached.

On the other hand, we may attempt to replicate the behavior *indirectly*. We might use an analogue device such as an electrical current to model a mechanical system. We might use a scaled-down model like a scaled model of a jet aircraft in a wind tunnel. Or we might attempt to replicate a behavior on a digital computer—for instance, simulating the global effects of the interactions of population, use of resources, and pollution, or simulating the operation of an elevator system during the morning rush hour.

The distinction between the various model types as depicted in Figure 2-4 is made solely for ease of discussion. For example, the distinction between experiments and simulations is based on whether the observations are obtained directly (experiments) or indirectly (simulations). In practical models this distinction is not nearly so sharp; one master model may employ several models as submodels, including selections from existing models, simulations, and experiments. Nevertheless, it is informative to contrast these types of models and compare their various capabilities for portraying the real world.

To that end, consider the following properties of a model:

Fidelity: The preciseness of a model's representation of reality
Cost: The total cost of the modeling process
Flexibility: The ability to change and control conditions affecting the
  model as required data is gathered

It is useful to know the degree to which a given model possesses each of these characteristics. However, since specific models vary greatly, even within the classes identified in Figure 2-4, the best we can hope for is a comparison in the ranges of performance between the classes of models for each of the characteristics. The comparisons are depicted in Figure 2-5, where the ordinate axis denotes the degree of effectiveness of each class.

Let's summarize the results shown in Figure 2-5. First, consider the characteristic of fidelity. One would expect observations made directly in the real world to demonstrate the greatest fidelity, even though some testing bias and measurement error may be present. The experimental models show the next greatest fidelity, since behavior is being observed directly in a more controlled environment, such as a laboratory. Since simulations are a step further from the real world, and because they introduce indirect observation (such as constructing and analyzing a scaled model), simulations suffer

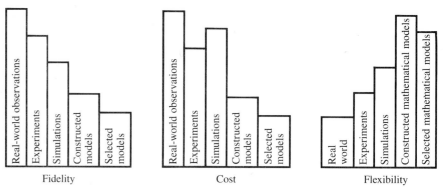

FIGURE 2-5 Comparisons between the model types.

from further loss in fidelity. Whenever a mathematical model is constructed, real-world conditions are simplified, resulting in more loss of fidelity. Finally, any selected model is based on additional simplifications that are not even tailored to the specific problem, and these simplifications imply still further loss in fidelity.

Next consider cost. Generally we would expect any selected mathematical model to be the least costly. Constructed mathematical models bear an additional cost of tailoring the simplifications to the phenomenon being studied. Experiments are usually expensive to set up and operate. Likewise, simulations employ indirect devices that are often expensive to develop, and simulations commonly involve large amounts of computer space and time for their operation.

Finally consider the characteristic of flexibility. Constructed mathematical models are generally the most flexible because different assumptions and conditions can be chosen relatively easily. Selected models are less flexible because they are developed under specific assumptions; nevertheless, specific conditions often can be varied over wide ranges. Simulations usually entail the development of some other indirect device in order to alter assumptions and conditions appreciably. Experiments are even less flexible, since some factors are very difficult to control beyond specific ranges. Observations of real-world behavior have little flexibility because the observer is limited to the specific conditions that pertain at the time of the observation. Moreover, other conditions might be highly improbable, or impossible, to create. It is important you understand that our discussion is only qualitative in nature, and that there are many exceptions to these generalizations.

The Construction of Models

In the preceding discussion we viewed modeling as a process and considered briefly the form of the model itself. Now let's focus attention on the construction of mathematical models. We begin by presenting an outline of a procedure that is helpful in constructing models. In the next section we will illustrate the various steps in the procedure by discussing several real-world examples.

Step 1. Identify the problem: What is it you would like to do or find out? Typically this is a very difficult step because people often have great difficulty in deciding what must be done. In real-life situations no one simply hands us a mathematical problem to solve. Usually we have to sort through large amounts of data and identify some particular aspect of the situation we wish to study. Moreover, we must be sufficiently precise (ultimately) in the formulation of the problem to allow for translation into mathematical symbology of the verbal statements describing it. This translation is accomplished through the next steps.

Step 2. Make assumptions: Generally you cannot hope to capture in a usable mathematical model all of the factors influencing the problem that has been identified. The task is simplified by reducing the number of factors under consideration. Then relationships between the remaining variables must be determined. Again, the complexity of the problem can be reduced by assuming relatively simple relationships. Thus the assumptions fall into two main categories:

a. Classification of the variables: What things influence the behavior you identified in Step 1? List these things as variables. The variables the model seeks to explain are the **dependent variables** and there may be several of these. The remaining variables are the **independent variables.** Each variable is classified as either dependent or independent, or you may choose to neglect it altogether. For example, in the nuclear arms race model the strength of the friendly missile force is a dependent variable that varies according to the number of enemy missiles, the enemy and friendly weapon technologies, and so forth. These latter variables are independent variables from the point of view of the friendly country.

dependent variables
independent variables

You may choose to neglect some of the independent variables for either of two reasons. First, the effect of the variable may be relatively small compared to other factors involved in the behavior. For example, in the nuclear arms race model we neglected such factors as the state of the economy, political considerations, and so forth, in order to reach a qualitative assessment of the behavior. You may also neglect a factor that affects the various alternatives in about the same way, even though it may have a very important influence on the behavior under investigation. For example, consider the problem of determining the optimal shape for a lecture hall where readability of a chalkboard or overhead projection is a dominant criterion. Lighting is certainly a crucial factor, but it would affect all possible shapes in about the same way. We can simplify the analysis considerably by neglecting such a variable, possibly incorporating it later in a separate, more refined model.

b. Determination of the interrelationships among the variables selected for study: Before you can hypothesize a relationship between the variables, you generally must make some additional simplifications. The problem may be

submodels

sufficiently complex so that you cannot see a relationship among all the variables initially. In such cases it may be possible to study **submodels.** That is, you study one or more of the independent variables separately. Eventually you will connect the submodels together. For instance, in the nuclear arms race model we hypothesized that the growth of the friendly arsenal depends on its current size, the size of the enemy arsenal, and the respective weapon technologies. We then proposed a submodel for the weapon technologies by assuming that the weapons could not be retargeted in flight. This assumption yielded a relatively simple submodel to incorporate into the master model. In later chapters you will study various techniques, such as proportionality, which will aid in hypothesizing relationships among the variables.

Step 3. Solve or interpret the model: Now put together all the submodels to see what the model is telling you. For example, the assumptions made in the nuclear arms race led to a graphical model that could be interpreted for various values of the independent variables and changes in the assumptions. In some cases the model may consist of mathematical equations or inequalities that must be solved in order to find out the information you are seeking. Often a problem statement requires a "best" or *optimal solution* to the model. Models of this type are discussed in Chapter 5.

While it was possible to construct a master model in the nuclear arms race model, often you will find that you are not quite ready to complete this step. Or you may end up with a model so unwieldy you cannot solve or interpret it. In such situations you might return to Step 2 and make additional simplifying assumptions. Sometimes you will even want to return to Step 1 to redefine the problem. This point will be amplified in the following discussion.

Step 4. Verify the model: Before you use the model, you must test it out. There are several questions you should ask before designing these tests and collecting data—a process that can be expensive and time consuming. First, does the model answer the problem you identified in Step 1, or did you stray from the key issue as you constructed the model? Second, is the model usable in a practical sense; that is, can you really gather the data necessary to operate the model? Third, does the model make common sense? In the nuclear arms race model, for instance, did we make a mathematical error in Step 3, or a faulty assumption in Step 2?

Once the commonsense tests are passed, you will want to test many models using actual data obtained from empirical observations. You need to be careful to design the test in such a way as to include observations over the *same range* of values of the various independent variables you expect to encounter when actually using the model. The assumptions you made in Step 2 may be reasonable over a restricted range of the independent variables, but very poor outside of those values. For instance, a frequently used interpretation of Newton's second law states that the net force acting on a body is equal to the mass of the body times its acceleration. This "law" is a reasonable "model" until the speed of the object approaches the speed of light.

Be very careful about the conclusions you draw from any tests. Just as you cannot prove a theorem simply by demonstrating many cases in which the theorem does hold, likewise, you cannot extrapolate broad generalizations from the particular evidence you gather about your model. A model does not become a law just because it is verified repeatedly in some specific instances. Rather you *corroborate the reasonableness* of your model through the data you collect.

Step 5. Implement the model: Of course your model is of no use just sitting in a filing cabinet. You will want to explain your model in terms that the decision makers and users can understand if it is ever to be of use to anyone. Further, unless the model is placed in a "user-friendly" mode, it will quickly fall into disuse. Expensive computer programs sometimes suffer such a demise. Often the inclusion of an additional step to facilitate the collection and input of the data necessary to operate the model determines its success or failure.

Step 6. Maintain the model: Remember that your model is derived from the specific problem you identified in Step 1 and from the assumptions you made in Step 2. Has the original problem changed in any way, or have some previously neglected factors become important? Does one of the submodels need to be adjusted? For instance, should we now consider the possibility of disarmament in the nuclear arms race model? Does the current economic situation dictate a cap on military expenditures? Is there a change in weapon technology that permits the retargeting of missiles in flight?

We summarize the steps for constructing mathematical models in Figure 2-6. We should not be too enamored with our work. Like any model, our procedure is an approximation process and therefore has its limitations. For example, the procedure seems to consist of discrete steps leading nicely to a usable result, but that's rarely the case in practice. Before offering an alternative procedure that emphasizes the iterative nature of the modeling process, let's discuss the advantages of the methodology depicted in Figure 2-6.

Step 1. Identify the problem.
Step 2. Make assumptions.
    a. Identify and classify the variables.
    b. Determine interrelationships between the variables and submodels.
Step 3. Solve the model.
Step 4. Verify the model.
    a. Does it address the problem?
    b. Does it make common sense?
    c. Test it with real-world data.
Step 5. Implement the model.
Step 6. Maintain the model.

FIGURE 2-6 The construction of a mathematical model.

The process shown in Figure 2-6 provides a methodology for progressively focusing on those aspects of the problem you wish to study. Furthermore, it demonstrates a rather curious blend of creativity with the scientific method used in the modeling process. The first two steps are more artistic or original in nature. They involve abstracting the essential features of the problem under study, neglecting any factors judged to be unimportant, and postulating relationships that are precise enough to help answer the questions posed by the problem yet simple enough to permit the completion of the remaining steps. While these steps admittedly involve a degree of craftsmanship, you will learn some scientific techniques you can apply to appraise the importance of a particular variable and the preciseness of an assumed relationship. Nevertheless, when generating numbers in Steps 3 and 4, you should remember that the process has been largely inexact and intuitive.

scientific method   Let's contrast the modeling process presented in Figure 2-6 with the scientific method. One version of the **scientific method** is as follows:

Step 1. Make some general observations of a phenomenon.
Step 2. Formulate a hypothesis about the phenomenon.
Step 3. Develop a method to test that hypothesis.
Step 4. Gather data to use in the test.
Step 5. Test the hypothesis using the data.
Step 6. Confirm or deny the hypothesis.

By design the mathematical modeling process and scientific method have some obvious similarities. For instance, both processes involve making assumptions or hypotheses, gathering real-world data, and testing or verification using that data. These similarities should not be surprising; while recognizing that part of the modeling process is an art, we do attempt to be scientific and objective whenever possible. Nevertheless, there are some subtle differences between the two procedures. One difference lies in the primary goal of the two processes. In the modeling process, assumptions are made in selecting which variables to include or neglect and in postulating the interrelationships among the included variables. The goal in the modeling process is to *hypothesize a model*, and like the scientific method, evidence is gathered to corroborate that model. However, unlike the scientific method, the objective is not to *confirm* or *deny* the model (we already know it is not precisely correct because of the simplifying assumptions we have made), but rather to test its *reasonableness*. We may decide that the model is quite satisfactory and useful, and elect to accept it. Or we may decide that the model needs to be refined or simplified. In extreme cases we may even need to redefine the problem, in a sense rejecting the model altogether. You will see in subsequent chapters that this decision process really constitutes the heart of mathematical modeling.

Figure 2-7 amplifies these ideas in viewing the modeling process, and attempts to display graphically its iterative nature. You begin by examining some system and identifying the particular behavior you wish to predict or explain. Next you identify the variables and simplifying assumptions, and

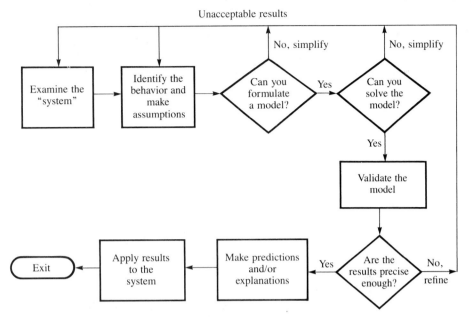

FIGURE 2-7 The iterative nature of model construction.

then you generate a model. You then attempt to validate the model with appropriate tests. If the results of the tests are satisfactory, you can use the model for its intended purpose. If the results are not satisfactory, there are several possibilities to pursue. You may decide the model needs to be refined by incorporating additional variables, or by restructuring a particular submodel. In some cases the test results may be so unsatisfactory that the original problem must be redefined because it turns out to be entirely too ambitious.

The process depicted in Figure 2-7 not only emphasizes the iterative nature of model construction, but also introduces the trade-offs between model simplification and model refinement. You will generally start with a rather simple model, progress through the modeling process, and then refine the model as the results of your validation procedures dictate. If you cannot come up with a model, or solve the one you have, you must *simplify*. You simplify a model by treating some variables as constants, by neglecting or aggregating some variables, by assuming simple relationships (such as linearity) in any submodels, or by restricting further the problem under investigation. On the other hand, if your results are not precise enough, you must *refine* the model. Refinement of a model is generally achieved in the opposite way: You introduce additional variables, assume more sophisticated relationships among the variables, or expand the scope of the problem. By trading off between simplification and refinement, you determine the generality, realism, and precision of your model. This trading-off process cannot be overemphasized and constitutes the "art of modeling."

robust

fragile
sensitivity

We complete this section by introducing several terms that are useful in describing models. A model is said to be **robust** when its conclusions do not depend on the precise satisfaction of the assumptions. On the other hand, a model is **fragile** if its conclusions do depend on the precise satisfaction of some set of conditions. The term **sensitivity** refers to the degree of change in a model's conclusions as some condition upon which they depend is varied; the greater the change, the more sensitive is the model to that condition.

2.1 PROBLEMS

In Problems 1 through 8, the scenarios are vaguely stated. From these, identify a problem you would like to study. What variables affect the behavior you have identified in the problem identification? Which variables are the most important? Remember, there are really no "right" answers.

1. The population growth of a single species.
2. A retail store intends to construct a new parking lot. How should the lot be illuminated?
3. A farmer wants to maximize the yield of a certain crop of food grown on his land. Has the farmer identified the correct problem? Discuss alternative objectives.
4. How would you design a lecture hall for a large class?
5. An object is to be dropped from a great height. How hard will it hit the ground?
6. How should a manufacturer of some product decide how many items of that product should be manufactured each year?
7. The United States Food and Drug Administration is interested in knowing if a new drug is effective in the control of a certain disease in the population.
8. How fast can a skier ski down a mountain slope?

For the scenarios presented in Problems 9 through 12, identify a problem worth studying and list the variables that affect the behavior you have identified. Which variables would be neglected completely? Which might be considered as constants initially? Can you identify any submodels you would want to study in detail? Identify any data you would want collected. (Problems 9 and 10 are from the 1980 *Nova* television program, "Living Machines.")

9. A botanist is interested in studying the shapes of leaves and the forces that mold them. He clips some leaves from the bottom of a white oak tree and finds the leaves to be rather broad, not very deeply indented. When he goes to the top of the tree, he gets very deeply indented leaves with hardly any broad expanse of blade.

10. Animals of different size work differently. The small ones talk in squeaky voices, their hearts beat faster, and they breathe more often than large animals. On the other hand, the skeleton of a larger animal is more robustly built than that of a small animal. The ratio of the diameter to the length in a larger animal is greater than it is in a smaller one. So there are regular distortions in the proportions of animals as the size increases from small to large.

11. A physicist is interested in studying properties of light. She wants to understand the path of a ray of light as it travels through the air into a smooth lake, particularly at the interface of the two different media.

12. A company with a fleet of trucks faces rising maintenance costs as the age and mileage of the trucks increase.

2.2 SOME ILLUSTRATIVE EXAMPLES

We now demonstrate the modeling process presented in the previous section with several illustrative examples. Special emphasis is placed here on identifying the problem and important variables. In subsequent chapters we will carry out the entire model-building process with these problems and use the techniques being studied there to suggest models that you can test.

Example 1 Vehicular Stopping Distance

Scenario   Consider the following rule of thumb often given in driver education classes:

> Allow one car length for every ten miles of speed under normal driving conditions, but more distance in adverse weather or road conditions. One way to accomplish this is to use the "Two-Second Rule" for measuring the correct following distance: If you stay two seconds behind the car in front, you have the correct distance no matter what your speed. To obtain that distance, watch the vehicle ahead of you pass some definite point on the highway, like a tar strip or overpass shadow. Then count to yourself "one thousand and one, one thousand and two"; that's two seconds. If you reach the mark before you finish saying those words, then you are following too close behind.

The preceding rule is implemented easily enough, but how good is it?

Problem identification   Our ultimate goal is to test this rule of thumb and suggest another rule if it fails. However, the statement of the problem "How good is the rule?" is rather vague. We need to be much more specific and spell out a problem, or ask a question, whose solution or answer will help us accomplish our goal while at the same time permitting a more exact mathematical analysis. Consider the following problem statement: *Predict the vehicle's total stopping distance as a function of its speed.*

Assumptions   We begin our analysis with a rather obvious model for total stopping distance:

total stopping distance = reaction distance + braking distance

By *reaction distance*, we mean the distance the vehicle travels from the instant the driver perceives a need to stop to the instant when the brakes are actually applied. *Braking distance* is the distance required for the brakes to bring the vehicle to a complete stop.

First let's develop a submodel for reaction distance. The reaction distance is a function of many variables, and we start by listing just two of these:

reaction distance = f(response time, speed)

We could continue developing the submodel in as much detail as we like. For instance, response time is influenced by both individual driving factors as well as by the vehicle operating system. *System time* is the time from which the driver touches the brake pedal until the brakes are mechanically applied. For modern cars we would probably neglect the influence of the system since it is quite small in comparison to the human factors. The portion of the response time determined by the driver depends on many things such as reflexes, alertness, visibility, and so forth. Since we are only developing a rule of thumb, we could just incorporate average values and conditions for these latter variables and so specify. Once all of the variables we deem important to the submodel have been identified, we can begin to determine interrelationships among them. We will suggest a submodel for reaction distance in Chapter 3.

Next consider the braking distance term. The weight and speed of the vehicle are certainly important factors to be taken into account. The efficiency of the brakes, the kind and condition of the tires, the road surface, and weather conditions are other legitimate factors. As before, we would most likely assume average values and conditions for these latter factors. Thus our initial submodel gives braking distance as a function of vehicular weight and speed:

braking distance = h(weight, speed)

In Chapter 3 we will also suggest and analyze a submodel for braking distance.

Finally, let's discuss briefly the last three steps in the modeling process for this problem. We would want to test our model against real-world data. Do the predictions afforded by the model agree with real driving situations? If not, we would want to assess some of our assumptions and perhaps restructure one (or both) of our submodels. If the model does predict real driving situations accurately, then does the rule of thumb stated in the opening discussion agree with the model? The answer to the latter question then gives an objective basis for answering "How good is the rule?" Whatever rule we come up with (in order to implement the model), it must be easily understood and easy to use if it is going to be effective. In this example, maintenance of the model does not seem to be a particular issue. Nevertheless, we would want to be sensitive to the effects on the model of such changes as power brakes or disc brakes, or a fundamental change in tire design, and so forth.

Example 2 Automobile Gasoline Mileage

Scenario During periods of concern when oil shortages and embargoes create an "energy crisis," there is always interest in how fuel economy varies with vehicular speed. We suspect that at very low speeds, when driving in low gears, automobiles convert power relatively inefficiently, and at very high speeds drag forces on the vehicle increase rapidly. It seems reasonable, then, to expect that automobiles have one or more speeds that yield optimum fuel mileage (the most miles per gallon of fuel). If this is so, fuel mileage would decrease beyond that optimum speed, but it would be beneficial to know just how this decrease takes place. Moreover, is the decrease significant? Consider the following excerpt from a newspaper article:

> Observe the 55 mile-an-hour national highway speed limit. For every five miles an hour over 50, there is a loss of one mile to the gallon. Insisting that drivers stay at the 55-mile-an-hour mark has cut fuel consumption 12 percent for Ryder Truck Lines of Jacksonville, Fla.—a savings of 631,000 gallons of fuel a year. The most fuel-efficient range for driving generally is considered to be between 35 and 45 miles an hour.*

Note especially the suggestion that there is a loss of 1 mile to the gallon for every 5 miles an hour over 50 mph. How good is this rule of thumb?

Problem identification *What is the relationship between the speed of a vehicle and its fuel mileage?* By answering that question, we can assess the accuracy of this rule of thumb.

Assumptions Let's consider the factors influencing fuel mileage. First, there are propulsion forces that drive the vehicle forward. These forces depend upon the power available from the type of fuel being burned, the engine's efficiency in converting that potential power, gear ratios, air temperature, and many other factors, including vehicular velocity. Next there are drag forces that tend to retard the vehicle's forward motion. The drag forces include frictional effects that depend upon the vehicle's weight, the type and condition of the tires, and the condition of the road surface. Air resistance is another drag force and depends upon the vehicular speed, vehicular surface area and shape, the wind, and air density. Another factor influencing fuel mileage relates to the driving habits of the driver. Does he or she drive at constant speeds, or is the driver constantly accelerating? Does he or she drive on level or mountainous terrain? Thus fuel mileage is a function of several factors, summarized in the following equation:

fuel mileage $= f$(propulsion forces, drag forces, driving habits, and so on)

* "Boost Fuel Economy," *Monterey Peninsula Herald*, May 16, 1982.

It is clear that the answer to the original problem will be quite detailed considering all the possible combinations of car types, drivers, and road conditions. Since such a study is far too ambitious to be undertaken here, we restrict the problem we are willing to address.

Restricted problem identification *For a particular individual driver, driving his or her car on a given day on a level highway at constant highway speeds near the optimum speed for fuel economy, provide a qualitative explanation of how fuel economy varies with small increases in speed.*

Under this restricted problem, environmental conditions, such as air temperature, air density, and road conditions, can be considered as constant. Since we have specified that the driver is driving his or her car, we have fixed the tire conditions, the shape and surface area of the vehicle, and fuel type. By restricting the highway driving speeds to be near the optimal speed, we obtain the simplifying assumptions of constant engine efficiency and constant gear ratio over small changes in vehicular velocity. Restricting a problem as originally posed is a very powerful technique for obtaining a manageable model. We will model the fuel economy problem under the stated restricted conditions in Chapter 3.

Example 3 The Assembly Line

Scenario There is increasing concern by both individual laborers and labor unions for both wages and job quality. On the other hand, corporations are feeling the competition from abroad. To increase domestic productivity, more assembly line operations have evolved. A question management faces is how to assign employees to the various jobs on the assembly line, realizing that the jobs are often quite tedious and repetitive, and demand various worker skill levels.

Problem identification The classical approach to making employee job assignments is to ensure maximum company profits. Typically, people are assigned to machines in such a way as to maximize the profit margin on the product being produced. In situations of fixed product demand, this approach amounts to minimizing the costs of production.

The difficulty with the classical approach is that it focuses on only one aspect of the problem—namely, short-term profits. In many situations the results of such assignment procedures include job dissatisfaction, excessive absenteeism, poor workmanship and product quality, and diminished worker efficiency and overall productivity. The net effect often is rising production costs coupled with reduced consumer demand. Thus while management attempts to maximize short-term profits, long-term profits actually suffer.

Although the maximization of profit is certainly a primary consideration, other factors, such as product quality and job satisfaction, must be taken into account. The precise reconciliation of the various factors into a problem statement depends on the particular situation. Let's assume that the firm has a contract and has the plant capacity to produce a fixed number of items per month. We then define the problem as follows: *Minimize production cost while meeting specified levels of demand, quality control, and job satisfaction.*

Assumptions Under this problem identification, production cost is a function of individual wages, each individual's productivity on each job, the expected number of defects each individual makes on each job per unit time, and the reduction in both quality and productivity as a function of the time an individual spends at a particular job assignment.

We will construct a model for the assembly line problem in Chapter 5. For now, let's ask whether or not we can obtain the needed data. Certainly the hourly wage presents no problem. In many instances we can build a history of an individual's productivity and workmanship by measuring the hourly output on each machine as well as the number of defects that are produced. The reduction in efficiency incurred by assigning an individual the same job for a prolonged period may be difficult to measure. Hopefully, after observing and interviewing several employees, we can establish guidelines to prevent excessive diminution in productivity while simultaneously maintaining job satisfaction. For example, we may determine that an individual should not be assigned to a particular tedious job for more than two hours each day.

The advantage of this approach is that it forces the modeler to consider many more aspects of the problem. *The mathematical modeler must guard against the tendency to ignore factors just because they are difficult to quantify.* In the preceding problem there may also be salutary effects from involving management in the data collection process. A lesson to be learned from this example is that a careful modeler considers all pertinent factors and then attempts to determine the sensitivity of the model to the various assumptions made.

Example 4 Morning Rush Hour

The next model stresses the importance of data collection.

Scenario You have been hired as a consultant by the manager of several skyscrapers used for offices in New York City. His or her clientele are particularly concerned with the poor elevator service they receive during the morning rush hour. They complain of long waits in the lobby and excessive stops by the elevator to discharge a few passengers. The manager is unwilling to construct additional elevators, being convinced the problem can be resolved through better scheduling.

Problem identification A number of formulations of the elevator problem are worth considering. Many people are very sensitive to the amount of time they spend waiting in queues, which suggests scheduling the elevators in such a way to minimize the *average waiting time* of a customer. Even then we need to be careful: drastic reduction of the service provided to floors used by relatively few customers could reduce the average waiting time yet significantly increase the *longest waiting time*. A similar argument applies to the amount of time spent in an elevator. If the vast majority of the people in an elevator require delivery to the top floor, the average travel time could be reduced by going to the top floor first and then delivering the remaining customers on the way back down. However, the longest delivery time might increase considerably. Moreover, consider the psychological effect on the customer who wants only to go to the second floor!

In this kind of problem, it is important to be sensitive to people's perceptions about the service being provided. How do the manager and customers measure a successful operation? It may well be based on the *length of the longest queue* or on the *number of complaints* received from dissatisfied customers. The point is that each of these criteria leads to a distinct mathematical model with generally different optimal solutions. Unless we consider such possibilities, we may succeed in formulating and solving a difficult mathematical problem yet fail to improve the manager's situation.

While any of the problem formulations might be appropriate in a given situation, let's consider the total delivery time for our problem identification: *Minimize the average total elevator delivery time of a morning's rush hour customers as measured by the difference between individual arrival time at the lobby and arrival time at the floor of destination.*

This formulation takes into account both the time spent in the lobby and the time spent in the elevator. A weighted combination of these two times might be more appropriate. While we have identified a particular problem to be solved here, the alternative problems mentioned earlier should make us sensitive to other criteria for formulating a problem, such as minimizing the longest total delivery time or the longest queue.

Assumptions Factors influencing the total delivery time include the building layout (location of offices, warehouses, factories, and the like), the distribution of starting times of the various agencies by the number of customers and also by the number of floors, and commuter schedules. In order to determine interrelationships among the variables, we need to gather additional information about them. We may determine that different agencies have distinct starting times, so the present poor service is caused by a few early arrivals or a few stragglers. In such cases we could use express elevators for high density floors to give them priority during their peak arrival periods. In other cases we might find that the arrival times are largely driven by commuter schedules, and we would want to investigate the predictability of the appropriate schedules. It may be determined that the distribution of arrivals and destinations approximate well-studied distribution functions. Then we could use properties of those distributions to estimate the effects of various schemes for routing the elevators on the average total delivery time.

Taking another approach, we may find it desirable to simulate the arrivals via a computer, and then test various schemes accordingly. Among the schemes worth considering are:

1. Assigning elevators to the even and odd floors.
2. Splitting the building into two or more groups of contiguous floors and assigning a different set of elevators to each group of floors.
3. Reserving express elevators for high density floors during peak periods.
4. Using express elevators for specified floors and local elevators in between those floors.

In Chapter 8 we present an algorithm simulating the elevator scheduling problem as our model.

Model verification How can we verify the solution to our model once it is obtained? One way is to sample customers during the rush hour and measure their total delivery time. Before and after implementing the solution to our model, we might be interested in gathering such statistics as average and longest time spent in the lobby, average and longest time in the elevator, and the length of the longest queue. However, the manager's clients may object to the harassment of such an approach. Then we would be forced to use more easily obtained measures such as queue length, running time versus stopped time of the elevators, and so forth. A word of caution is in order here. When using indirect measures such as those just mentioned, you may have a tendency to experiment with the solution to improve the indirect measure. But the improvement of an indirect measure may not result in a more satisfied customer. For example, you might work to minimize the longest queue, if that is the measure of effectiveness. As a result, customers may find themselves spending less time in lobbies, but more time in elevators, and they may still feel dissatisfied with the service.

2.2 PROBLEMS

1. Consider the assumptions in the example on vehicular stopping distance. Name some variables other than the two listed in the text that also influence the reaction distance. Once a model is constructed for this problem, how would you go about validating it? What kind of data would you collect? Is it possible to obtain the data? How would you use the model, once it is validated to your satisfaction, to test the "rule of thumb" stated in the scenario for the example?
2. In the automobile gasoline mileage example, discuss the difficulties associated with constructing a model for the original problem posed: "What is the relationship between the speed of a vehicle and its fuel mileage?" Discuss the relationship between the automobile gasoline mileage problem and the energy crisis model presented in Chapter 1.

3. Consider a new company that is just getting started in producing a single product in a competitive market situation. Discuss some of the short- and long-term goals the company might have as it enters into business. How do these goals affect employee job assignments? Would the company necessarily decide to maximize profits in the short run?

4. Discuss the differences between using a model to predict, versus to explain, a real-world system. Think of some situations in which you would like to explain a system; likewise, imagine others in which you would want to predict a system. Relate this question to the examples presented in the text.

5. Suppose the manager of the skyscraper in the elevator problem asks you (the consultant), "How long does it take to get to the top?" What questions would you begin to ask to formulate a problem for purposes of constructing a model?

   a. If the elevator is an express to the top, how might a graph of its speed versus time appear? Assume a constant acceleration and deceleration of the elevator of α ft per sec per sec.

   b. How long will it take for the elevator to reach its full speed of v ft per sec?

   c. What is the total distance traveled during both acceleration and deceleration?

   d. If D is the distance in feet to the top, answer the manager's question.

2.2 PROJECTS

1. Consider the taste of brewed coffee. What are some of the variables affecting taste? Which variables might be neglected initially? Suppose you hold all variables fixed except water temperature. Most coffeepots use boiled water in some manner to extract the flavor from the ground coffee. Do you think boiled water is optimal for producing the best flavor? How would you test this submodel? What data would you collect and how would you gather it?

2. A transportation company is considering transporting people between skyscrapers in New York City via helicopter. You are hired as a consultant to determine the number of helicopters needed. Identify an appropriate problem precisely. Use the model-building process to identify the data you would like to have in order to determine the relationships between the variables you select. You may want to consider redefining your problem as you proceed.

3. Consider wine making. Suggest some objectives a commercial producer might have. Consider taste as a submodel. What are some of the variables affecting taste? Which variables might be neglected initially? How would you relate the remaining variables? What data would be useful to determine the relationships?

4. Should a couple buy or rent a home? As the cost of a mortgage rises, intuitively it would seem that there is a point where it no longer "pays" to buy a house. What variables determine the *total cost* of a mortgage?

5. Consider the operation of a medical office. Records have to be kept on individual patients, and accounting procedures are a daily task. Should the office buy or lease a small computer system? Suggest objectives that might be considered. What variables would you consider? How would you relate the variables? What data would you like to have in order to determine the relationships between the variables you select? Why might solutions to this problem differ from office to office?

6. When should a person replace his or her vehicle? What factors should affect the decision? Which variables might be neglected initially? Identify the data you would like to have to determine the relationships among the variables you select.

7. How far can a person long jump? In the 1968 Olympic Games in Mexico City Bob Beamon of the United States increased the record by a remarkable 10%. List the variables that affect the length of the jump. Do you think the low air density of Mexico City accounts for the 10% difference?

8. Is college a *financially* sound investment? Income is forfeited for four years, and the cost of college is extremely high. What factors determine the total cost of a college education? How would you determine the circumstances necessary for the investment to be profitable?

MODELING USING PROPORTIONALITY

INTRODUCTION

In the previous chapter we presented an overview of the modeling process. Using several specific examples, we carefully worked through the initial steps of that process by identifying the problem and the variables influencing the particular behavior or phenomenon under investigation. In some instances we identified data that would be needed to help determine the relationship among the selected variables. However, we did not delve into discovering the nature of the relationship itself. In this chapter you will learn a technique of proportionality that helps to suggest the form of the relationship among the variables you have selected. The technique is also useful for testing the reasonableness of the assumptions underlying the model. In Chapter 4 you will see how proportionality can be used in fitting a model to data that has been collected.

3.1 PROPORTIONALITY AND GEOMETRIC SIMILARITY

proportional Two positive quantities x and y are said to be **proportional** (to one another) if one quantity is a constant positive multiple of the other; that is, if

$$y = kx$$

for some positive constant k. We write $y \propto x$ in that situation, and say that y "is proportional to" x. Thus,

$$y \propto x \quad \text{if and only if} \quad y = kx \quad \text{for some constant} \quad k > 0 \quad \text{(3.1)}$$

Of course, if $y \propto x$, then $x \propto y$ also, because the constant k in Equation (3.1) cannot be zero and $x = (1/k)y$. Other examples of proportionality relationships include the following:

$$y \propto x^2 \quad \text{if and only if} \quad y = k_1 x^2 \quad \text{for } k_1 \text{ a constant} \qquad \textbf{(3.2)}$$

$$y \propto \ln x \quad \text{if and only if} \quad y = k_2 \ln x \quad \text{for } k_2 \text{ a constant} \qquad \textbf{(3.3)}$$

$$y \propto e^x \quad \text{if and only if} \quad y = k_3 e^x \quad \text{for } k_3 \text{ a constant} \qquad \textbf{(3.4)}$$

Now let's explore a geometric interpretation of proportionality. In Model (3.1), $y = kx$ yields $k = y/x$. Thus k may be interpreted as the tangent of the angle θ depicted in Figure 3-1, and the relation $y \propto x$ defines a set of points along a line in the plane with angle of inclination θ.

Comparing the general form of a proportionality relationship $y = kx$ with the equation for a straight line $y = mx + b$, you can see that the graph of a proportionality relationship is a line (possibly extended) passing through the origin. If you plot the proportionality variables for the Models (3.2), (3.3), and (3.4), you obtain the straight-line graphs presented in Figure 3-2.

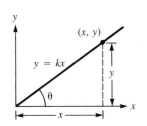

FIGURE 3-1 Geometrical interpretation of $y \propto x$.

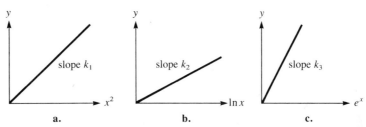

a.   **b.**   **c.**

FIGURE 3-2 Geometrical interpretation of Models (a) (3.2), (b) (3.3), and (c) (3.4).

It is important to note that just any straight line does not represent a proportionality relationship: the y-intercept *must* be zero so that the line passes through the origin. Failure to recognize this point can lead to erroneous results when using your model. For example, suppose you are interested in predicting the volume of water displaced by a boat as it is loaded with cargo. Since a floating object displaces a volume of water equal to its weight, you might be tempted to assume that the total volume y of displaced water is proportional to the weight x of the added cargo. However, there is a difficulty with that assumption because the unloaded boat already displaces a volume of water equal to its weight. Although the graph of total volume of displaced water versus weight of added cargo is given by a straight line, it is not given by a line passing through the origin (see Figure 3-3), so the proportionality argument is incorrect.

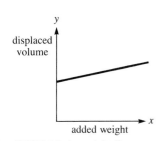

FIGURE 3-3 A straight-line relationship exists between displaced volume and total weight, but it is not a proportionality because the line fails to pass through the origin.

However, a proportionality relationship may be a reasonable simplifying assumption, depending on the size of the y-intercept and the slope of the line. The domain of the independent variable also can be significant since the relative error

$$\frac{y_a - y_p}{y_a}$$

is greater for smaller values of x. These features are depicted in Figure 3-4. If the slope is nearly zero, proportionality may be a poor assumption because the initial displacement dwarfs the effect of the added weight. For example, there would be virtually no effect in placing 400 lb on an aircraft carrier already weighing many tons. On the other hand, if the initial displacement is relatively small and the slope is large, the effect of the initial displacement is dwarfed quickly, and proportionality will be a good simplifying assumption.

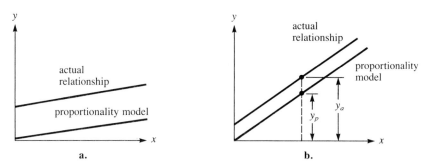

FIGURE 3-4 Proportionality as a simplifying assumption.

Geometric similarity is a concept related to proportionality and can be useful to simplify the mathematical modeling process. The definition is as **geometrically similar** follows: Two objects are said to be **geometrically similar** if there is a 1–1 correspondence between points of the objects such that the ratio of distances between corresponding points is constant for all possible pairs of points.

For example, consider the two boxes depicted in Figure 3-5. Let l denote the distance between the points A and B in Figure 3-5a, and l' the distance between the corresponding points A' and B' in Figure 3-5b. Other corresponding points in the two figures, and the associated distances between the points, are marked in the same way. In order that the boxes be geo-

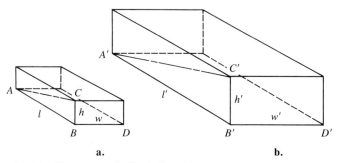

FIGURE 3-5 Two geometrically similar objects.

metrically similar, it must be true that

$$\frac{l}{l'} = \frac{w}{w'} = \frac{h}{h'} = k \quad \text{for some constant } k > 0$$

Let's interpret the last result geometrically. In Figure 3-5, consider the triangles ABC and $A'B'C'$. If the two boxes are geometrically similar, so must these two triangles be similar. This same argument can be applied to any corresponding pair of triangles, such as CBD and $C'B'D'$. Thus, *corresponding angles are equal for objects that are geometrically similar.* In other words, the shape is the same for two geometrically similar objects, and one object is simply an enlarged copy of the other. You may think of geometrically similar objects as scaled replicas of one another, as in an architectural drawing where all the dimensions are simply scaled by some constant factor k.

One advantage that results when two objects are geometrically similar is a simplification in certain computations, such as volume and surface area. For the boxes depicted in Figure 3-5, consider the following argument for the ratio of their volumes V and V':

$$\frac{V}{V'} = \frac{lwh}{l'w'h'} = k^3 \tag{3.5}$$

Similarly, the ratio of their total surface areas S and S' is given by:

$$\frac{S}{S'} = \frac{2lh + 2wh + 2wl}{2l'h' + 2w'h' + 2w'l'} = k^2 \tag{3.6}$$

characteristic dimension

Not only are these ratios immediately known once the scaling factor k has been specified, but the surface area and volume may be expressed as a proportionality in terms of some selected **characteristic dimension.** For example, since $l/l' = k$, we have

$$\frac{S}{S'} = k^2 = \frac{l^2}{l'^2}$$

Therefore,

$$\frac{S}{l^2} = \frac{S'}{l'^2} = \text{constant}$$

holds for any two geometrically similar objects. That is, surface area is always proportional to the square of the characteristic dimension length:

$$S \propto l^2$$

Likewise, volume is proportional to the length cubed:

$$V \propto l^3$$

Thus, if you are interested in some function depending on an object's length, surface area, and volume, for example,

$$y = f(l, S, V)$$

you could express all the function arguments in terms of some selected characteristic dimension, such as length, giving

$$y = g(l, l^2, l^3)$$

To illustrate these ideas, suppose you are interested in the terminal velocity of a raindrop falling from a motionless cloud. Assume that the atmospheric drag on the raindrop is proportional to its surface area S times the square of its speed v. The mass m of the raindrop is proportional to the volume (assuming constant density) so that

$$S \propto l^2 \quad \text{and} \quad m \propto V \propto l^3$$

implies that

$$S \propto m^{2/3}$$

Now, the downward force on the raindrop is mg, and the force of the air resistance is kSv^2 (where g is the constant acceleration due to gravity and k is some proportionality constant). Then Newton's second law gives

$$\frac{m \, dv}{dt} = mg - kSv^2$$

When the raindrop reaches its terminal velocity v_T, dv/dt is zero for all remaining time, yielding

$$mg - kSv_T^2 = 0$$

or

$$v_T^2 = \frac{mg}{kS} \propto m^{1/3}$$

Therefore, we conclude that the terminal velocity v_T of the raindrop is proportional to $m^{1/6}$.

In the next section we will apply these concepts of proportionality and geometric similarity to the modeling process.

3.1 PROBLEMS

1. Explain graphically the proportionality $y \propto u/v$.
2. Verify Equations (3.5) and (3.6) in the text.
3. If a spring is stretched 0.37 in. by a 14-lb force, what stretch will be produced by a 9-lb force? by a 22-lb force? Assume Hooke's law, which asserts that the distance stretched is proportional to the force applied.
4. If an architectural drawing is scaled so that 0.75 in. represents 4 ft, what length represents 27 ft?

5. Suppose two maps of the same country on different scales are drawn on tracing paper and superimposed, with possibly one of the maps being turned over before being superimposed on the other. Show there is just one place that is represented by the same spot on both maps.

6. For a falling object subject to an atmospheric drag proportional to Sv^2, show that the kinetic energy per unit area that must be converted into other energy when the object hits the ground is proportional to $m^{2/3}$. Assume the object reaches its terminal velocity.

7. The following table displays the winning lifts for weight lifting at the 1976 Montreal Olympic Games.

Bodyweight class		Total winning lifts (lb)		
	Max. weight (lb)	Snatch	Jerk	Total weight
Flyweight	114.5	231.5	303.1	534.6
Bantamweight	123.5	259.0	319.7	578.7
Featherweight	132.5	275.6	352.7	628.3
Lightweight	149.0	297.6	380.3	677.9
Middleweight	165.5	319.7	418.9	738.5
Light-heavyweight	182.0	358.3	446.4	804.7
Middle-heavyweight	198.5	374.8	468.5	843.3
Heavyweight	242.5	385.8	496.0	881.8

a. Assume that the lift L is proportional to the cross-sectional area A of the weight lifter's muscle. Find a proportionality relationship between L and body weight W.

b. Graphically compare your result with the data in the table by plotting ln L versus ln W.

c. Formulate a rule for handicapping a lift that compensates for the various body weights so that an "overall" winner can be selected for the Montreal Games.

8. Lumbercutters wish to use a readily available measurement to estimate the number of board feet of lumber in a tree. Let's assume they measure the diameter of the tree in inches at waist height. Initially assume that the trees are all the same height. Develop a model that predicts board feet as a function of diameter in inches, assuming that the cross-sectional areas of the trees are geometrically similar. Develop a second model that assumes that the trees are geometrically similar—that is, the height is proportional to the diameter. Test both models against the following data. Which is the better model?

x	17	19	20	23	25	28	32	38	39	41
y	19	25	32	57	71	113	123	252	259	294

The variable x is the diameter of a ponderosa pine in inches and y is the number of board feet divided by 10.

9. Are the hearts of mammals geometrically similar? Use the following data to support or refute your argument.

Animal	Heart weight (g)	Length of cavity of left ventricle
Mouse	0.13	0.55
Rat	0.64	1.0
Rabbit	5.8	2.2
Dog	102	4.0
Sheep	210	6.5
Ox	2030	12.0
Horse	3900	16.0

Data from A. J. Clark, *Comparative Physiology of the Heart* (New York: Macmillan, 1927), p. 84.

10. Warm-blooded animals use large quantities of energy to maintain body temperature because of heat loss through the body surface. In fact, biologists believe that the primary energy drain on a *resting* warm-blooded animal is the maintenance of body temperature.

Animal	Body weight (g)	Pulse rate
Mammals		
Vesperugo pipistrellus	4	660
Mouse	25	670
Rat	200	420
Guinea pig	300	300
Rabbit	2,000	205
Hare	3,000	64
Dog (a)	5,000	120
Dog (b)	30,000	85
Sheep	50,000	70
Man	70,000	72
Horse	450,000	38
Ox	500,000	40
Elephant	3,000,000	48
Birds		
Canary	20	1000
Pigeon	300	185
Crow	341	378
Buzzard	658	300
Hen	2,000	312
Domestic duck	2,300	240
Wild duck	1,100	190
Turkey	8,750	193
Ostrich	71,000	60–70

Data from A. J. Clark, *Comparative Physiology of the Heart* (New York: Macmillan, 1927), p. 99.

a. Construct a model relating blood flow through the heart to body weight. Assume that the amount of energy available is proportional to the blood flow through the lungs, which is the source of oxygen. Assuming the least amount of blood needed is circulated, the amount of available energy will equal the amount used.

b. The tabled data (p. 56) relate the weights of some mammals to their heart rate in beats per minute. Construct a model that relates heart rate to body weight. Discuss the assumptions of your model. Use the data to check your modeling. Consider the mammals, the small birds (< 1000 g), and the large birds (> 1000 g) on separate graphs.

MODEL

3.2 A BASS FISHING DERBY

Consider a sport fishing club that for conservation purposes wishes to encourage its membership to release their fish immediately after catching them. On the other hand, the club wishes to make awards based on the total weight of fish that are caught: honorary membership in "The 1000 Pound Club," "Greatest Total Weight Caught During a Derby Award," and so forth. But how does someone fishing determine the weight of a fish he or she has caught? You might suggest that each individual carry a small portable scale. However, portable weight scales tend to be inconvenient and inaccurate, especially for smaller fish.

Problem identification   Thus we can identify a problem as follows: *Predict the weight of a fish in terms of some easily measurable dimensions.*

Assumptions   Many factors that affect the weight of a fish can easily be identified. Different species have different shapes and different average weights per unit volume (weight density) based on the proportions and densities of meat, bone, and so forth. Gender also plays an important role, especially during the spawning season. The various seasons themselves probably have a considerable effect on weight.

Since a general rule for sport fishing is sought, let's initially restrict attention to a single species of fish, say bass, and assume that within the species the average weight density is constant. Later on, it may be desirable to refine our model if the results prove unsatisfactory or if it is determined that considerable variability in density does exist. Furthermore, let's also neglect gender and season. Thus, initially, we will predict weight as a function of size (volume) and constant average weight density.

Assuming that all bass are geometrically similar, the volume of any single bass is proportional to the cube of some characteristic dimension. Note that we are not assuming any particular shape, but only that the bass are "scaled models" of one another. The basic shape can be quite irregular so long as the ratio between corresponding pairs of points in two distinct bass remains constant for all possible pairs of points. The idea is illustrated in Figure 3-6.

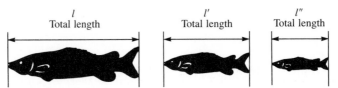

FIGURE 3-6 Fish that are geometrically similar are simply scaled models of one another.

Now choose the length l of the fish as the characteristic dimension. This choice is depicted in Figure 3-6. Thus, the volume of a bass satisfies the proportionality

$$V \propto l^3$$

Since weight W is volume times average density and a constant average density is being assumed, it follows immediately that

$$W \propto l^3$$

Model verification   Let's test our model. Consider the following data collected during a fishing derby*:

Length, l(in.)	14.5	12.5	17.25	14.5	12.625	17.75	14.125	12.625
Weight, W(oz)	27	17	41	26	17	49	23	16

If our model is correct, then the graph of W versus l^3 should be a straight line passing through the origin. The graph showing an approximating straight line is presented in Figure 3-7. (Note that the judgment here is qualitative. In Chapter 4 we develop analytic methods to determine a "best fitting" model for collected data.)

Let's accept the model, at least for further testing, based on the small amount of data presented so far. Since the data point $(14.5^3, 26)$ happens

* Data collected by CPT Murray Williams, West Point Fishing Club.

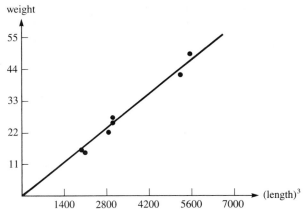

FIGURE 3-7 If the model is valid, the graph of W versus l^3 should be a straight line passing through the origin.

to lie along the line we have drawn in Figure 3-7, we can estimate the slope of the line as $26/3049 = 0.00853$ yielding the model

$$W = 0.00853 \, l^3 \qquad (3.7)$$

Of course if we had drawn our line a little differently, we would have obtained a slightly different slope. In Chapter 4 you will be asked to show analytically that the coefficient that minimizes the sum of the squared deviations between the model $W = Kl^3$ and the given data points is $K = 0.008437$. A graph of Model (3.7) is presented in Figure 3-8, showing also a plot of the original data points.

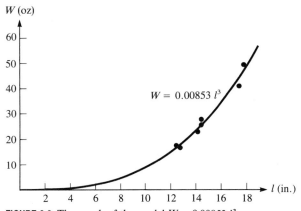

FIGURE 3-8 The graph of the model $W = 0.00853 \, l^3$.

Model (3.7) provides a convenient rule of thumb. For example, from Figure 3-8 we might estimate that a 12-in. bass weighs approximately 1 lb. This means that an 18-in. bass should weigh approximately $(1.5)^3 = 3.4$ lb and a 24-in. bass approximately $2^3 = 8$ lb. For the fishing derby, a card converting the length of a caught bass to its weight in ounces or pounds could be given to each fisher, or a cloth tape or a retractable metal tape could be marked with the conversion scales if the use of the rule became popular enough. A conversion scale for Model (3.7) is as follows:

Length (in.)	12	13	14	15	16	17	18	19	20	21	22	23	24	25	26
Weight (oz)	15	19	23	29	35	42	50	59	68	79	91	104	118	133	150
Weight (lb)	0.9	1.2	1.5	1.8	2.2	2.6	3.1	3.7	4.3	4.9	5.7	6.5	7.4	8.3	9.4

Even though our rule seems reasonable based on the limited data we have obtained, a fisher may not like the rule because it doesn't reward the catching of a "fat" fish: the model treats fat and skinny fish alike. Let's address this dissatisfaction. Instead of assuming that the fish are geometrically similar, assume that only their cross-sectional areas are similar. This does not imply any particular shape for the cross section, only that the definition of geometric similarity is satisfied. For example, consider the ellipses depicted in Figure 3-9. From the definition of geometric similarity, if the ellipses are geometrically similar, then the ratio of the distance $\overline{P_1P_2}$ and the corresponding distance $\overline{P_1'P_2'}$ must be the same as the ratio between $\overline{P_3P_4}$ and its corresponding distance $\overline{P_3'P_4'}$. That is,

$$\overline{P_1P_2}:\overline{P_1'P_2'} = \overline{P_3P_4}:\overline{P_3'P_4'}$$

or

$$\frac{a_1}{a_2} = \frac{b_1}{b_2} \tag{3.8}$$

shape factor

Now a characteristic dimension such as d in Figure 3-9, and a factor such as $r = a/b$, which is the ratio of the major axis to the minor axis, can be used to describe any two-dimensional ellipse. This factor is called a **shape factor.** Shape factors are very important in similitude as you shall see in

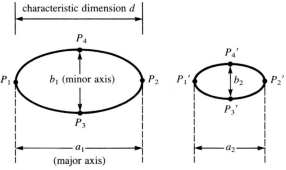

FIGURE 3-9 An ellipse is completely described by the shape factor $r = a/b$ and a convenient characteristic dimension d. Two geometrically similar ellipses have the same shape factor and are "scaled" by the ratio of their characteristic dimensions.

Section 7.5. From Equation (3.8), $a_1/b_1 = a_2/b_2$ so that geometrically similar ellipses have identical shape factors. Thus, once the shape factor for a given ellipse is known, all geometrically similar ellipses can be described by giving the characteristic dimension d. Likewise, once the shape of the cross section of a bass is described, all bass with geometrically similar cross sections are described by giving the chosen characteristic dimension because they have the same shape factors. We assume that the cross sections of the bass are geometrically similar and choose the characteristic dimension l shown in Figure 3-6.

Now assume that the major portion of the weight of the fish is due to the main body. Thus the head and tail contribute relatively little to the total weight. Constant terms can be added later if our model proves worthy of refinement. Next assume that the main body is of varying cross-sectional area. Then the volume can be found by multiplying the average cross-sectional area A_{avg} by the "effective length" l_{eff}:

$$V \approx l_{\text{eff}}(A_{\text{avg}})$$

But how shall the effective length l_{eff} and average cross-sectional area A_{avg} be measured? Have the fishing contestants measure the length of the fish l as before, and assume the proportionality $l_{\text{eff}} \propto l$. To estimate the average cross-sectional area, have each fisher take a cloth measuring-tape and measure the circumference of the fish at its widest point. Call this measurement the *girth*, g. Assume the average cross-sectional area is proportional to the square of the girth. Combining these two proportionality assumptions gives

$$V \propto lg^2$$

Finally, assuming constant density, $W \propto V$ as before so that

$$W = klg^2 \tag{3.9}$$

for some positive constant k.

A number of assumptions have been made, so let's get an initial test of our model. Consider again the following data.

Length, l (in.)	14.5	12.5	17.25	14.5	12.625	17.75	14.125	12.625
Girth, g (in.)	9.75	8.375	11.0	9.75	8.5	12.5	9.0	8.5
Weight, W (oz)	27	17	41	26	17	49	23	16

Since our model suggests a proportionality exists between W and lg^2, we consider a plot of W versus lg^2. This plot is depicted in Figure 3-10. Since the plotted data lies approximately along a straight line passing through the origin, the proportionality assumption seems reasonable. Now the point corresponding to the 41-oz fish happens to lie along the line shown in Figure 3-10 so the slope can be estimated as

$$\frac{41}{(17.25)(11)^2} \approx 0.0196$$

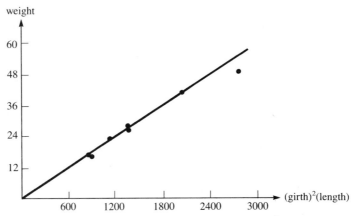

FIGURE 3-10 Testing the proportionality between W and lg^2.

This computation leads to the model

$$W = 0.0196 \, lg^2 \qquad \textbf{(3.10)}$$

In Chapter 4 you will show analytically that choosing the slope in such a way that the sum of the squared deviations from the given data points is minimized leads to the model

$$W = 0.0187 \, lg^2$$

A fisher would probably be happier with the new rule (3.10), since doubling the girth leads to a fourfold increase in the weight of the fish. However, the model appears more inconvenient to apply. Since $1/0.0196 \approx 50.9$, we could round this coefficient to 50 and have a fisher apply one of the simpler rules:

$$W = \frac{lg^2}{50} \qquad \text{for } W \text{ in ounces and } l, g \text{ measured in inches}$$

$$W = \frac{lg^2}{800} \qquad \text{for } W \text{ in pounds and } l, g \text{ measured in inches}$$

However, the application of either of the preceding rules would probably require the fisher to record the length and girth of each fish and then compute its weight on a four-function calculator. Or perhaps he or she could be given a two-dimensional card showing the correct weights for different values of length and girth. Another possibility could be to design a "slide rule" device, as we shall now explain. Since

$$W = \frac{l}{50} (g^2) \quad \text{(approximately)}$$

it follows that

$$\ln W = \ln \frac{l}{50} + 2 \ln g$$

Thus W can be computed by adding together two logarithms and then taking the inverse logarithm. This idea is the same as that underlying the slide rule. The design and use of a slide rule device to compute Model (3.10) is shown and explained in Figure 3-11.

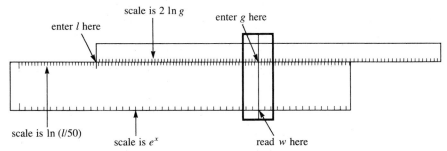

FIGURE 3-11 A "slide rule" device to compute the weight of a bass.

While the slide rule device might make some older engineers and slide rule companies happy, the fishing competitors probably would prefer a simple plastic disk on which the girth and length measurements can be entered in such a way that the weight of the bass appears "magically" in a window. You are asked to design such a disk in the following problem section.

3.2 PROBLEMS

1. Design a plastic disk to perform the calculations given by Model (3.10).
2. Which of the two models, (3.7) or (3.10), do you think is better? Why? Discuss the models qualitatively. In Chapter 4 you will be asked to compare the two models analytically.
3. Under what circumstances, if any, will Models (3.7) and (3.10) coincide? Explain fully.
4. Consider the models $W \propto l^2 g$ and $W \propto g^3$. Interpret each of these models geometrically. Explain how these two models differ from Models (3.7) and (3.10), respectively. Under what circumstances, if any, would the four models coincide? Which model do you think would do the best job of predicting W? Why? In Chapter 4 you will be asked to compare the four models analytically.
5. **a.** Let $A(x)$ denote a typical cross-sectional area of a bass, $0 \leqslant x \leqslant l$, where l denotes the length of the fish. Use the mean value theorem from calculus to show that the volume V of the fish is given by

$$V = l \cdot \bar{A}$$

where $\bar{A}$ is the average value of $A(x)$.

b. Assuming that $\bar{A}$ is proportional to the square of the girth g and that the weight density for the bass is constant, establish that

$$W \propto lg^2$$

MODEL

3.3 VEHICULAR STOPPING DISTANCE

Consider again the scenario posed in Example 1 of Section 2.2. Recall the rule of thumb that allows one car length for every 10 mph of speed. It was also stated in the scenario that this rule is the same as allowing for 2 sec between cars. However, the rules are in fact different from one another (at least for most cars). In order that the rules be the same, at 10 mph both should allow one car length:

$$1 \text{ car length} = \text{distance} = \frac{\text{speed in ft}}{\text{sec}} (2 \text{ sec})$$

$$= \left(\frac{10 \text{ mi}}{\text{hr}}\right)\left(\frac{5{,}280 \text{ ft}}{\text{mi}}\right)\left(\frac{1 \text{ hr}}{3600 \text{ sec}}\right)(2 \text{ sec})$$

$$= 29.33 \text{ ft}$$

This is absurd for an average car length of 15 ft.

Let's interpret the one-car-length rule geometrically. If we assume a car length of 15 ft and plot this rule, we obtain the graph shown in Figure 3-12, which shows that the distance allowed by the rule is proportional to the velocity. In fact, if we plot the speed in feet per second, the constant of proportionality has the units "seconds" and represents the total reaction time in order for the equation $D = kv$ to make sense. Moreover, in the case of a 15-ft car, we obtain the constant of proportionality as follows:

$$k = \frac{15 \text{ ft}}{10 \text{ mph}} = \frac{15 \text{ ft}}{52{,}800 \text{ ft}/3{,}600 \text{ sec}} = \frac{90}{80} \text{ sec}$$

In our previous discussion of this problem, we presented the model:

total stopping distance = reaction distance + braking distance

Let's consider the submodels for reaction distance and braking distance.

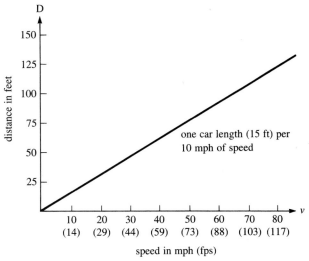

FIGURE 3-12 Geometrical interpretation of the one-car-length rule of thumb.

Recall that

$$\text{reaction distance} = f(\text{response time, speed})$$

Now assume that the vehicle continues at constant speed from the time the driver determines the need to stop until the brakes are applied. Under this assumption, reaction distance d_r is simply the product of response time t_r and velocity v:

$$d_r = t_r v \tag{3.11}$$

To test the submodel (3.11), plot measured reaction distance versus velocity. If the resultant graph approximates a straight line through the origin, you could estimate the slope t_r and feel fairly confident in the submodel. Alternatively, you could test a group of drivers representative of the assumptions made in Example 1 of Section 2.2 and estimate t_r directly.

Next, consider the braking distance:

$$\text{braking distance} = h(\text{weight, speed})$$

Suppose there is a panic stop and that the maximum brake force F is applied throughout the stop. The brakes are basically an energy dissipating device; that is, the brakes do work on the vehicle producing a change in the velocity that results in a loss of kinetic energy. Now, the work done is the force F times the braking distance d_b. This work must equal the change in kinetic energy which, in this situation, is simply $0.5mv^2$. Thus, we have

$$\text{Work done} = Fd_b = 0.5mv^2 \tag{3.12}$$

Next, we consider how the force F relates to the mass of the car. A reasonable design criterion would be to build cars in such a way that the maximum deceleration is constant when the maximum brake force is applied regardless of the mass of the car. This assumption means that the panic deceleration of a large car, like a Cadillac, is the same as that of a small car, like a Honda, owing to the design of the braking system. Moreover, constant deceleration occurs throughout the panic stop. From Newton's second law, $F = ma$, it follows that the force F is proportional to the mass. Combining this result with Equation (3.12) gives the proportionality relation

$$d_b \propto v^2 \qquad\qquad \textbf{(3.13)}$$

At this point we might want to design a test for the two submodels, or we could test the submodels against the data provided by the U.S. Bureau of Public Roads given in Table 3-1.

TABLE 3-1 Observed reaction and braking distances

Speed (mph)	Driver reaction distance (ft)	Braking distance* (ft)		Total stopping distance (ft)	
20	22	18–22	(20)	40–44	(42)
25	28	25–31	(28)	53–59	(56)
30	33	36–45	(40.5)	69–78	(73.5)
35	39	47–58	(52.5)	86–97	(91.5)
40	44	64–80	(72)	108–124	(116)
45	50	82–103	(92.5)	132–153	(142.5)
50	55	105–131	(118)	160–186	(173)
55	61	132–165	(148.5)	193–226	(209.5)
60	66	162–202	(182)	228–268	(248)
65	72	196–245	(220.5)	268–317	(292.5)
70	77	237–295	(266)	314–372	(343)
75	83	283–353	(318)	366–436	(401)
80	88	334–418	(376)	422–506	(464)

* Interval given includes 85% of the observations based on tests conducted by the U.S. Bureau of Public Roads. Figures in parentheses represent average values.

Figure 3-13 depicts the plot of driver reaction distance against velocity, using the data in Table 3-1. The graph is a straight line of approximate slope 1.1 passing through the origin: our results are *too good!* Since we always expect some deviation in experimental results, we should be suspicious. In fact, the results of Table 3-1 are based on the submodel (3.11), where an average response time of 3/4 sec was obtained independently. So we might later decide to design another test for the submodel.

To test the submodel for braking distance, we plot the observed braking distance recorded in Table 3-1 against v^2, as shown in Figure 3-14. Proportionality seems to be a reasonable assumption at the lower speeds, although it does seem to be less convincing at the higher speeds. By graphically fitting a straight line to the data, we estimate the slope and obtain the submodel:

$$d_b = 0.054v^2 \qquad\qquad \textbf{(3.14)}$$

You will learn how to fit the model to the data analytically in Chapter 4.

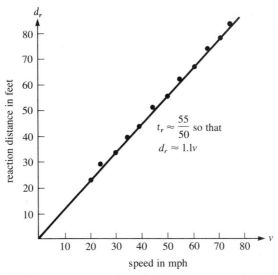

FIGURE 3-13 Proportionality of reaction distance and speed.

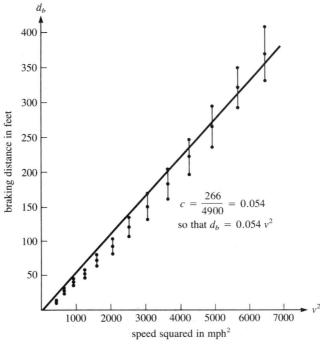

FIGURE 3-14 Proportionality of braking distance and the speed squared.

Connecting the two submodels, (3.11) and (3.14), we obtain the following model for the total stopping distance d:

$$d = 1.1v + 0.054v^2 \qquad\qquad \textbf{(3.15)}$$

The predictions of Model (3.15) and the actual observed stopping distances recorded in Table 3-1 are plotted in Figure 3-15. Considering the grossness of the assumptions and the inaccuracies of the data, the model seems to agree fairly reasonably with the observations up to about 70 mph. The rule of thumb of one 15-ft car length for every 10 mph of speed is also plotted in Figure 3-15. You can see that the rule significantly underestimates the total stopping distance at speeds exceeding 40 mph.

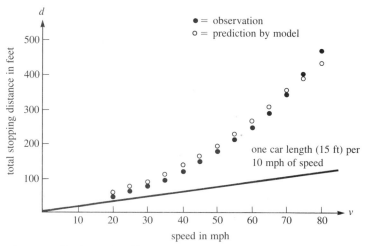

FIGURE 3-15 Total stopping distance.

Let's suggest an alternate rule of thumb. It must be easy to understand and to use. Assume that the driver of the trailing vehicle must be fully stopped by the time he or she reaches the point occupied by the lead vehicle at the exact time of the observation. Thus the driver must trail the lead vehicle by the total stopping distance, either based on our model (3.15) or on the observed data in Table 3-1. The maximum stopping distance can readily be converted to a trailing time. The results of these computations for the observed distances, in which 85% of the drivers were able to stop, are given in Table 3-2 (p. 69). The computations suggest the following rule of thumb.

Speed (mph)	Guideline (sec)
0–10	1
10–40	2
40–60	3
60–75	4

TABLE 3-2 Time required to allow the proper stopping distance

Speed (mph) (fps)	Stopping distance* (ft)	Trailing time required for maximum stopping distance (sec)
20 (29.3)	42 (44)[†]	1.5
25 (36.7)	56 (59)	1.6
30 (44.0)	73.5 (78)	1.8
35 (51.3)	91.5 (97)	1.9
40 (58.7)	116 (124)	2.1
45 (66.0)	142.5 (153)	2.3
50 (73.3)	173 (186)	2.5
55 (80.7)	209.5 (226)	2.8
60 (88.0)	248 (268)	3.0
65 (95.3)	292.5 (317)	3.3
70 (102.7)	343 (372)	3.6
75 (110.0)	401 (436)	4.0
80 (117.3)	464 (506)	4.3

* Includes 85% of the observations based on tests conducted by the U.S. Bureau of Public Roads.
† Figures in parentheses under stopping distance represent maximum values and are used to calculate trailing times.

This alternate rule of thumb is plotted in Figure 3-16. An alternative to using the rule of thumb might be to convince manufacturers to modify existing speedometers to compute stopping distance and time for the car's speed v based on our formula (3.15). We will revisit the braking system problem in Section 9.3 with a model based on the derivative.

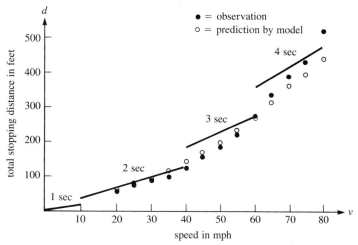

FIGURE 3-16 Total stopping distance and alternate rule of thumb. The plotted observations are the maximum values from Table 3.2.

3.3 PROBLEMS

1. How would you design a test to determine the average response time? Design a test to determine average reaction distance. Discuss the difference between the two statistics. If you were going to use the results of this test to predict the total stopping distance, would you want the true average?

2. For the submodel concerning braking distance, how would you design brake systems so that the maximum deceleration is constant for all vehicles regardless of their mass? Consider the surface area of the brake pads and the capacity of the hydraulic system to apply a force.

3.3 PROJECTS

1. Suppose after years of experience your auto company has designed an "optimum braking system" for its prestigious full-sized car. That is, the distance required to brake the car is the best in its weight class, and the occupants feel that the system is very smooth. Your firm has decided to build cars in the lighter weight classes. Discuss how you would "scale" the braking system of your present car in order to have the same performance in the smaller versions. Be sure to consider the hydraulic system and the size of the brake pads. Would a simple geometric similarity suffice? Let's suppose that the wheels are scaled in such a manner that the pressure (at rest) on the tires is constant in all car models. Would the brake pads seem proportionately larger or smaller in the scaled-down cars?

3.4 AUTOMOBILE GASOLINE MILEAGE

In Example 2 of Section 2.2 we posed the following problem:

For a particular individual driver, driving her car on a given day on a level highway at constant highway speeds near the optimum speed for fuel economy, provide a qualitative explanation of how fuel economy varies with small increases in speed.

The newspaper article from which this problem was derived also gave the rule of thumb that for every 5 mph over 50 mph, there is a loss of 1 mile to the gallon. Let's graph this rule of thumb. If you plot the loss in miles per gallon against the speed minus 50, the graph is a straight line passing through the origin as depicted in Figure 3-17. Let's see if this linear graph is qualitatively correct.

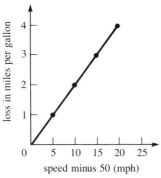

FIGURE 3-17 For every 5 mph over 50 mph, there is a loss of 1 mile to the gallon.

Since the automobile is to be driven at constant speeds, the acceleration is zero. Newton's second law then yields that the resultant force must be zero, or that the forces of propulsion and resistance must be in equilibrium. That is,

$$F_p = F_r$$

where F_p represents the propulsion force, and F_r the resisting force.

First, consider the force of propulsion. Each gallon of gasoline contains an amount of energy, say K. If C_r represents the amount of fuel burned per unit time, then $C_r K$ represents the power available to the car. Assuming a constant rate of power conversion, it follows that the converted power is proportional to $C_r K$. Since power is the product of force and velocity for a constant force, this argument yields the proportionality relation:

$$F_p \propto \frac{C_r K}{v}$$

By further assuming a constant fuel rating K, this last proportionality simplifies to

$$F_p \propto \frac{C_r}{v} \tag{3.16}$$

Next, consider the resisting force. Since we have restricted the problem to highway speeds, it is reasonable to assume that the frictional forces are small when compared to the drag forces caused by air resistance. At highway speeds, one sensible submodel for these drag forces is

$$F_r \propto Sv^2$$

where S is the cross-sectional area perpendicular to the direction of the moving car. (This assumption is commonly used in engineering for moderate sizes of S.) Since S is constant in our restricted problem, it follows that

$$F_r \propto v^2$$

Application of the condition $F_p = F_r$ and the proportionality (3.16) then yields

$$\frac{C_r}{v} \propto v^2$$

or

$$C_r \propto v^3 \tag{3.17}$$

The proportionality (3.17) gives the qualitative information that the fuel consumption rate should increase as the cube of the velocity. However, fuel consumption rate is not an especially good indicator of fuel efficiency: although the proportionality (3.17) says that the car is using more fuel per unit time at higher speeds, the car is also traveling further. Therefore, we define gasoline mileage as follows:

$$\text{mileage} = \frac{\text{distance}}{\text{consumption}}$$

Substitution of vt for distance, and $C_r t$ for consumption then gives the proportionality,

$$\text{mileage} = \frac{v}{C_r} \propto v^{-2} \tag{3.18}$$

Thus, gasoline mileage is inversely proportional to the square of the velocity.

Model (3.18) provides some useful qualitative information to assist in explaining automobile fuel consumption. For one thing, we should be very suspicious of the rule of thumb suggesting a linear graph as depicted in Figure 3-17. However, be very careful about the conclusions you do draw. While the exponential relationship in (3.18) appears impressive, it is valid only over a restricted range of speeds. Over that restricted range the relationship could be nearly linear, depending on the size of the constant of proportionality. Moreover, don't forget that we have ignored completely many factors in our analysis, and we have assumed that several important factors are constant. Thus our model is quite fragile, and its use is limited to a qualitative explanation over a restricted range of speeds. But then, that's precisely how we identified our problem.

3.4 PROBLEMS

1. In the automobile gasoline mileage example, suppose you plot miles per gallon against speed as a graphical representation of the rule of thumb (instead of the graph depicted in Figure 3-17). Explain why it would be difficult to deduce a proportionality relationship from that graph.

2. In the automobile gasoline mileage example, assume that the drag forces are proportional to Sv, where S is the cross-sectional area perpendicular to the direction of the moving car and v is its speed. What conclusions can you draw? Discuss the factors that might influence the choice of Sv^2 over Sv for the drag forces submodel. How could you test the submodel?

3. Discuss several factors that were completely ignored in our analysis of the gasoline mileage problem.

4. An object is sliding down a ramp inclined at an angle of θ radians and attains a terminal velocity before reaching the bottom. Assume that the drag force caused by the air is proportional to Sv^2, where S is the cross-sectional area perpendicular to the direction of motion and v is the speed. Further assume that the sliding friction between the object and the ramp is proportional to the normal weight of the object. Determine a relationship between terminal velocity and the mass of the object. If two different boxes, weighing 600 and 800 lb, are pushed down the ramp, find a relationship between their terminal velocities.

5. Assume that under certain conditions the heat loss of an object is proportional to the exposed surface area. Relate the heat loss of a cubic object with side length 6 in. to one with a side length of 12 in. Now, consider two irregularly shaped objects, such as two submarines. Relate the heat loss of a 70-ft submarine with a 7-ft scaled model. Suppose you are interested in the amount of energy needed to maintain a constant internal temperature in the submarine. Relate the energy needed in the actual submarine to that required by the scaled model. Specify the assumptions you have made.

6. Consider the situation of two warm-blooded adult animals essentially at rest and under the same conditions (as in a zoo). Assume the animals maintain the same body temperature, and that the energy available to maintain this temperature is proportional to the amount of food provided to them. Challenge this last assumption. If you are willing to assume that the animals are geometrically similar, relate the relative amounts of food necessary to maintain their body temperatures to their lengths and volumes (Hint: see Problem 5). List any assumptions you have made. What additional assumptions are necessary to relate the amount of food necessary with their body weights?

MODEL

3.5 ELAPSED TIME OF A TAPE RECORDER

Most stereo cassette decks and reel-to-reel audiotape recording devices are equipped with a numerical counter. This tape counter provides a numerical reference point, which can be used to index a recorded cassette or tape for playback purposes. The counter is incremented as the tape advances over the read/write heads. In many instances it is desired to relate the number displayed by the counter to time; for example, to find the playing time of a selection or to determine its tape count.

Does the counter on your tape recorder operate in such a manner that the elapsed time is proportional to the displayed count? For example, suppose during the playing of a tape requiring 90 minutes to play, your counter goes from 0 to 1600. Does it take 45 minutes for the counter to reach 800? When your counter reaches 1400, does enough playing time still remain to record another album or particular selection? Can you construct a model that relates the counter reading to the amount of time that has elapsed?

Problem identification   *For a particular cassette deck or tape recorder equipped with a tape counter, relate the counter to the amount of playing time that has elapsed.*

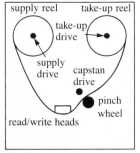

FIGURE 3-18 Schematic of a tape recorder.

Assumptions   At first you might think the counter number is proportional to the elapsed time. However, consider the tape recorder depicted in Figure 3-18. When the recorder is playing or recording, it appears at first that the take-up drive is pulling the tape across the read/write heads. If this were so, it would be reasonable to assume that the take-up drive turns at constant angular velocity. This assumption implies a relatively low linear speed of the tape initially across the read/write heads when the take-up reel is nearly empty, but a relatively high speed when the take-up reel is nearly full. (Why is this true?) Consequently, if we splice a length of recorded tape onto our original tape, it will sound different depending on whether the splice occurs at the beginning, middle, or end of the original tape. Since this situation does not correspond with actual experience in splicing tapes, we reject the hypothesis of constant angular velocity of the take-up drive during the recording or playback of a tape.

The preceding discussion suggests that during the play or recording of a tape the linear speed of the tape across the read/write heads is constant. Thus our first assumption is that the linear speed is a constant s. How is this assumption implemented mechanically? Inspection of a tape recorder reveals that a capstan drive pulls the tape across the read/write heads whenever the capstan is engaged. During record/play the take-up reel merely picks up the slack in the tape. Consequently, the take-up drive turns relatively fast when it is nearly empty and relatively slow when it is nearly full. You will notice this phenomenon if you observe a tape recorder in action.

Now consider how the counter might work. A simple procedure would be to run a belt around a pulley attached to one of the drives (the supply, take-up, or the capstan drive) and around a pulley attached to a drive turning some gears to advance the counter. This situation is illustrated in Figure 3-19. If the counter system is designed in this way, the reading on the counter will be proportional to the number of turns made by the appropriate drive. (Why?)

Which of the three drives actually operates the counter mechanism? If it's the capstan, then since its angular velocity is proportional to the speed of the tape, the number of revolutions of the drive will be proportional to the length of tape played. However, when the tape is wound in fast forward, or rewound in fast reverse, the capstan drive is disengaged and yet the counter continues to operate. (In fact, that's precisely how we search for a selection known to be located at some particular counter reading.) Thus the counter must be operated by either the supply or the take-up drive. We need further testing to see which drive actually does the job. Hence our submodel for the counter mechanism is

$$c \propto \frac{\theta}{2\pi}$$

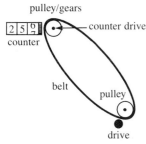

pulley/gears

2 5 9
counter

counter drive

belt
pulley

drive

FIGURE 3-19 A schematic showing the operation of the counter mechanism.

where c is the current counter reading and $\theta/2\pi$ is the number of revolutions of either the supply or take-up drive (depending on which one operates the counter). For the sake of definiteness in the presentation to follow we assume the counter is operated by the take-up drive.

Next, we relate the turning angle θ of the take-up reel to the length l of the tape played. When the take-up drive turns through an angle θ measured in radians, the amount l of tape played is given by $l = r\theta$, where r is the radius of the tape. This situation is depicted in Figure 3-20. A complication of the problem is that the radius r varies with the angle θ. But how, exactly, does r vary with θ?

For each revolution of the take-up drive, let's assume that the radius increases by some constant amount. This amount would be close to the thickness of the tape (less when the tape is stretched, more when it is wrapped loosely). We would expect the tape to be wrapped tighter initially when the radius of the tape on the take-up reel is small, but that the wrapping would become progressively looser as the radius of the tape on the reel increases. Nevertheless, to simplify the model, we assume the radius of the tape

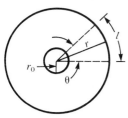

r_0
r
l
θ

FIGURE 3-20 When the take-up reel turns through an angle of θ radians, a length l of tape is played, where $l = r\theta$.

wrapped on the take-up reel increases *uniformly* with the number of revolutions of the take-up reel. Since the take-up reel has a hub of some radius r_0, this assumption implies that

$$r = r_0 + k_1\theta$$

for some constant k_1. Notice that this last equation implies that Δr is proportional to $\Delta\theta$.

We summarize our conclusions. The counter number c is a function of the linear speed s of the tape across the read/write heads, the thickness of the tape, the uniformity of the wrapping on the take-up reel, the counts per revolution existing between the counter and the take-up reel, and the initial radius of the take-up reel. Our assumptions have produced the following submodels:

a. s is constant
b. $c \propto \theta/2\pi$, where θ is the turning angle of the take-up reel measured in radians
c. $r = r_0 + k_1\theta$ for some constant k_1, where r_0 denotes the radius of the hub and r is the radius of the played out tape on the take-up reel

Now let's construct our model. As the take-up drive rotates through an angle of θ radians, the radius r varies and is given by $r = r_0 + k_1\theta$. By definition, the *average value* of r during such a rotation is

$$\bar{r} = \frac{1}{\theta}\int_0^\theta (r_0 + k_1\theta)\,d\theta$$

$$= \frac{1}{\theta}\left(r_0\theta + \frac{k_1\theta^2}{2}\right)$$

(See Problem 2 for an alternate derivation of $\bar{r}$ using only a proportionality argument.) Moreover, by the mean value theorem the length l of tape played during this rotation is given by $l = \bar{r}\theta$, so that

$$l = r_0\theta + \frac{k_1\theta^2}{2}$$

Next we know $l = st$, and since the linear speed s is a constant, this result implies

$$t = \frac{r_0}{s}\theta + \frac{k_1}{2s}\theta^2$$

Finally, the assumption $c \propto \theta/2\pi$ leads to the equation

$$t = k_2 c + k_3 c^2 \tag{3.19}$$

where k_2 and k_3 are constants. Once these constants are determined, the quadratic equation (3.19) can be solved for c in terms of the playing time if that is desired. We will determine the constants k_2 and k_3 experimentally in Section 4.4.

In Chapter 9 we will derive Model (3.19) by a method using the derivative.

Testing the submodels   The three submodels upon which our model depends can be tested easily. The assumption of constant linear speed of the tape across the read/write heads turns out to be the case. In fact, several tape recorders offer a selection of more than one speed. This submodel can be tested in the following way. Mark a specific length of tape at the beginning, at the middle, and at the end of the tape on the reel. Then record the amount of time it takes to play the length of marked tape in each of the three positions. The speed can then be computed and checked to see if it is constant. Our experimental results compare favorably with the manufacturer's claim of $3\frac{3}{4}$ in. per sec to within an error of 0.8%.

The submodel $c \propto \theta/2\pi$ can be checked by comparing the counter reading with the number of revolutions of the supply and take-up reels. Our data is summarized in Figures 3-21 and 3-22. For the tape recorder tested, the take-up drive operates the counter and the constant of proportionality is $\frac{11}{10}$. That is, the counter increases by 11 for every 10 revolutions of the take-up drive.

The hub of the take-up reel is easily measured. The assumption of uniform wrapping is checked by measuring an initial radius r_i and a final radius r_f for a given number of revolutions and then computing the ratio

$$h_{\text{eff}} = \frac{r_f - r_i}{\text{number of revolutions}} \tag{3.20}$$

The ratio represents the increase in radius per revolution or the "effective thickness" h_{eff} of the tape. If h_{eff} is the same at the beginning, middle, and

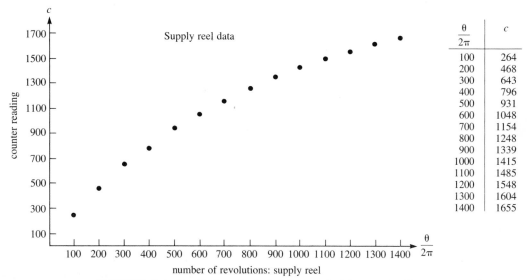

FIGURE 3-21 Rejecting the hypothesis that the supply drive operates the counter mechanism.

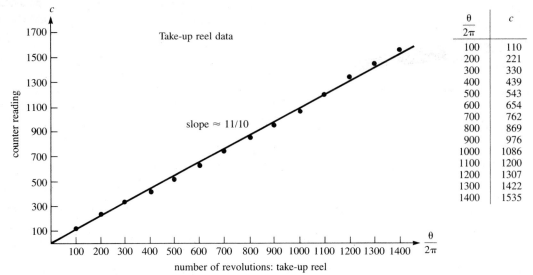

FIGURE 3-22 Accepting the hypothesis that the take-up drive operates the counter mechanism.

end of the tape, then the assumption of uniform wrapping is satisfied. For the tape recorder we tested, the effective thickness was computed to be 0.001503, which compares favorably with the manufacturer's claim of 0.0015 for the thickness of the tape. Thus, the uniformity of wrapping assumption seems reasonable. (A word of caution is in order. The thickness measurements must be taken while the recorder is operated in the record/play mode. If fast forward is used, the increased torque causes stretching of the tape and a distortion of the tape thickness. For the recorder we tested, an "effective thickness" as low as 0.00143 was computed in the fast forward mode.)

Since the submodels seem to be reasonably satisfied, we will determine the constants k_2 and k_3 in Chapter 4 where we consider model fitting. For the model developed in this section, it is sufficient to satisfy the inherent assumptions. However, for the analytic model to be developed in Problem 1 and again in Chapter 9, more information is needed (see Problem 2).

3.5 PROBLEMS

1. Using the assumptions of the model to relate the elapsed time of a tape recorder with the counter reading, show that

$$t = \frac{\pi}{ms}\left[2r_0c + \left(\frac{h_{\text{eff}}}{m}\right)c^2\right]$$

where r_0: radius of the hub of take-up reel
s: linear speed of the tape recorder in playback mode
m: constant of proportionality for $c = m(\theta/2\pi)$

h_{eff}: effective thickness of the tape
c: counter reading

2. Show that the total wrapped thickness of the tape played out on the take-up reel equals $(h_{eff}/2\pi)\theta$, where θ is the total number of radians of rotation through which the take-up reel turns. Thus the total wrapped thickness is proportional to θ. Use this result to conclude that the average radius on the take-up reel is

$$\bar{r} = r_0 + \frac{h_{eff}}{4\pi}\theta$$

Note: The above calculation for r is valid because of the proportionality relationship of the wrapped thickness to θ. Our presentation in the text, using the mean value theorem, is valid for more general relationships.

3. Answer the following questions for your tape recorder.
   a. Is it the supply or take-up drive that operates the counter? (The model you construct depends on which drive operates the counter.) What is the constant of proportionality m relating the counter and the appropriate drive $[c = m(\theta/2\pi)]$?
   b. What is the constant linear speed s of the tape across the read/write heads in the record/play mode?
   c. What is the "effective thickness" h_{eff} of the tape as it is wrapped on the supply or take-up reel?
   d. What is the radius of the hub r_0 of your take-up reel?

4. Using the data you collected in Problem 3 and the model you constructed in Problem 1, predict the elapsed time as a function of the counter reading for your tape recorder. Test it against data that you collect. How well does the model do? Explain any discrepancies.

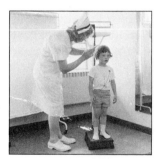

MODEL

3.6 BODY WEIGHT VERSUS HEIGHT*

A question of interest to almost all Americans is "How much should I weigh?" A rule of thumb often given to people desiring to run a marathon is 2 lb of body weight per inch of height, but shorter marathoners seem to

* Optional section.

have a much easier time meeting this rule than the taller ones. Tables have been designed to suggest weights for different purposes: doctors are concerned about a reasonable weight for health purposes, and many Americans seek weight standards based on physical appearance. Moreover, some organizations, such as the Army, are concerned about physical conditioning and define an upper weight allowance of acceptability. Quite often these weight tables are categorized in some manner. For example, consider Table 3-3, which gives upper weight limits of acceptability for males between the ages of 17 and 21. (The table has no further delineators such as bone structure.)

TABLE 3-3 Weight versus height for males age 17–21

Height (in.)	Weight (lb)	Height (in.)	Weight (lb)
60	132	71	185
61	136	72	190
62	141	73	195
63	145	74	201
64	150	75	206
65	155	76	212
66	160	77	218
67	165	78	223
68	170	79	229
69	175	80	234
70	180		

If the differences between successive weight entries in Table 3-3 are computed to determine how much weight is allowed for each additional inch of height, it will be seen that throughout a large portion of the table a constant 5 lb per inch is allowed (with some fours and sixes appearing at the lower and upper ends of the scales, respectively). Certainly this last rule is more liberal than the previous "2 lb per inch" rule recommended for the marathoners, but just how reasonable a constant weight per height rule is remains to be seen. In this section we examine qualitatively how weight and height should vary.

Body weight depends on a number of factors, some of which have already been mentioned. In addition to height, bone density could be a factor. Is there a significant variation in bone density, or is it essentially constant? What about the relative volume occupied by the bones? Is the volume essentially constant, or are there "heavy, medium, and light" bone structures? And what about a body density factor? How can differences in the densities of bone, muscle, and fat be accounted for? Do these densities vary? Is body density typically a function of age and gender in the sense that the relative composition of muscle, bone, and fat vary as a person becomes older? Are there different compositions of muscle and fat between males and females at the same ages?

Let's define the problem so that bone density is considered constant (by accepting an upper limit), and predict weight as a function of height, gender, age, and body density. The purpose or basis of the weight table must also be specified, so we base the table on physical appearance.

Problem identification   We now identify the problem as follows: *For various heights, genders, and age groups, determine upper weight limits that represent maximum levels of acceptability based on physical appearance.*

Assumptions   Now assumptions about body density are needed if we are to predict successfully weight as a function of height. As one simplifying assumption, suppose that some parts of the body are composed of an "inner core" of one density and an "outer core" of a different density. Assume too that the inner core is composed primarily of bones and muscle material and that the outer core is primarily a fatty material, giving rise to the different densities (see Figure 3-23). We next construct submodels to explain how the weight of each core might vary with height.

How does body weight vary with height? To begin, assume that for adults certain parts of the body, such as the head, have the same volume and density for different people. Thus the weight of an adult is given by

$$W = k_1 + W_{in} + W_{out} \qquad (3.21)$$

inner core has one density

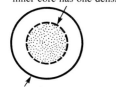

·outer core has another density

FIGURE 3-23 Assume that parts of the body are composed of an inner and outer core of distinct densities.

where $k_1 > 0$ is the constant weight of those parts having the same volume and density for different individuals, and W_{in} and W_{out} are the weights of the inner and outer cores, respectively.

Next, let's turn our attention to the inner core. How does the volume of the extremities and trunk vary with height? We know that people are not geometrically similar because they don't appear to be scaled models of one another, having different shapes and different relative proportions of the trunk and extremities. However, we are concerned with an "upper weight limit" based on physical appearance by definition of the problem at hand. Even though this may be somewhat subjective, it would seem reasonable that whatever image might be visualized as an upper limit standard of acceptability for a 74-in. person would be a scaled image of a 65-in. person. Thus, for purposes of our problem, geometric similarity of individuals is a reasonable assumption. Note that no particular shape is being assumed, but only that the ratio of distances between corresponding points in individuals is the same. Under this assumption the volume of each component we are considering is proportional to the cube of a characteristic dimension, which we select to be height h. Hence the sum of the components must be proportional to the cube of the height, or

$$V_{in} \propto h^3 \qquad (3.22)$$

Now what should be assumed about the average weight density of the inner core? Assuming that the inner core is composed of muscle and bone, each of which has a different density, what percentage of the total volume of the inner core is occupied by the bones? If bone diameter is assumed to be proportional to the height, then the total volume occupied by the bones is proportional to the cube of the height. This implies that the percentage of the total volume of the inner core occupied by the bones in geometrically similar individuals is constant. It follows from the ensuing argument that the average weight density ρ_{in} is constant as well. For example, consider the average weight density ρ_{avg} of a volume V consisting of two components V_1 and V_2, each with a distinct density ρ_1 and ρ_2. Then

$$V = V_1 + V_2$$

and

$$\rho_{avg}V = W = \rho_1 V_1 + \rho_2 V_2$$

yield

$$\rho_{avg} = \rho_1 \frac{V_1}{V} + \rho_2 \frac{V_2}{V}$$

Therefore as long as the ratios V_1/V and V_2/V do not change, the average weight density ρ_{avg} is constant. Application of this result to the inner core implies that the average weight density ρ_{in} is constant yielding

$$W_{in} = V_{in}\rho_{in} \propto h^3$$

or

$$W_{in} = k_2 h^3 \quad \text{for } k_2 > 0 \tag{3.23}$$

Note that the preceding submodel includes any case of materials with densities different than muscle and bone (such as tendons, ligaments, organs, and so forth) as long as the percentage of the total volume of the inner core occupied by those materials is constant.

Now consider the outer core of fatty material. Since the table is to be based on personal appearance, it can be argued that the thickness of the outer core should be constant regardless of the height (see Problem 3). If τ represents this thickness, then the weight of the outer core is

$$W_{out} = \tau \rho_{out} S_{out}$$

where S_{out} is the surface area of the outer core and ρ_{out} is the density of the outer core. Again assuming that the subjects are geometrically similar, it follows that the surface area is proportional to the square of the height. If the density of the outer core of fatty material is assumed to be constant for all individuals, then we have

$$W_{out} \propto h^2$$

It may be argued, however, that taller people can carry a greater thickness for the fatty layer. If it is assumed that the thickness of the outer core is proportional to the height, then

$$W_{out} \propto h^3$$

Allowing both of these assumptions to reside in a single submodel gives

$$W_{\text{out}} = k_3 h^2 + k_4 h^3, \quad \text{where } k_3, k_4 \geq 0 \tag{3.24}$$

Here the constants k_3 and k_4 are allowed to assume a zero value.

Summing the submodels represented by Equations (3.21), (3.23), and (3.24) to determine a model for weight yields

$$W = k_1 + k_3 h^2 + k_5 h^3 \quad \text{for } k_1, k_5 > 0 \text{ and } k_3 \geq 0 \tag{3.25}$$

where $k_5 = k_2 + k_4$. Note that the model (3.25) suggests variations of a higher order than the first power of h. If the model is valid, then taller people will indeed have a difficult time satisfying the "linear" rules of thumb given earlier. However, at the moment our judgment can only be qualitative since we have not verified our submodels. Some ideas on how to test the model are discussed in the problem set. Furthermore, we have not placed any relative significance on the higher-order terms occuring in (3.25). In the study of statistics, regression techniques are given to provide insight to the significance of each term in the model.

Model interpretation   Let's interpret the rules of thumb given earlier, which allowed a constant weight increase for each additional inch of height, in terms of our submodels. Consider the amount of weight attributable to an increase in the length of the trunk. Since the total allowable weight increase per inch is assumed constant by the given rules, the portion allowed for the trunk increase may also be assumed constant. To allow a constant weight increase, the trunk must increase in length while maintaining *the same cross-sectional area*. This implies, for example, that the waist size remains constant. Let's suppose that a 30-in. waist is judged the upper limit acceptable for the sake of personal appearance in a male with a height of 66 in. in the 17–21 category. The 2 lb per inch rule would allow a 30-in. waist for a male with a height of 72 in. as well. On the other hand, the model based on geometric similarity suggests that all distances between corresponding points should increase by the same ratio. Thus, the male with a height of 72 in. should have a waist of 30(72/66), or approximately 32.7 in., in order to be proportioned similarly. Comparing the two models, we get the following data.

Height (in.)	Linear models (in., waist measure)	Geometric similarity model (in., waist measure)
66	30	30.0
72	30	32.7
78	30	35.5
84	30	38.2

Now we can see why tall marathoners who follow the 2 lb per inch rule appear *very thin*.

3.6 PROBLEMS

1. Describe in detail the data you would like to obtain in order to test the various submodels supporting Model (3.25). How would you go about collecting the data?

2. Tests now exist to measure the percent of body fat. Assume that such tests are accurate and that a large amount of carefully collected data is available. You may also specify any other statistics, such as waist size and height, that you would like collected. Explain how the data could be arranged to check the assumptions underlying the submodels in this section. For example, suppose the data for males between 17 and 21 with constant body fat and height are examined. Explain how the assumption of constant density of the inner core could be checked.

3. A popular measure of physical condition and personal appearance is the "pinch test." To administer this test, you measure the thickness of the "outer core" at selected locations in the body by pinching. Where and how should the pinch be made? What thickness of pinch should be allowed? Should the pinch thickness be allowed to vary with height?

PART TWO

MODEL FITTING AND
MODELS REQUIRING OPTIMIZATION

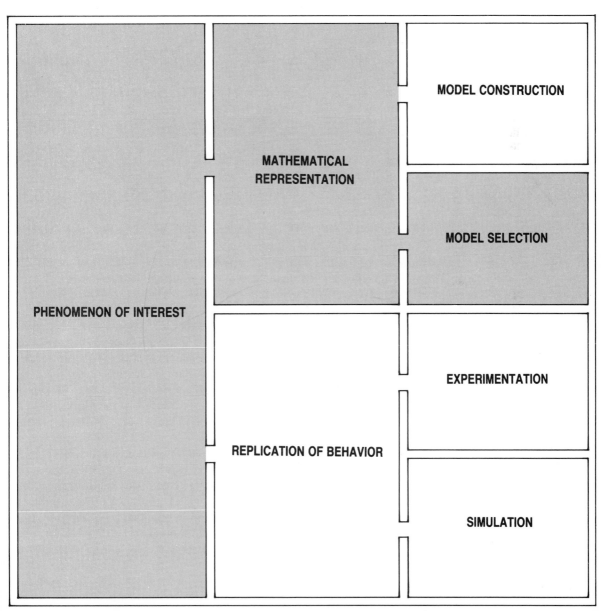

MODEL FITTING

INTRODUCTION

In the mathematical modeling process we encounter situations that cause us to analyze data for different purposes. You have already seen how our assumptions can lead to a model of a particular type. For example, in Chapter 3, when we analyzed the distance required to bring a car to a safe stop once the brakes are applied, our assumptions led to a submodel of the form

$$d_b = Cv^2$$

where d_b is the distance required to stop the car, v is the velocity of the car at the time the brakes are applied, and C is some arbitrary constant of proportionality. At this point we can collect and analyze sufficient data to determine if the assumptions are reasonable. If they are, we want to determine the constant C that selects the particular member from the family $y = Cv^2$ corresponding to the braking distance submodel.

We may encounter situations in which there are different assumptions leading to different submodels. For example, when studying the motion of a projectile through a medium such as air, we can make different assumptions about the nature of a drag force, such as the drag force being proportional to v or to v^2. We might even choose to neglect the drag force completely. As another example, when we are determining how fuel consumption varies with automobile velocity, our different assumptions about the drag force can lead to models that predict mileage varies as $C_1 v^{-1}$ or as $C_2 v^{-2}$ (see Problem 2 in Section 3.4). The resulting problem can be thought of in the following way: First, use some collected data to choose C_1 and C_2 in a way that selects the curve from each family that "best fits" the data, and then choose whichever resultant model is most appropriate for the particular situation under investigation.

A different case arises when the problem is so complex as to prevent the formulation of a model explaining the situation. For instance, if the submodels involve partial differential equations that are not solvable in closed form, there is little hope for constructing a master model that can be solved and analyzed without the aid of a computer. Or there may be so many significant variables involved that one would not even attempt to construct an explicative model. In such cases experiments may have to be conducted to investigate the behavior of the dependent variable(s). Then the experimental data can be used to predict the values of the dependent variable(s) for selected values of the independent variable(s) within the range of the data points.

The preceding discussion identifies three possible tasks when analyzing a collection of data points:

1. Fitting a selected model type or types to the data.
2. Choosing the most appropriate model from competing types that have been fitted. For example, we may need to determine whether the best-fitting exponential model is a better model than the best-fitting polynomial model.
3. Making predictions from the collected data.

In the first two tasks a model or competing models exist that seem to *explain* the behavior being observed. We address these two cases in this chapter under the general heading of *model fitting*. However, in the third case, a model does not exist to explain the behavior being observed. Rather, there exists a collection of data points that can be used to *predict* the behavior within some range of interest. In essence we wish to construct an *empirical model* based on the collected data. In Part III of the text we study such empirical model construction under the general heading of *interpolation*. It is important to understand both the philosophical and the mathematical distinctions between model fitting and interpolation.

The Relationship Between Model Fitting and Interpolation

Let's analyze the three tasks identified in the preceding paragraph to determine what must be done in each case. In Task 1 the precise meaning of "best" model must be identified and the resulting mathematical problem resolved. In Task 2 a criterion is needed for comparing models of different types. In Task 3 criteria must be established for determining how to make predictions in between the observed data points.

Note the difference in the modeler's attitude in each of these situations. In the two model-fitting tasks a relationship of a particular type is strongly suspected, and the modeler is willing to accept some deviation between the model and the collected data points in order to have a model that satisfactorily *explains* the situation under investigation. In fact, the modeler expects errors to be present in both the model and the data. On the other hand, when interpolating, the modeler is strongly guided by the data that have been carefully collected and analyzed, and a curve is sought that captures the trend of the data in order to *predict* in between the data points. Thus the modeler generally attaches little explicative significance to the interpolating curves. In all situations the modeler may ultimately want to make predictions from the model. However, the modeler tends to emphasize the *proposed models* over the data when model fitting, whereas he or she places greater confidence in the *collected data* when interpolating and attaches less signifi- cance to the form of the model. In a sense, explicative models are "theory" driven while predictive models are "data" driven.

Let's illustrate the preceding ideas with an example. Suppose we are at- tempting to relate two variables, y and x, and have gathered the data plotted in Figure 4-1. If the modeler is going to make predictions based solely upon the data in the figure, he or she might use a technique such as *spline inter- polation* (which you will study in Chapter 6) in order to pass a smooth polynomial through the points (see Figure 4-2). Note that in Figure 4-2 the interpolating curve passes through the data points and captures the trend of the behavior over the range of observations.

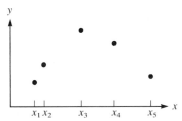

FIGURE 4-1 Observations relating the variables y and x.

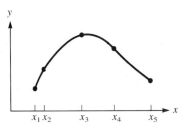

FIGURE 4-2 Interpolating the data using a smooth polynomial.

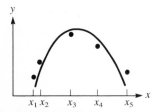

FIGURE 4-3 Fitting a parabola $y = C_1 x^2 + C_2 x + C_3$ to the data points.

However, suppose that in studying the particular behavior depicted in Figure 4-1 the modeler makes assumptions leading to the expectation of a quadratic model, or parabola, of the form $y = C_1 x^2 + C_2 x + C_3$. In this case the data of Figure 4-1 would be used to determine the arbitrary constants C_1, C_2, and C_3 in order to select the "best" parabola (see Figure 4-3). The fact that the parabola may deviate from some or all of the data points would be of no concern. Note the difference in the values of the predictions made by the curves in Figures 4-2 and 4-3 in the vicinity of the values x_1 and x_5.

A modeler may find it necessary both to fit a model and to interpolate in the same problem. The best-fitting model of a given type may prove to be unwieldy, or even impossible, for subsequent analysis involving operations like integration or differentiation. In such situations the model may be replaced with an interpolating curve (such as a polynomial) that is more readily differentiated or integrated. For example, a step function used to model a square wave might be replaced by a trigonometric approximation to facilitate subsequent analysis. In these instances the modeler desires the interpolating curve to approximate closely the essential characteristics of the func-

approximation   tion it replaces. This type of interpolation is usually called **approximation** and is typically addressed in introductory numerical analysis courses.

Sources of Error in the Modeling Process

Before discussing criteria upon which to base curve-fitting and interpolation decisions, we need to examine the modeling process in order to ascertain where errors can arise. If error considerations are neglected, undue confidence may be placed in intermediate results, causing faulty decisions in subsequent steps. Our goal is to ensure that all parts of the modeling process are computationally compatible and to consider the effects of cumulative errors likely to exist from previous steps.

For purposes of easy reference we classify errors under the following category scheme:

1. Formulation error
2. Truncation error
3. Round-off error
4. Measurement error

formulation error   A **formulation error** is an error resulting from the assumption that certain variables are negligible, or from simplifications arising in describing interrelationships among the variables in the various submodels. For example, when we determined a submodel for braking distance in Chapter 3, we completely neglected road friction, and we assumed a very simple relationship for the nature of the drag force due to air resistance. Formulation errors are present in even the best models.

truncation errors   **Truncation errors** are those errors attributable to the numerical method used to solve a mathematical problem. For example, we may find it necessary to approximate $\sin x$ with a polynomial representation obtained from the power series

$$\sin x = x - \frac{x^3}{3!} + \frac{x^5}{5!} - \cdots$$

An error will be introduced when the series is truncated to produce the polynomial.

round-off error

Round-off error refers to any error caused by using a finite digit machine for computation. Since all numbers cannot be represented exactly using only finite representations, we must always expect round-off errors to be present. For example, consider a calculator or computer that uses 8-digit arithmetic. Then the number 1/3 is represented by .33333333 so that 3 times 1/3 is the number .99999999 rather than the actual value 1. The error of 10^{-8} is due to round-off. The ideal real number 1/3 is an *infinite* string of decimal digits .33333..., but any calculator or computer can do arithmetic only with numbers having finite precision. When many arithmetic operations are performed in succession, each with its own round-off, the accumulated effect of round-off can significantly alter the numbers that are supposed to be the answer. Round-off is just one of the things we have to live with—*and be aware of*—when we use computing machines.

measurement errors

Measurement errors are caused by imprecision in the data collection. This imprecision may include such diverse things as human errors in recording or reporting the data, or the actual physical limitations of the laboratory equipment. For example, considerable measurement error would be expected in the data reflecting the response distance and the braking distance in the braking distance problem.

4.1 FITTING MODELS TO DATA GRAPHICALLY

Assume the modeler has made certain assumptions leading to a model of a particular type. The model generally contains one or more parameters, and sufficient data must be gathered to determine them. Let's consider the problem of data collection.

The determination of how many data points to collect involves a trade-off between the cost of obtaining them and the accuracy required of the model. As a minimum, the modeler needs at least as many data points as there are arbitrary constants in the model curve. Additional points are required to determine any arbitrary constants involved with some technique of "best fit" that is being employed. The *range* over which the model is to be used determines the end points of the interval for the independent variable(s). The *spacing* of the data points within that interval is also important because any part of the interval over which the model must fit particularly well can be weighted by using unequal spacing. We may choose to take more data points where maximum use of the model is expected, or we may collect more data points where we anticipate abrupt changes in the dependent variable(s).

Even if the experiment has been carefully designed and the experiments meticulously conducted, the modeler needs to appraise the accuracy of the data before attempting to fit the model. How were the data collected? What is the accuracy of the measuring devices used in the collection process? Do any points appear suspicious? Following such an appraisal and elimination (or replacement) of spurious data, it is useful to think of each data

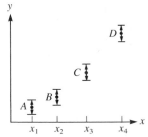

FIGURE 4-4 Each data point is thought of as an interval of confidence.

absolute deviations

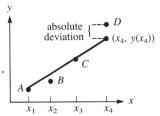

FIGURE 4-5 Minimizing the sum of the absolute deviations from the fitted line.

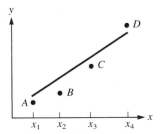

FIGURE 4-6 Minimizing the largest absolute deviation from the fitted line.

point as an interval of relative confidence rather than as a single point. This idea is shown in Figure 4-4. The length of each interval should be commensurate with the appraisal of the errors present in the data collection process.

Visual Model Fitting with the Original Data

Suppose it is desired to fit the model $y = ax + b$ to the data shown in Figure 4-4. How might the constants a and b be chosen to determine the line that "best fits" the data? Generally, when more than two data points exist, all of them cannot be expected to lie exactly along a single straight line, even if such a line accurately models the relationship between the two variables, x and y. That is, ordinarily there will be some vertical discrepancy between some of the data points and any particular line being considered. We refer to these vertical discrepancies as **absolute deviations** (see Figure 4-5). For the "best-fitting" line we might try to minimize the sum of these absolute deviations leading to the model depicted in Figure 4-5. While success may be achieved in minimizing the sum of the absolute deviations, the absolute deviation from individual points may be quite large. For example, consider point D in Figure 4-5. If the modeler has confidence in the accuracy of this data point, there would be concern for the predictions made from the fitted line near the point. As an alternative, suppose a line is selected that minimizes the largest deviation from any point. Applying this criterion to the data points might give the line shown in Figure 4-6.

Although these visual methods for fitting a line to data points may appear imprecise, the methods are often quite *compatible* with the accuracy of the modeling process itself. The grossness of the assumptions and the imprecision involved in the data collection may not warrant a more sophisticated analysis. In such situations the blind application of one of the analytic methods to be presented in Section 4.2 may lead to models far less appropriate than one obtained graphically. Furthermore, a visual inspection of the model fitted graphically to the data immediately gives an impression of *how good* the fit is and *where* it appears to fit well. Unfortunately, these important considerations are often overlooked in problems with large amounts of data analytically fitted via computer codes. Since the model-fitting portion of the modeling process seems to be more precise and analytic than some of the other steps, there is a tendency to place undue faith in the numerical computations.

Transforming the Data

Most of us are limited visually to fitting only lines. So how can we graphically fit other curves as models? Suppose, for example, that a relationship of the form $y = Ce^x$ is suspected for some submodel and the data shown in Table 4-1 have been collected (see p. 92).

The model states that y is proportional to e^x. Thus, if we plot y versus e^x, we should obtain approximately a straight line. The situation is depicted

TABLE 4-1 Collected data

x	1	2	3	4
y	8.1	22.1	60.1	165

in Figure 4-7. Since the plotted data points do lie approximately along a line that projects through the origin, we conclude that the assumed proportionality is reasonable. From the figure the slope of the line is approximated as

$$C = \frac{165 - 60.1}{54.6 - 20.1} \approx 3.0$$

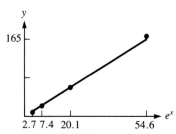

FIGURE 4-7 Plot of y versus e^x for the data given in Table 4-1.

Now let's consider an alternate technique that is useful in a variety of problems. Take the logarithm of each side of the equation $y = Ce^x$ to obtain

$$\ln y = \ln C + x$$

Note that this expression is an equation of a line in the variables $\ln y$ and x. The number $\ln C$ is the intercept when $x = 0$. The transformed data are shown in Table 4-2 and plotted in Figure 4-8. "Semi-log paper" or a computer is useful when plotting large amounts of data.

TABLE 4-2 The transformed data from Table 4-1

x	1	2	3	4
$\ln y$	2.1	3.1	4.1	5.1

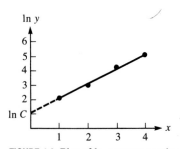

FIGURE 4-8 Plot of $\ln y$ versus x using Table 4-2.

From Figure 4-8 we can determine that the intercept ln C is approximately 1.1, giving $C = e^{1.1} \approx 3.0$ as before.

A similar transformation can be performed on a variety of other curves to produce linear relationships among the resulting transformed variables. For example, if $y = x^a$, then

$$\ln y = a \ln x$$

is a linear relationship in the transformed variables ln y and ln x. Here "log-log paper" or a computer is useful when plotting large amounts of data.

Let's pause and make an important observation. Suppose we do invoke a transformation and plot ln y versus x, as in Figure 4-8, and find the line that successfully minimizes the sum of the absolute deviations of the transformed data points. The line then determines ln C, which in turn produces the proportionality constant C. Although it is not obvious, the resulting model $y = Ce^x$ is *not* the member of the family of exponential curves of the form ke^x that minimizes the sum of the absolute deviations from the original data points (when we plot y versus x). This important idea will be demonstrated both graphically and analytically in the ensuing discussion. When transformations of the form $y = \ln x$ are made, the distance concept is distorted. While a fit that is compatible with the inherent limitations of a graphical analysis may be obtained, the modeler must be aware of this distortion and *verify the model using the graph from which it is intended to make predictions or conclusions—namely, the y versus x graph in the original data rather than the graph of the transformed variables.*

We now present an example illustrating how a transformation may distort distance in the xy-plane. Consider the data plotted in Figure 4-9 and assume the data are expected to fit a model of the form $y = Ce^{1/x}$. Using a logarithmic transformation as before, we find

$$\ln y = \frac{1}{x} + \ln C$$

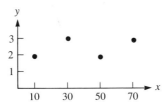

FIGURE 4-9 A plot of some collected data points.

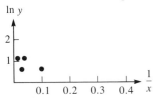

FIGURE 4-10 A plot of the transformed data points.

A plot of the points ln y versus $1/x$ based on the original data is shown in Figure 4-10. Note from the figure how the transformation distorts the distances between the original data points and squeezes them all together. Consequently, if a straight line is made to fit the transformed data plotted in Figure 4-10, the absolute deviations appear relatively small (that is, small computed on the Figure 4-10 scale rather than on the Figure 4-9 scale). However, if you were to plot the fitted model $y = Ce^{1/x}$ to the data in Figure 4-9, you would see that it fits the data relatively poorly as shown in Figure 4-11.

From the preceding example it can be seen that if a modeler is not careful when using transformations, he or she can be tricked into selecting a relatively poor model. This realization becomes especially important when comparing alternative models. Very serious errors can be introduced when selecting the best model unless all the comparisons are made with the original data (plotted in Figure 4-9 in our example). Otherwise, the choice of "best" model may be determined by a peculiarity of the transformation rather than on the merits of the model and how well it fits the original data. While the danger of making transformations is evident in this graphical illustration, a modeler may be fooled if he or she is not especially observant since many computer codes fit models by first making a transformation. If the modeler intends to use indicators, such as the sum of the absolute deviations, to make decisions about the adequacy of a particular submodel or choose among competing submodels, the modeler must first ascertain how those indicators were computed.

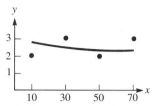

FIGURE 4-11 A plot of the curve $y = Ce^{1/x}$ based on the value $\ln C \approx 0.9$ from Figure 4-10.

4.1 PROBLEMS

1. The model in Figure 4-2 would normally be used to predict behavior between x_1 and x_5. What would be the danger of using the model to predict y for values of x less than x_1 or greater than x_5? Suppose we are modeling the trajectory of a thrown baseball.

2. The following table gives the elongation e in inches per inch (in./in.) for a given stress S on a steel wire measured in pounds per square inch (lb/in.^{2}). Test the model $e = c_1 S$ by plotting the data. Estimate c_1 graphically.

$S(\times 10^{-3})$	5	10	20	30	40	50	60	70	80	90	100
$e(\times 10^5)$	0	19	57	94	134	173	216	256	297	343	390

3. In the following data, x is the diameter of a ponderosa pine in inches measured at breast height and y is a measure of volume—number of board feet divided by 10. Test the model $y = ax^b$ by plotting the transformed data. If the model seems reasonable, estimate the parameters a and b of the model graphically. Compare your results with the models you proposed in Problem 8 in Section 3.1.

x	17	19	20	22	23	25	28	31	32	33	36	37	38	39	41
y	19	25	32	51	57	71	113	141	123	187	192	205	252	259	294

4. In the following data, V represents a mean walking velocity and P represents the population size. We wish to know if we can predict the popula-

tion size P by observing how fast people walk. We discuss the scenario in Section 6.1. Plot the data. What kind of a relationship is suggested? Test the following models by plotting the appropriate transformed data:
a. $P = aV^b$ **b.** $P = a \ln V$
If either model seems appropriate, estimate the parameters graphically.

V	2.27	2.76	3.27	3.31	3.70	3.85	4.31	4.39	4.42
P	2500	365	23700	5491	14000	78200	70700	138000	304500

V	4.81	4.90	5.05	5.21	5.62	5.88
P	341948	49375	2602000	867023	1340000	1092759

5. The following data represent the growth of a population of fruit flies over a six-week period. Test the following models by plotting an appropriate set of data. Estimate the parameters of the models:
a. $P = c_1 t$ **b.** $P = ae^{bt}$

t (days)	7	14	21	28	35	42
P (number of observed flies)	8	41	133	250	280	297

6. The following data represent (hypothetical) energy consumption normalized to the year 1900. Plot the data. Test the model $Q = ae^{bx}$ by plotting the transformed data. Estimate the parameters of the model graphically.

x	Year	Consumption Q
0	1900	1.00
10	1910	2.01
20	1920	4.06
30	1930	8.17
40	1940	16.44
50	1950	33.12
60	1960	66.69
70	1970	134.29
80	1980	270.43
90	1990	544.57
100	2000	1,096.63

7. In 1601 the German astronomer Johannes Kepler became director of the Prague Observatory. Kepler had been helping Tycho Brahe in collecting 13 years of observations on the relative motion of the planet Mars. By 1609 Kepler had formulated his first two laws:
    i. Each planet moves on an ellipse with the sun at one focus.
    ii. For each planet, the line from the sun to the planet sweeps out equal areas in equal times.

Kepler spent many years verifying these laws and formulating a third law, which relates the orbital periods and mean distances from the sun.

a. Plot the period time T versus the mean distance r using the updated observational data as follows:

Planet	Period (days)	Mean distance from sun (km $\times 10^{-6}$)
Mercury	88	57.9
Venus	225	108.2
Earth	365	149.6
Mars	687	227.9
Jupiter	4329	778.3
Saturn	10,753	1427
Uranus	30,660	2870
Neptune	60,150	4497
Pluto	90,670	5907

b. Assuming a relationship of the form

$$T = Cr^a$$

determine the parameters C and a by plotting $\ln T$ versus $\ln r$. Does the model seem reasonable? Can you now formulate Kepler's third law?

4.2 ANALYTIC METHODS OF MODEL FITTING

In this section we investigate several criteria for fitting curves to a collection of data points. Each criterion gives a way for selecting the "best" curve from a given family in the sense that the curve most accurately represents the data according to the criterion. We also discuss how the various criteria are related.

The Chebyshev Approximation Criterion

In the preceding section we graphically fit lines to a given collection of data points. One of the "best fit" criteria used was to minimize the largest distance from the line to any corresponding data point. Let's analyze this geometric construction. Given a collection of m data points (x_i, y_i), $i = 1, 2, \ldots,$ m, fit the collection to that line $y = ax + b$, determined by the parameters a and b, that minimizes the distance between any data point (x_i, y_i) and its corresponding point on the line $(x_i, ax_i + b)$. That is, the largest absolute deviation $|y_i - y(x_i)|$ is minimized over the entire collection of data points. Now let's generalize this criterion.

Given some function type $y = f(x)$ and a collection of m data points (x_i, y_i), minimize the largest absolute deviation $|y_i - f(x_i)|$ over the entire collection. That is, determine the parameters of the function type $y = f(x)$ that minimizes the number

$$\text{Maximum} \, |y_i - f(x_i)| \qquad i = 1, 2, \ldots, m \qquad (4.1)$$

Chebyshev approximation criterion
This important criterion is often called the **Chebyshev approximation criterion.** The difficulty with the Chebyshev criterion is that it is often complicated to apply in practice, at least using only the elementary calculus. The optimization problems that result from applying the criterion may require advanced mathematical procedures or numerical algorithms requiring the use of a computer.

For example, suppose you want to measure the line segments AB, BC, and AC represented in Figure 4-12. Assume your measurements yield the estimates $AB = 13$, $BC = 7$, and $AC = 19$. As you should expect in any physical measuring process, discrepancy results. In this situation, the values of AB and BC add up to 20 rather than the estimated $AC = 19$. Let's resolve the discrepancy of 1 unit using the Chebyshev criterion. That is, we will assign values to the three line segments in such a way that the largest absolute deviation between any corresponding pair of assigned and observed values is minimized. Assume the same degree of confidence in each measurement so that each measurement has equal weight. In that case the discrepancy should be distributed equally across each segment, resulting in the predictions $AB = 12\frac{2}{3}$, $BC = 6\frac{2}{3}$, and $AC = 19\frac{1}{3}$. Thus, each absolute deviation is $1/3$. Convince yourself that reducing any of these deviations causes one of the other deviations to increase. (Remember that $AB + BC$ must equal AC.) Let's formulate this problem symbolically.

Let x_1 represent the true value of the length of the segment AB, and x_2 the true value for BC. For ease of our presentation, let r_1, r_2, and r_3 represent the discrepancies between the true and measured values as follows:

$$x_1 - 13 = r_1 \qquad \text{(Line Segment } AB\text{)}$$
$$x_2 - 7 = r_2 \qquad \text{(Line Segment } BC\text{)}$$
$$x_1 + x_2 - 19 = r_3 \qquad \text{(Line Segment } AC\text{)}$$

FIGURE 4-12 The line segment AC is divided into two segments AB and BC.

residuals
The numbers r_1, r_2, and r_3 are called **residuals.**

If the Chebyshev approximation criterion is applied, values would be assigned to x_1 and x_2 in such a way as to minimize the largest of the three numbers $|r_1|, |r_2|, |r_3|$. If we call that largest number r, then we want to

$$\text{Minimize } r$$

Subject to the three conditions:

$$|r_1| \leqslant r \quad \text{or} \quad -r \leqslant r_1 \leqslant r$$
$$|r_2| \leqslant r \quad \text{or} \quad -r \leqslant r_2 \leqslant r$$
$$|r_3| \leqslant r \quad \text{or} \quad -r \leqslant r_3 \leqslant r$$

Each of these conditions can be replaced by two inequalities. For example, $|r_1| \leqslant r$ can be replaced by $r - r_1 \geqslant 0$ and $r + r_1 \geqslant 0$. If this is done for each condition, the problem can be stated in the form of a classical mathematical problem:

$$\text{Minimize } r$$

Subject to:

$$
\begin{array}{lll}
r - x_1 & + 13 \geqslant 0 & (r - r_1 \geqslant 0) \\
r + x_1 & - 13 \geqslant 0 & (r + r_1 \geqslant 0) \\
r & - x_2 + 7 \geqslant 0 & (r - r_2 \geqslant 0) \\
r & + x_2 - 7 \geqslant 0 & (r + r_2 \geqslant 0) \\
r - x_1 & - x_2 + 19 \geqslant 0 & (r - r_3 \geqslant 0) \\
r + x_1 & + x_2 - 19 \geqslant 0 & (r + r_3 \geqslant 0)
\end{array}
$$

linear program   This problem is called a **linear program.** We will discuss linear programs further in Chapter 5. Even large linear programs can be solved by computer implementation of an algorithm known as "the Simplex Method." In the preceding line segment example, the Simplex Method yields a minimum value of $r = 1/3$, and $x_1 = 12\frac{2}{3}$ and $x_2 = 6\frac{2}{3}$.

We now generalize this procedure. Given some function type $y = f(x)$, whose parameters are to be determined, and a collection of m data points (x_i, y_i), define the residuals $r_i = y_i - f(x_i)$. If r represents the largest absolute value of these residuals, then the problem is to

$$\text{Minimize } r$$

Subject to:

$$
\left.\begin{array}{l}
r - r_i \geqslant 0 \\
r + r_i \geqslant 0
\end{array}\right\} \text{ for } i = 1, 2, \ldots, m
$$

Although we discuss linear programs in Chapter 5, we should note here that the model resulting from this procedure is not always a linear program; for example, consider fitting the function $f(x) = \sin kx$. It should also be noted that many computer codes of the simplex algorithm require using variables that are allowed to assume only nonnegative values. This requirement can be accomplished with a simple substitution (see Problem 5).

As you will see, alternative criteria lead to optimization problems that often can be resolved more conveniently. Primarily for this reason, the Chebyshev criterion is not used often for fitting a curve to a finite collection of data points. However, its application should be considered whenever minimizing the largest absolute deviation is important. (We consider several applications of the criterion in Chapter 5.) Furthermore, the principle underlying the Chebyshev criterion is extremely important when you are replacing a function defined over an interval by another function and the largest difference between the two functions over the interval must be minimized. This topic is studied in approximation theory and is typically covered in introductory numerical analysis.

Minimizing the Sum of the Absolute Deviations

When we were graphically fitting lines to the data in Section 4.1, one of our criteria minimized the total sum of the absolute deviations between the data points and their corresponding points on the fitted line. This criterion can be generalized: given some function type $y = f(x)$ and a collection of m data points (x_i, y_i), minimize the sum of the absolute deviations $|y_i - f(x_i)|$. That is, determine the parameters of the function type $y = f(x)$ to minimize

$$\sum_{i=1}^{m} |y_i - f(x_i)| \tag{4.2}$$

If we let $R_i = |y_i - f(x_i)|$, $i = 1, 2, \ldots, m$ represent each absolute deviation, then the preceding criterion (4.2) can be interpreted as minimizing the length of the line formed by adding together the numbers R_i. This is illustrated for the case $m = 2$ in Figure 4-13.

Although we geometrically applied this criterion in Section 4.1 when the function type $y = f(x)$ is a line, the general criterion presents severe problems. In order to solve this optimization problem using the calculus, we need to differentiate the sum (4.2) with respect to the parameters of $f(x)$ in order to find the critical points. However, the various derivatives of the sum fail to be continuous because of the presence of the absolute values, so we will not pursue this criterion any further. In Chapter 5 we consider other applications of the criterion and discuss briefly how solutions to the problem can be approximated numerically.

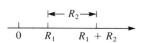

FIGURE 4-13 A geometrical interpretation of minimizing the sum of the absolute deviations.

The Least-Squares Criterion

least-squares criterion   Currently the most frequently used curve-fitting criterion is the **least-squares criterion.** Using the same notation as shown earlier, the problem is to determine the parameters of the function type $y = f(x)$ in order to minimize the sum

$$\sum_{i=1}^{m} [y_i - f(x_i)]^2 \tag{4.3}$$

Part of the popularity of this criterion stems from the ease with which the resulting optimization problem can be solved using only the calculus of several variables. However, relatively recent advances in mathematical programming techniques (such as the Simplex Method for solving many applications of the Chebyshev criterion), and advances in numerical methods for approximating solutions to the criterion (4.2), promise to dissipate this advantage. The justification for the use of the least-squares method increases when considering probabilistic arguments that assume the errors are distributed randomly. However, we will not discuss such probabilistic arguments in this text.

We now give a geometric interpretation of the least-squares criterion. Consider the case of three data points and let $R_i = |y_i - f(x_i)|$ denote the

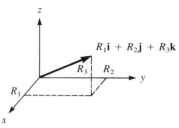

FIGURE 4-14 A geometrical interpretation of the least-squares criterion.

absolute deviation between the observed and predicted values for $i = 1, 2, 3$. You can think of the R_i as the scalar components of a "deviation vector" as depicted in Figure 4-14. Thus the vector $\mathbf{R} = R_1\mathbf{i} + R_2\mathbf{j} + R_3\mathbf{k}$ represents the resultant deviation between the observed and predicted values. The magnitude of the deviation vector is given by

$$|\mathbf{R}| = \sqrt{R_1{}^2 + R_2{}^2 + R_3{}^2}$$

In order to minimize $|\mathbf{R}|$ we can minimize $|\mathbf{R}|^2$ (see Problem 1). Thus the least-squares problem is to determine the parameters of the function type $y = f(x)$ such that

$$|\mathbf{R}|^2 = \sum_{i=1}^{3} R_i{}^2 = \sum_{i=1}^{3} [y_i - f(x_i)]^2$$

is minimized. That is, we may interpret the least-squares criterion as minimizing the magnitude of the vector whose coordinates represent the absolute deviations between the observed and predicted values.

Relating the Criteria

The geometric interpretations of the three curve-fitting criteria help in providing a qualitative description comparing the criteria. Minimizing the sum of the absolute deviations tends to treat each data point with equal weight and to average the deviations. The Chebyshev criterion gives more weight to a single point potentially having a large deviation. The least-squares criterion is somewhere in between as far as weighting individual points with significant deviations is concerned. But let's be more precise. Because the Chebyshev and least-squares criteria are the most convenient to apply analytically, we now derive a method for relating the deviations resulting from using these two criteria.

Suppose the Chebyshev criterion is applied and the resulting optimization problem solved to yield the function $f_1(x)$. The absolute deviations resulting from the fit are defined as follows:

$$|y_i - f_1(x_i)| = c_i, \qquad i = 1, 2, \ldots, m$$

Now, define c_{max} as the largest of the absolute deviations c_i. There is a special significance attached to c_{max}. Since the parameters of the function $f_1(x)$ are determined so as to minimize the value of c_{max}, it is the minimal largest absolute deviation obtainable.

On the other hand, suppose the least-squares criterion is applied and the resulting optimization problem solved to yield the function $f_2(x)$. The absolute deviations resulting from this fit are then given by

$$|y_i - f_2(x_i)| = d_i, \qquad i = 1, 2, \ldots, m$$

Define d_{max} as the largest of the absolute deviations d_i. At this point it can only be said that d_{max} is at least as large as c_{max} because of the special significance of the latter as discussed previously. However, let's attempt to relate d_{max} and c_{max} more precisely.

The special significance the least-squares criterion attaches to the d_i is that the sum of their squares is the smallest such sum obtainable. Thus it must be true that

$$d_1{}^2 + d_2{}^2 + \cdots + d_m{}^2 \leqslant c_1{}^2 + c_2{}^2 + \cdots + c_m{}^2$$

Since $c_i \leqslant c_{max}$ for every i, these inequalities imply

$$d_1{}^2 + d_2{}^2 + \cdots + d_m{}^2 \leqslant mc_{max}^2$$

or

$$\sqrt{\frac{d_1{}^2 + d_2{}^2 + \cdots + d_m{}^2}{m}} \leqslant c_{max}$$

For ease of discussion define

$$D = \sqrt{\frac{d_1{}^2 + d_2{}^2 + \cdots + d_m{}^2}{m}}$$

Thus,

$$D \leqslant c_{max} \leqslant d_{max}$$

This last relationship is very revealing. Suppose it is more convenient to apply the least-squares criterion in a particular situation, but there is concern about the largest absolute deviation c_{max} that may result. If we compute D, a lower bound on c_{max} is obtained, and d_{max} gives an upper bound. Thus, if there is considerable difference between D and d_{max}, the modeler should consider applying the Chebyshev criterion.

4.2 PROBLEMS

1. Using elementary calculus, show that the minimum and maximum points for $y = f(x)$ occur among the minimum and maximum points for $y = f^2(x)$. Assuming $f(x) \geqslant 0$, why can we minimize $f(x)$ by minimizing $f^2(x)$?

2. For each of the following data sets, formulate the mathematical model that minimizes the largest deviation between the data and the line $y = ax + b$. If a computer is available, solve for the estimates of a and b.

a.

x	1.0	2.3	3.7	4.2	6.1	7.0
y	3.6	3.0	3.2	5.1	5.3	6.8

b.

x	29.1	48.2	72.7	92.0	118	140	165	199
y	0.0493	0.0821	0.123	0.154	0.197	0.234	0.274	0.328

c.

x	2.5	3.0	3.5	4.0	4.5	5.0	5.5
y	4.32	4.83	5.27	5.74	6.26	6.79	7.23

3. For the data below, formulate the mathematical model that minimizes the largest deviation between the data and the model $y = c_1x^2 + c_2x + c_3$. If a computer code is available, solve for the estimates of c_1, c_2, and c_3.

x	0.1	0.2	0.3	0.4	0.5
y	0.06	0.12	0.36	0.65	0.95

4. For the following data, formulate the mathematical model that minimizes the largest deviation between the data and the model $P = ae^{bt}$. If a computer code is available, solve for the estimates of a and b.

t	7	14	21	28	35	42
P	8	41	133	250	280	297

5. Suppose the variable x_1 can assume any real value. Show that the following substitution using nonnegative variables x_2 and x_3 permits x_1 to assume any real value:

$$x_1 = x_2 - x_3 \quad \text{where } x_1 \text{ is unconstrained}$$
$$\text{and} \quad x_2 \geqslant 0 \quad \text{and} \quad x_3 \geqslant 0$$

Thus if a computer code allows only nonnegative variables, the substitution allows for solving the linear program in the variables x_2 and x_3, and then recovering the value of the variable x_1.

4.3 APPLYING THE LEAST-SQUARES CRITERION

Suppose our assumptions lead us to expect a model of a certain type and that data have been collected and analyzed. In this section the least-squares criterion is applied to estimate the parameters for several types of curves.

Fitting a Straight Line

Suppose a model of the form $y = Ax + B$ is expected and it's been decided to use the m data points (x_i, y_i), $i = 1, 2, \ldots, m$ to estimate A and B. Denote the least-squares estimate of $y = Ax + B$ by $f(x) = ax + b$. Applying the least-squares criterion (4.3) to this situation requires the minimization of

$$S = \sum_{i=1}^{m} [y_i - f(x_i)]^2 = \sum_{i=1}^{m} (y_i - ax_i - b)^2$$

A necessary condition for optimality is that the two partial derivatives $\partial S/\partial a$ and $\partial S/\partial b$ equal zero yielding the equations

$$\frac{\partial S}{\partial a} = -2 \sum_{i=1}^{m} (y_i - ax_i - b)x_i = 0$$

$$\frac{\partial S}{\partial b} = -2 \sum_{i=1}^{m} (y_i - ax_i - b) = 0$$

These equations can be rewritten to give

$$\left.\begin{aligned} a \sum_{i=1}^{m} x_i^2 + b \sum_{i=1}^{m} x_i &= \sum_{i=1}^{m} x_i y_i \\ a \sum_{i=1}^{m} x_i + mb &= \sum_{i=1}^{m} y_i \end{aligned}\right\} \tag{4.4}$$

The preceding equations can be solved for a and b once all the values for x_i and y_i are substituted into them. The solutions for the parameters a and b are easily obtained by elimination and are found to be (see Problem 1)

$$a = \frac{m \sum x_i y_i - \sum x_i \sum y_i}{m \sum x_i^2 - (\sum x_i)^2}, \text{ the slope} \tag{4.5}$$

and

$$b = \frac{\sum x_i^2 \sum y_i - \sum x_i y_i \sum x_i}{m \sum x_i^2 - (\sum x_i)^2}, \text{ the intercept} \tag{4.6}$$

Computer codes are easily written to compute these values for a and b for any collection of data points. Equations (4.4) are called the "normal equations."

Fitting a Power Curve

Next let's use the least-squares criterion to fit a curve of the form $y = Ax^n$, where n is fixed, to a given collection of data points. Call the least-squares estimate of the model $f(x) = ax^n$. Application of the criterion then requires the minimization of

$$S = \sum_{i=1}^{m} [y_i - f(x_i)]^2 = \sum_{i=1}^{m} [y_i - ax_i^n]^2$$

A necessary condition for optimality is that the derivative dS/da equals zero giving the equation:

$$\frac{dS}{da} = -2 \sum_{i=1}^{m} x_i^n [y_i - ax_i^n] = 0$$

Solving the equation for a yields

$$a = \frac{\sum x_i^n y_i}{\sum x_i^{2n}} \tag{4.7}$$

Remember, the number n is *fixed* in Equation (4.7).

Of course, the least-squares criterion can be applied to other models as well. The limitation in applying the method lies in calculating the various derivatives required in the optimization process, setting these derivatives equal to zero, and solving the resulting equations for the parameters in the model type.

For example, let's fit $y = Ax^2$ to the following collection of data and predict the value of y when $x = 2.25$:

TABLE 4-3 Data collected to fit $y = Ax^2$

x	0.5	1.0	1.5	2.0	2.5
y	0.7	3.4	7.2	12.4	20.1

In this case the least-squares estimate a is given by

$$a = \frac{\sum x_i^2 y_i}{\sum x_i^4}$$

We compute $\sum x_i^4 = 61.1875$, $\sum x_i^2 y_i = 195.0$ to yield $a = 3.1869$ (to four decimal places). This computation gives the least-squares approximate model

$$y = 3.1869x^2$$

When $x = 2.25$, the predicted value for y is 16.1337.

Transformed Least-Squares Fit

While the least-squares criterion appears easy to apply in theory, in practice it may be difficult. For example, consider fitting the model $y = Ae^{Bx}$ using the least-squares criterion. Call the least-squares estimate of the model $f(x) = ae^{bx}$. Application of the criterion then requires the minimization of

$$S = \sum_{i=1}^{m} [y_i - f(x_i)]^2 = \sum_{i=1}^{m} [y_i - ae^{bx_i}]^2$$

A necessary condition for optimality is that $\partial S/\partial a = \partial S/\partial b = 0$. Formulate the conditions and convince yourself that solving the resulting system of nonlinear equations would not be easy. Many simple models result in derivatives that are very complex or in systems of equations that are difficult to solve. For this reason we use transformations that allow us to *approximate* the least-squares model.

In graphically fitting lines to data in Section 4.1, we sometimes found it convenient to transform the data first and then fit a line to the transformed points. For example, in graphically fitting $y = Ce^x$, we found it convenient to plot $\ln y$ versus x and then fit a line to the transformed data. The same idea can be used with the least-squares criterion in order to simplify the computational aspects of the process. In particular, if a convenient substitution can be found so that the problem takes the form $Y = AX + B$ in the transformed variables X and Y, then Equations (4.4) can be used to fit a line to the transformed variables. We illustrate the technique with the example that we just worked out.

Suppose we wish to fit the power curve $y = Ax^N$ to a collection of data points. Let's denote the estimate of A by α and the estimate of N by n. Taking the logarithm of both sides of the equation $y = \alpha x^n$ yields

$$\ln y = \ln \alpha + n \ln x \qquad \textbf{(4.8)}$$

Note that in plotting the variables $\ln y$ versus $\ln x$, Equation (4.8) yields a straight line. On that graph, $\ln \alpha$ is the intercept when $\ln x = 0$ and the slope of the line is n. Using Equations (4.5) and (4.6) to solve for the slope n and intercept $\ln \alpha$ with the transformed variables and $m = 5$ data points, we have

$$n = \frac{5 \sum (\ln x_i)(\ln y_i) - (\sum \ln x_i)(\sum \ln y_i)}{5 \sum (\ln x_i)^2 - (\sum \ln x_i)^2}$$

$$\ln \alpha = \frac{\sum (\ln x_i)^2 \sum \ln y_i - \sum (\ln x_i)(\ln y_i) \sum \ln x_i}{5 \sum (\ln x_i)^2 - (\sum \ln x_i)^2}$$

For the data displayed in Table 4-3 we get $\sum \ln x_i = 1.3217558$, $\sum \ln y_i = 8.359597801$, $\sum (\ln x_i)^2 = 1.9648967$, and $\sum (\ln x_i)(\ln y_i) = 5.542315175$ yielding $n = 2.062809314$ and $\ln \alpha = 1.126613508$ or $\alpha = 3.085190815$. Thus our least-squares best fit of Equation (4.8) is (rounded to four decimal places)

$$y = 3.0852 x^{2.0628}$$

This model predicts $y = 16.4348$ when $x = 2.25$. Note, however, that this model fails to be a quadratic like the one we fit previously.

Suppose we still wish to fit a *quadratic* $y = Ax^2$ to the collection of data. Denote the estimate of A by a_1 to distinguish this constant from the constants a and α computed previously. Taking the logarithm of both sides of the equation $y = a_1 x^2$ yields

$$\ln y = \ln a_1 + 2 \ln x$$

In this situation the graph of $\ln y$ versus $\ln x$ is a straight line of slope 2 and intercept $\ln a_1$. Using the second equation in (4.4) to compute the intercept, we have

$$2 \sum \ln x_i + 5 \ln a_1 = \sum \ln y_i$$

For the data displayed in Table 4-3 we get $\sum \ln x_i = 1.3217558$ and $\sum \ln y_i = 8.359597801$. Therefore this last equation gives $\ln a_1 = 1.14321724$ or $a_1 = 3.136844129$, yielding the least-squares best fit (rounded to four decimal places)

$$y = 3.1368x^2$$

The model predicts $y = 15.8801$ when $x = 2.25$, which differs significantly from the value 16.1337 predicted by the first quadratic $y = 3.1869x^2$ obtained as the least-squares best fit of $y = Ax^2$ without transforming the data. We compare these two quadratic models (as well as a third model) in the next section.

The preceding example illustrates two facts. First, if an equation can be transformed to yield an equation of a straight line in the transformed variables, then Equations (4.4) can be used directly to solve for the slope and intercept of the transformed graph. Second, the least-squares best fit of the transformed equation *does not* coincide with the least-squares best fit of the original equation. The reason for this discrepancy is that the resulting optimization problems are different. In the case of the original problem we are finding the curve that minimizes the sum of the squares of the deviations using the original data, whereas in the case of the transformed problem we are minimizing the sum of the squares of the deviations using the *transformed* variables.

4.3 PROBLEMS

1. Solve the equations given by (4.4) to obtain the values of the parameters a and b given by (4.5) and (4.6), respectively.
2. Use Equations (4.5) and (4.6) to estimate the coefficients of the line $y = ax + b$ such that the sum of the squared deviations between the line and the following data points is minimized.

a.

x	1.0	2.3	3.7	4.2	6.1	7.0
y	3.6	3.0	3.2	5.1	5.3	6.8

b.

x	29.1	48.2	72.7	92.0	118	140	165	199
y	0.0493	0.0821	0.123	0.154	0.197	0.234	0.274	0.328

c.

x	2.5	3.0	3.5	4.0	4.5	5.0	5.5
y	4.32	4.83	5.27	5.74	6.26	6.79	7.23

For each problem, compute D and d_{max} to bound c_{max}. Compare the results with your solutions in Problem 2 in Section 4.2.

3. Derive the equations that minimize the sum of the squared deviations between a set of data points and the quadratic model $y = c_1 x^2 + c_2 x + c_3$. Use the equations to find estimates of c_1, c_2, and c_3 for the following set of data:

x	0.1	0.2	0.3	0.4	0.5
y	0.06	0.12	0.36	0.65	0.95

Compute D and d_{max} to bound c_{max}. Compare the results with your solution to Problem 3 in Section 4.2.

4. Make an appropriate transformation to fit the model $P = ae^{bt}$ using Equations (4.4). Estimate a and b.

t	7	14	21	28	35	42
P	8	41	133	250	280	297

5. Examine closely the system of equations that result when you fit the quadratic in Problem 3. Suppose $c_2 = 0$. What would be the corresponding system of equations? Repeat for the cases $c_1 = 0$ and $c_3 = 0$. Can you suggest a system of equations for a cubic? Check your result. Explain how you would generalize the system of Equations (4.4) to fit any polynomial. Explain what you would do if one or more of the coefficients in the polynomial is zero.

6. A rule of thumb given for computing a person's weight is as follows: for a female, multiply the height in inches by 3.5 and subtract 108; for a male, multiply the height in inches by 4.0 and subtract 128. If the person is "small bone-structured" adjust this computation by subtracting 10%; for a "large bone-structured" person, add 10%. No adjustment is made for an average-size person. Gather some data on the weight versus height of people of differing age, size, and sex. Using Equations (4.4), fit a straight line to your data for males and another straight line to your data for females. What are the slopes and intercepts of those lines? How do the results compare with the rule of thumb?

4.3 PROJECTS

1. Complete the requirements of the module "Curve Fitting via the Criterion of Least Squares," by John W. Alexander, Jr., UMAP 321. This unit provides an easy introduction to correlations, scatter diagrams (polynomial, logarithmic, and exponential scatters), and lines and curves of regression. Students construct scatter diagrams, choose appropriate functions to fit specific data, and use a computer program to fit curves. Recommended for students who wish an introduction to statistical measures of correlation.

4.3 FURTHER READING

Burden, Richard L., J. Douglas Faires, and Albert C. Reynolds. *Numerical Analysis,* 2nd ed. Boston: Prindle, Weber, and Schmidt, 1978.

Hamming, R. W. *Numerical Methods for Scientists and Engineers.* New York: McGraw-Hill, 1973.

Cheney, E. Ward, and David Kincaid. *Numerical Mathematics and Computing.* Monterey, Calif.: Brooks/Cole, 1984.

Stanton, Ralph G. *Numerical Methods for Science and Engineering.* Englewood Cliffs, N.J.: Prentice-Hall, 1961.

Stiefel, Edward L. *An Introduction to Numerical Mathematics.* New York: Academic Press, 1963.

4.4 CHOOSING A BEST MODEL: SOME EXAMPLES

Let's consider the adequacy of the various models of the form $y = Ax^2$ that we fit using the least-squares and transformed least-squares criteria in the previous section. Using the least-squares criterion, we obtained the model $y = 3.1869x^2$. An obvious way of evaluating how well the model fits the data is to compute the deviations between the model and the actual data. If we compute the sum of the squares of the deviations, we can bound c_{max} as well. For the model $y = 3.1869x^2$ and the data given in Table 4-3, we compute the deviations shown in Table 4-4.

TABLE 4-4 Deviations between the data in Table 4-3 and the fitted model $y = 3.1869x^2$

x_i	0.5	1.0	1.5	2.0	2.5
$y_i - y(x_i)$	−0.0967	0.2131	0.02948	−0.3476	0.181875

From Table 4-4 we compute the sum of the squares of the deviations as 0.20954, so that $D = (0.20954/5)^{1/2} = 0.204714$. Since the largest absolute deviation is 0.3476 when $x = 2.0$, c_{max} can be bounded as follows:

$$D = 0.204714 \leqslant c_{max} \leqslant 0.3476 = d_{max}$$

Let's find c_{max}. Since there are five data points, the mathematical problem is to minimize the largest of the five numbers $|r_i| = |y_i - y(x_i)|$. Calling that largest number r, we want to minimize r subject to $r \geqslant |r_i|$, $i = 1, 2, 3, 4, 5$. Denote our model by $y(x) = a_2x^2$. Then substitution of the observed data points in Table 4-3 into the inequalities $r \geqslant r_i$ and $r \geqslant -r_i$ for each $i = 1, 2, 3, 4, 5$ yields the following linear program:

$$\text{Minimize } r$$

Subject to:

$$r - r_1 = r - \ \ (0.7 - 0.25a_2) \geqslant 0$$
$$r + r_1 = r + \ \ (0.7 - 0.25a_2) \geqslant 0$$
$$r - r_2 = r - \ \ (3.4 - \ \ \ \ \ a_2) \geqslant 0$$
$$r + r_2 = r + \ \ (3.4 - \ \ \ \ \ a_2) \geqslant 0$$
$$r - r_3 = r - \ \ (7.2 - 2.25a_2) \geqslant 0$$
$$r + r_3 = r + \ \ (7.2 - 2.25a_2) \geqslant 0$$
$$r - r_4 = r - (12.4 - \ \ \ \ 4a_2) \geqslant 0$$
$$r + r_4 = r + (12.4 - \ \ \ \ 4a_2) \geqslant 0$$
$$r - r_5 = r - (20.1 - 6.25a_2) \geqslant 0$$
$$r + r_5 = r + (20.1 - 6.25a_2) \geqslant 0$$

The solution of the preceding linear program yields $r = 0.28293$, and $a_2 = 3.17073$. Thus we have reduced our largest deviation from $d_{max} = 0.3476$ to $c_{max} = 0.28293$. Note that we can reduce the largest deviation no further than 0.28293 for the model type $y = Ax^2$.

Now we have determined three estimates of the parameter A for the model type $y = Ax^2$. Which estimate is best? For each model we can readily compute the deviations from each data point as recorded in Table 4-5:

TABLE 4-5 Summary of the deviations for each model $y = Ax^2$

x_i	y_i	$y_i - 3.1869x_i^2$	$y_i - 3.1368x_i^2$	$y_i - 3.17073x_i^2$
0.5	0.7	-0.0967	-0.0842	-0.0927
1.0	3.4	0.2131	0.2632	0.2293
1.5	7.2	0.029475	0.1422	0.0659
2.0	12.4	-0.3476	-0.1472	-0.2829
2.5	20.1	0.181875	0.4950	0.28293

For each of the three models we can compute the sum of the squares of the deviations and the maximum absolute deviation. The results are shown in Table 4-6.

TABLE 4-6 Summary of the results for the three models

Criterion	Model	$\sum [y_i - y(x_i)]^2$	Max $|y_i - y(x_i)|$
Least-squares	$y = 3.1869x^2$	0.2095	0.3476
Transformed least-squares	$y = 3.1368x^2$	0.3633	0.4950
Chebyshev	$y = 3.17073x^2$	0.2256	0.28293

As you would expect, each model has something to commend it. However, notice the increase in the sum of the squares of the deviations in the transformed least-squares model. It is tempting to apply a simple rule such as "choose the model with the smallest sum of squares," or "choose the model with the smallest absolute deviation." (Other statistical indicators of goodness of fit exist as well. For example, see *Probability and Statistics in Engineering and Management Science*, by William W. Hines and Douglas C. Montgomery, Wiley, New York, 1972.) These indicators are very useful for eliminating obviously poor models, but there is no easy answer to the question "which model is best?" The model with the smallest absolute deviation or the smallest sum of squares may fit very poorly over the range where you most intend to use it. Furthermore, as you will see in Chapter 6, models can easily be constructed which pass through each data point thereby yielding a zero sum of squares and zero maximum deviation. So we need to answer the question "Which model is best?" on a case-by-case basis, taking into account such things as the purpose of the model, the precision demanded by the scenario, the accuracy of the data, and the range of values for the independent variable over which the model will be used.

When choosing among models or judging the adequacy of a model, we may find it tempting to rely on the value of the "best fit" criterion being used. For example, it is tempting to choose the model that has the smallest sum of squared deviations for the given data set or conclude that a sum of squared deviations less than a predetermined value indicates a good fit. However, in isolation these indicators may be very misleading. For example, consider the data displayed in Figure 4-15. In each of the four cases the model $y = x$ results in exactly the same deviations. Without the benefit of the graphs, therefore, we might conclude that in each case the model fits the data "about the same." However, as the graphs show, there is a significant variation in each model's ability *to capture the trend of the data*. The fol-

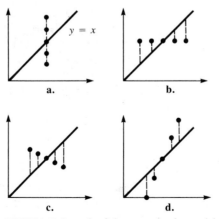

FIGURE 4-15 In each of these graphs the model $y = x$ has the same sum of squared deviations.

lowing examples illustrate how the various indicators may be used to help in reaching a decision on the adequacy of a particular model. Normally, a graphical plot is of great benefit.

Example 1 Vehicular Stopping Distance

Let's reconsider the problem of predicting a motor vehicle's stopping distance as a function of its speed. (This problem was addressed in Sections 2.2 and 3.3.) In Section 3.3 the submodel that reaction distance d_r is proportional to the velocity v was tested graphically and the constant of proportionality was estimated to be 1.1. Similarly, the submodel predicting a proportionality between braking distance d_b and the square of the velocity was tested. We found reasonable agreement with the submodel and estimated the proportionality constant to be 0.054. Hence the model for stopping distance was given by

$$d = 1.1v + 0.054v^2 \qquad (4.9)$$

We now fit these submodels analytically and compare the various fits.

To fit the model $d_r = Av$ using the least-squares criterion, we use the formula from (4.7):

$$A = \frac{\sum x_i y_i}{\sum x_i^2}$$

where y_i denotes the driver reaction distance and x_i denotes the speed at each data point. For the 13 data points given in Table 3-1 of Chapter 3, we compute $\sum x_i y_i = 40905$ and $\sum x_i^2 = 37050$, giving $A = 1.104049$.

For the model type $d_b = Bv^2$, we use the formula:

$$B = \frac{\sum x_i^2 y_i}{\sum x_i^4}$$

where y_i denotes the average braking distance and x_i denotes the speed at each data point. For the 13 data points given in Table 3-1 of Chapter 3, we compute $\sum x_i^2 y_i = 8258350$ and $\sum x_i^4 = 152343750$, giving $B = 0.054209$. Since the data are relatively imprecise and the modeling done qualitatively, we round the coefficients to obtain the model:

$$d = 1.104v + 0.0542v^2 \qquad (4.10)$$

Model (4.10) does not differ significantly from that obtained graphically in Chapter 3.

Next, let's analyze how well the model fits. We can readily compute the deviations between the observed data points in Table 3-2 and the values predicted by Models (4.9) and (4.10). The deviations are summarized in Table 4-7. The fits of both models are very similar. The largest absolute deviation for Model (4.9) is 30.4 and for Model (4.10) it is 28.8. Note that both models overestimate the stopping distance up to 70 mph, and then they begin to

TABLE 4-7 Deviations from the observed data points and Models (4.9) and (4.10)

Speed	Graphical model (4.9)	Least-squares model (4.10)
20	1.6	1.76
25	5.25	5.475
30	8.1	8.4
35	13.15	13.535
40	14.4	14.88
45	16.35	16.935
50	17	17.7
55	14.35	15.175
60	12.4	13.36
65	7.15	8.255
70	−1.4	−0.14
75	−14.75	−13.325
80	−30.4	−28.8

underestimate the stopping distance. We should point out that a better fitting model would be obtained by directly fitting the data for total stopping distance to

$$d = k_1 v + k_2 v^2$$

instead of fitting the submodels individually as we did. The advantage of fitting the submodels individually and then testing each submodel is that we can measure how well they explain the behavior.

A plot of the proposed model(s) and the observed data points is useful to determine how well the model fits the data. Model (4.10) and the observations are plotted in Figure 4-16. It is evident from the figure that a definite trend exists in the data and that Model (4.10) does a reasonable job of capturing that trend, especially at the lower speeds.

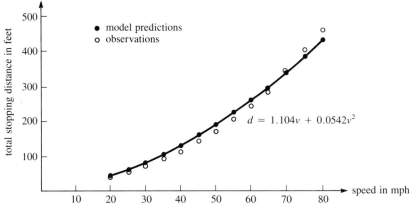

FIGURE 4-16 A plot of the proposed model and the observed data points provides a visual check on the adequacy of the model.

A powerful technique for quickly determining where the model is breaking down is to plot the deviations (residuals) as a function of the independent variable(s). For Model (4.10) a plot of the deviations is given in Figure 4-17 showing that the model is indeed reasonable up to 70 mph. However, beyond 70 mph there is a breakdown in the model's ability to predict the observed behavior.

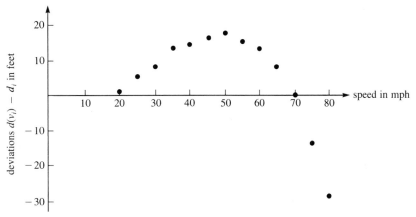

FIGURE 4-17 A plot of the deviations (residuals) reveals those regions where the model does not fit well.

Let's examine Figure 4-17 more closely. Note that while the deviations up to 70 mph are relatively small, they are all positive! If the model fully explains the behavior, the deviations should not only be small, but they should be positive and negative as well. Why? However, in Figure 4-17 we note a definite pattern in the nature of the deviations, which might cause us to reexamine the model and/or the data. The nature of the pattern in the deviations can give us clues on how to refine the model further. In this case the imprecision in the data collection process probably does not warrant further model refinement.

Example 2 The Elapsed Time of a Tape Recorder

In Section 3.5 we identified the problem of relating a counter on a tape recorder to the amount of playing time that had elapsed. The following model was constructed to explain the behavior:

$$t = k_2 c + k_3 c^2$$

where t is the elapsed time, c is the counter reading, and k_2 and k_3 are constants that must be determined. We now determine k_2 and k_3 for several tape recorders.

Let's begin by determining the equations for the parameters to fit the curve $y = ax + bx^2$ using the least-squares criterion. Thus the problem is to minimize

$$S = \sum [y_i - ax_i - bx_i^2]^2$$

The necessary conditions for the minimum are $\partial S/\partial a = 0$ and $\partial S/\partial b = 0$, which yield the two equations:

$$\left.\begin{array}{l}(\sum x_i^2)a + (\sum x_i^3)b = \sum x_i y_i \\ (\sum x_i^3)a + (\sum x_i^4)b = \sum x_i^2 y_i\end{array}\right\} \qquad \textbf{(4.11)}$$

Now let's consider fitting the model $y = ac + bc^2$, where a and b represent the least-squares estimates of k_2 and k_3 to the following set of data collected from a cassette recorder:*

c_i	100	200	300	400	500	600	700	800
t_i (sec)	205	430	677	945	1233	1542	1872	2224

Computing the coefficients in (4.11), where $x_i = c_i$ and $y_i = t_i$, yields the following system of equations:

$$(2.04 \times 10^6)a + (1.296 \times 10^9)\,b = 5.3189 \ \times 10^6$$
$$(1.296 \times 10^9)a + (8.772 \times 10^{11})\,b = 3.43539 \times 10^9$$

Solution of this system yields

$$a = 1.94293$$
$$b = 0.001045773$$

which gives the least-squares model:

$$t = 1.94293c + 0.001045773c^2$$

Next, let's see how well the model fits the data points. The results are summarized as follows:

c_i	100	200	300	400	500	600	700	800
t_i	205	430	677	945	1233	1542	1872	2224
$t(c_i)$	204.75	430.72	676.99	944.50	1232.91	1542.24	1872.48	2223.64

Note how well the model agrees with the data. The largest deviation is 0.72, which is probably about as accurate as the seconds were measured.

At this point we would want to corroborate our success with data from a second recorder. We collected the following set of data using a reel-to-reel recorder:

c	200	400	600	800	1000	1200	1400	1600
t (sec)	397	873	1425	2055	2764	3550	4413	5355

* Data collected by Barbara Henneike and Richard Staats.

Computing the appropriate coefficients in (4.11) using these data yields the system

$$(8.16 \times 10^6)a + (1.0368 \times 10^{10})b = 2.46978 \times 10^7$$
$$(1.0368 \times 10^{10})a + (1.40352 \times 10^{13})b = 3.221804 \times 10^{10}$$

Solving for a and b gives

$$a = 1.79213$$
$$b = 0.00097164$$

These results yield the model:

$$t = 1.79213c + 0.00097164c^2$$

We can compare the values predicted by the model with the observed data points as follows:

c_i	200	400	600	800	1000	1200	1400	1600
t_i	397	873	1425	2055	2764	3550	4413	5355
$t_i - t(c_i)$	.29	.69	.07	.55	.23	.28	.40	.19
% Error	.08	.08	0	.05	.01	.01	.01	0

Note that it is useful to examine the percent error $|t_i - t(c_i)|/t_i$ since a large deviation may actually represent a very good prediction. Specifically, in the example, the absolute deviation of 0.40 when $c = 1400$ represents a much smaller percentage error than does the absolute deviation of 0.29 when $c = 200$. Computing D and d_{max} for the least-squares estimate, we find

$$D = 0.386 \leqslant c_{max} \leqslant d_{max} = 0.69$$

with a sum of the squares of the deviations equal to 1.195.

Testing Your Model

Let's conclude with an important observation. In the preceding examples the data points used to fit the model are the same as those used to test the model. This situation is often necessary because a limited amount of data is available. However, instances will occur when you will want to collect additional data in order to conduct an independent test of the model. In other words, the model is fit using one set of data but tested against a second set of data. Even if a clear trend in the data seems to exist, the independent data set gives a better estimate to the size of the expected deviations, since the parameters of the model were not determined by minimizing the deviations with respect to this second set. For instance, in the reel-to-reel tape recorder model, we could have recorded the elapsed time at counter readings of 100, 300, 500, 700, 900, 1100, 1300, and 1500 and then used those observations for a test instead of the times at the counter readings 200, 400, 600, 800, 1000, 1200, 1400, and 1600.

4.4 PROBLEMS

For each of the following problems, find a model using the least-squares criterion either on the data or the transformed data (as appropriate). Compare your results with the graphical fits obtained in Problems 4.1 by computing the deviations, maximum absolute deviation, and the sum of the squared deviations for each model. Find a bound on c_{max} if the model was fit using the least-squares criterion.

1. Problem 3 in Section 4.1
2. Problem 4a in Section 4.1
3. Problem 4b in Section 4.1
4. Problem 5a in Section 4.1
5. Problem 2 in Section 4.1
6. Problem 6 in Section 4.1
7. **a.** In the following data, W represents the weight of a fish (bass) and l represents its length. Fit the model $W = kl^3$ to the data using the least-squares criterion.

Length, l (in.)	14.5	12.5	17.25	14.5	12.625	17.75	14.125	12.625
Weight, W (oz)	27	17	41	26	17	49	23	16

   b. In the following data, g represents the girth of a fish. Fit the model $W = klg^2$ to the data using the least-squares criterion.

Length, l (in.)	14.5	12.5	17.25	14.5	12.625	17.75	14.125	12.625
Girth, g (in.)	9.75	8.375	11.0	9.75	8.5	12.5	9.0	8.5
Weight, W (oz)	27	17	41	26	17	49	23	16

   c. Which of the two models fits the data better? Justify fully. Which model do you prefer? Why?
8. Use the data presented in Problem 7b to fit the models $W = cg^3$ and $W = kgl^2$. Interpret these models. Compute appropriate indicators and determine which model is "best." Explain.

4.4 PROJECTS

1. Write a computer program that finds the least-squares estimates of the coefficients in the following models:
   a. $y = ax^2 + bx + c$
   b. $y = ax^n$
2. Write a computer program that computes the deviation from the data points and any model that the user enters. Assuming that the model

was fitted using the least-squares criterion, compute D and d_{max}. Output each data point, the deviation from each data point, D, d_{max}, and the sum of the squared deviations.

3. Write a computer program that uses Equations (4.4) and the appropriate transformed data to estimate the parameters of the following models:

  a. $y = bx^a$
  b. $y = be^{ax}$
  c. $y = a \ln x + b$
  d. $y = ax^2$
  e. $y = ax^3$

MODELS REQUIRING OPTIMIZATION

INTRODUCTION

In the previous chapter we discussed the problem of fitting tested and accepted models to collected data. Several curve-fitting criteria were identified, which led to distinct mathematical models that we then attempted to solve. In particular, the criteria of minimizing the largest absolute deviation and minimizing the sum of the absolute deviations led to mathematical models that were generally quite difficult to solve. Partly because of the difficulty in solving those mathematical models, the alternative criterion of minimizing the sum of the squared deviations was studied.

There are other situations in mathematical modeling that require us to determine a "best" or "optimal" solution. It may be the problem of determining the *maximum* profit a firm can make (as in Chapter 1) or that of finding the *minimum* sum of absolute deviations between a fitted model and a set of data points (as in Chapter 4). The process of finding the "best" **optimization** solution to such problems is known as **optimization.** As you will see in this chapter, the solution of a model requiring optimization can be very difficult to obtain. The study of optimization constitutes a large and interesting field of mathematics in which extensive research is currently being conducted. While many optimization problems can be solved by a direct application of elementary calculus, others require the application of specialized mathematics, which is best studied in a separate course or even a sequence of courses. In this chapter we first present an overview of the field of optimization. Later in the chapter, some scenarios are presented that naturally give rise to models requiring optimization. Because of the objectives of this introductory text, we limit any detailed discussion of the model solutions to those that can be handled by a direct application of the elementary calculus.

In Sections 5.1 and 5.2 we provide a general classification of optimization problems, and we address situations that lead to models illustrating many types of optimization problems. In both of these sections the emphasis is on *model formulation*. This emphasis will allow you additional practice on the first several steps of the modeling process while simultaneously providing a preview of the kinds of problems you will learn to solve in advanced mathematics courses.

In Section 5.3 we address a special class of problems that can be solved using only elementary calculus. In the illustrative problem of that section we develop a model for determining an optimal "inventory strategy." That problem is concerned with deciding in what quantities and how often goods should be ordered in order to minimize the total cost of carrying an inventory. The restrictions on the various submodels are rather severe, so the sensitivity of the solutions to the assumptions is examined. (A "simulation model" with relaxed assumptions is constructed in Section 8.2.) The emphasis in Section 5.3 is on *model solution* and *model sensitivity*.

Section 5.4 is an optional section, where we address an "assignment" problem. In the illustrative example we develop a model for the assignment of people to jobs in an assembly line. This problem was first addressed in Section 2.2. In the problem section you are asked to formulate models for problems like scheduling final exams at your school and determining the optimal way of transporting goods from distribution centers to markets.

Section 5.5 is also an optional section, in which *graphical optimization* is presented. The illustrative problem addresses the management of a fishing industry and is concerned with whether a free market can lead to a satisfactory solution for fishers, consumers, and ecologists alike, or whether some type of government intervention is necessary. The graphical analysis in the example provides a qualitative preview for the type of analytic models we will develop in Chapters 9 and 10 using differential equations.

The projects in this chapter allow for a more detailed study of the optimization topics we discuss. For instance, students who wish can study linear programming or elementary ideas in the calculus of variations using the UMAP modules referenced in the various project sections.

5.1 CLASSIFYING OPTIMIZATION PROBLEMS

In order to provide a framework for discussing a class of optimization problems, we offer a basic model for such problems. Then problems are classified according to various characteristics of the basic model that are possessed by the particular problem. We discuss, too, variations from the basic model itself. The basic model is

$$\text{Optimize } f_j(\mathbf{X}) \text{ for } j \text{ in } J \qquad (5.1)$$

Subject to:

$$g_i(\mathbf{X}) \begin{Bmatrix} \geq \\ = \\ \leq \end{Bmatrix} b_i \text{ for } i \text{ in } I$$

Now let's explain the rather intimidating notation. To optimize means to maximize or minimize. The subscript j indicates that there may be one or more functions to optimize. The functions are distinguished by their integer subscripts that belong to a finite set J. We seek the vector $\mathbf{X}_0$ giving the optimal value for each of the functions $f_j(\mathbf{X})$. The various components of the vector $\mathbf{X}$ are called the **decision variables** of the model, while the functions $f_j(\mathbf{X})$ to be optimized are called the **objective functions.** By "subject to," we connote that there may be certain "side" conditions that must be met. For example, if the objective is to minimize the cost of producing a particular product, it might be specified that all contractual obligations for the product be met as side conditions. Side conditions are typically called **constraints.** The integer subscript i indicates that there may be one or more constraint relationships that must be satisfied. A constraint may be an equality (such as precisely meeting the demand for a product) or inequality (such as not exceeding budgetary limitations, or providing the minimal nutritional requirements in a diet problem). Finally, each constant b_i represents the level that the associated constraint function $g_i(\mathbf{X})$ must achieve and, because of the way optimization problems are typically written, is often called the **right-hand side** in the model. Thus the solution vector $\mathbf{X}_0$ must optimize each of the objective functions $f_j(\mathbf{X})$ and simultaneously satisfy each constraint relationship. We now consider one simplistic problem illustrating the basic model ideas.

decision variables
objective functions

constraints

right-hand side

An Example: Determining a Production Schedule

A carpenter makes tables and bookcases for a net unit profit that he estimates as $25 and $30, respectively. He is trying to determine how many of each piece of furniture he should make each week. He has up to 600 board ft of lumber to devote weekly to the project and up to 40 hours of labor. He can use the lumber and labor productively elsewhere if they are not used in the production of tables and bookcases. He estimates that it requires 20 board ft of lumber and 5 hours of labor to complete a table, and 30 board ft of lumber and 4 hours of labor for a bookcase. He also estimates that he can sell all the tables and bookcases that are produced. Moreover, he has signed contracts to deliver four tables and two bookcases every week. The carpenter wishes to determine a weekly production schedule for tables and bookcases that maximizes his profits.

In this example, the problem is to decide how many tables and bookcases to make every week. Consequently, the decision variables represent the quantities of tables and bookcases to be made. The objective function represents the net weekly profit to be realized from selling the furniture. The constraints are given by the restrictions on the quantities to be produced, which are posed by the availability of lumber, labor, and contractual obligations. Finally, the specified levels of these constraints represent the right-hand side. We will construct a mathematical model for this scenario after discussing the problem in general terms.

There are various ways of classifying optimization problems. These classifications are not meant to be mutually exclusive, but to describe certain mathematical characteristics possessed by the problem under investigation. We now describe several of these classifications.

unconstrained
constrained
An optimization problem is said to be **unconstrained** if there are no constraints, and **constrained** if one or more side conditions are present. The production schedule problem just described illustrates a constrained problem.

linear program
An optimization problem is said to be a **linear program** if it satisfies the following properties:*

1. There is a unique objective function.
2. Whenever a decision variable appears in either the objective function or one of the constraint functions, it must appear only as a power term with an exponent of 1, possibly multiplied by a constant.
3. No term in the objective function or in any of the constraints can contain products of the decision variables.
4. The coefficients of the decision variables in the objective function and each constraint are constants.
5. The decision variables are permitted to assume fractional as well as integer values.

These properties ensure, among other things, that the effect of any decision variable is *proportional* to its value. For instance, in the production schedule example, if the number of tables produced is doubled, then the profit due to their sale is doubled and so are the amounts of labor and lumber required to produce them. Let's examine each property more closely.

multiobjective
goal programs
Property 1 limits the problem to a single objective function. Problems with more than one objective function are called **multiobjective,** or **goal programs.** (We give an example of a multiobjective problem in the next section.) Properties 2 and 3 are self-explanatory, and any optimization problem that fails

nonlinear
to satisfy either one of them is said to be **nonlinear.** Property 4 is quite restrictive for many scenarios you might wish to model. In the production schedule example, for instance, it would be required to know the unit selling price and unit cost of each item precisely, as well as the exact number of board feet and labor required to produce each item. But quite often it is impossible to predict precisely the required values in advance (consider trying to predict the market price of corn), or the coefficients represent average values with rather large deviations from the actual values occurring in practice. The coefficients may be time-dependent as well. Time-dependent problems in

dynamic programs
a certain class are called **dynamic programs.** (We give an example in the next section.) If the coefficients are not constant but instead are probabilistic in

stochastic program
nature, the problem is classified as a **stochastic program.** Finally, if one or

---

* Although a program satisfying properties 1–5 is linear even if it contains variables that are allowed to assume negative values, the Simplex Method requires using nonnegative variables only. A simple substitution allows this (see Problem 5 in Section 4.2). Many computer codes will make this substitution for you.

integer program
mixed integer program

more of the decision variables is restricted to integer values (hence violating Property 5), the resulting problem is called an **integer program** (or a **mixed integer program** if the integer restriction applies only to a subset of the decision variables). In the production schedule example, it makes sense to allow fractional numbers for the tables and bookcases in determining a weekly schedule. (Why?) However, it probably would not make sense to produce a fractional table or bookcase in modeling the carpenter's total production. We next consider the production schedule example in more detail.

Constrained Optimization: A Linear Program Model

While Properties 1 through 5 are precisely satisfied only rarely in practice, there are many instances where reasonable assumptions lead to their satisfaction. In the production schedule example, a constant profit of $25 per table implies a constant unit selling price and a constant unit cost of material, labor, overhead, and the like. Since the 5 hours of labor required for producing each table is probably at best an average estimate, the requirement of constant coefficients in Property 4 may not be satisfied. Nevertheless, these average values may be an acceptable simplification, especially if it is possible to see how sensitive the optimal production schedule is to variations from these values. Moreover, the carpenter probably has more objectives than merely maximizing total profit, violating Property 1. If we are willing to accept the given coefficients as constants and a single objective function, then we can model the production schedule scenario as follows:

Let x_1 denote the number of tables to be produced weekly, and let x_2 denote the number of bookcases. Then the model becomes

$$\text{Maximize } 25x_1 + 30x_2 \tag{5.2}$$

Subject to:

$$
\begin{aligned}
20x_1 + 30x_2 &\leqslant 600 && \text{(Lumber)} \\
5x_1 + 4x_2 &\leqslant 40 && \text{(Labor)} \\
x_1 &\geqslant 4 && \text{(Contract)} \\
x_2 &\geqslant 2 && \text{(Contract)}
\end{aligned}
$$

Simplex Method

sensitivity analysis

Since Properties 1 through 5 are satisfied, the preceding model is a linear program. Even very large linear programs can be solved readily using a technique called the **Simplex Method,** which was created by George Dantzig in 1947. A powerful advantage of the Simplex Method is the relative ease with which it allows a **sensitivity analysis.** That is, not only is it possible to find an optimal solution (if one exists) to the linear program, but it is also possible to determine how sensitive the solution is to variations in the coefficients in the objective function, the constraint functions, or the right-hand side. If the selling price of the tables is only an estimate, we can determine how much the selling price of a table would have to increase or decrease before the optimal production schedule changes. Similarly, it is possible to determine the change in the profit resulting from a unit change in the amount of labor or lumber available, and the range of values over which that change per unit is valid. A sensitivity analysis not only is useful in checking the assumptions, but it is also of great assistance in determining what price to pay for additional resources (such as labor or lumber in our example) in order to increase the profit.

The assumptions required of a linear program become immensely important when one considers that the linear program is the only constrained problem for which solutions can generally be found for large-scale problems in reasonable amounts of time, even with the aid of very powerful computers. As you will see, even small unconstrained nonlinear problems, modeling rather simple scenarios, are often computationally intractable. As for constrained nonlinear programs, only those satisfying specialized conditions can be solved in general. Furthermore, in spite of the large amount of research being conducted, finding good numerical approximations to many nonlinear problems is still very difficult. Primarily for these reasons, many modelers strongly consider the possibility of meeting the assumptions required by a linear program before considering a more refined, and perhaps intractable, model.

5.1 PROJECTS

Complete the requirements in the referenced UMAP module or monograph.

1. "Methods of Linear Programming for Students of Mathematics and Sciences," by J. N. Boyd and P. N. Raychowdhury, UMAP 605. This unit provides an introduction to linear programming and is written for the natural sciences student. Familiarity with elementary matrix theory and determinants is required.
2. "Linear Programming in Two Dimensions, I and II," by Nancy S. Rosenberg, UMAP 453 and 454. These units describe in detail a graphical method for solving linear programming problems in two dimensions, and they show how the changes in the parameters of a linear programming problem affect its optimal solution.
3. Complete the requirements in the UMAP monograph, *Modeling Tomorrow's Energy System*, by T. Owen Carroll. This monograph deals with the energy resource allocation problem in the United States and describes models actually used by the Department of Energy. A linear programming optimization model is developed to examine optimal supply–demand allocations. Since economic activity drives the demand for energy, an expanded input–output model of the nation's economy, containing explicit representation of energy use, is coupled with the optimization model. Long-term effects are also discussed. Student prerequisites include an introduction to linear programming and matrix algebra.

5.2 FORMULATION OF OPTIMIZATION MODELS

In this section we discuss several classes of optimization models other than the linear program. First, let's review the criteria for fitting models to data, studied in Chapter 4. The criterion of minimizing the largest absolute deviation from the fitted model to any corresponding data point led to the optimization problem formulated in (4.1). Verify that the optimization problem represented by that formulation is a linear program (and hence can be solved by the Simplex Method). Minimizing the largest absolute deviation may also arise in a variety of physical problems. For instance, consider the problem of blending various types of available foods containing distinct levels of certain nutrients to obtain a diet that meets specified nutrition levels. It may be just as important not to exceed the specified levels as not to fall short of them. In such a case it may be desirable to find the combination of foods available that minimizes the largest absolute deviation from any specified nutrition level, perhaps weighting the deviations to reflect the relative importance of the various nutrients. Another physical problem occurs when a factory discharges water into a river. It would do no good to satisfy precisely all but one of the environmental standards, especially if excessive deviation in either direction from any established standard threatens wildlife. A better approach might be to blend the components of the discharge in such a way that the largest deviation from any standard is minimized.

Unconstrained Optimization Problems

Another criterion considered for fitting a model to data points is minimizing the sum of the absolute deviations. For the model $y = f(x)$, if $y(x_i)$ represents the function evaluated at $x = x_i$ and (x_i, y_i) denotes the corresponding data point for $i = 1, 2, \ldots, m$ points, then this criterion can be formulated as follows: Find the parameters of the model $y = f(x)$ in order to

$$\text{Minimize } \sum_{i=1}^{m} |y_i - y(x_i)| \qquad (5.3)$$

This last condition illustrates an unconstrained optimization problem. Since the derivative of the function being minimized fails to be continuous (because of the presence of the absolute value), it is impossible to solve this problem with a straightforward application of the elementary calculus.

Finally, the criterion of minimizing the sum of the squared deviations was considered. Using the same notation as earlier, the problem is to find the parameters of the model $y = f(x)$ in order to

$$\text{Minimize } \sum_{i=1}^{m} [y_i - y(x_i)]^2 \qquad (5.4)$$

As you saw in Chapter 4, this unconstrained optimization problem can be solved by direct application of the calculus. However, for many problems the computations may be very difficult, or even impossible, to carry out.

An Example: Locating a Plant's Headquarters

Consider now an unconstrained optimization problem that arises in a physical scenario. Suppose we are attempting to locate a plant's headquarters in such a way as to minimize the distance traveled to subsidiary plants. Assume that all plants are visited with the same frequency and neglect vertical distances, giving a two-dimensional model. These assumptions might be reasonable where travel is by private aircraft and each plant and the headquarters have an airstrip, or where the travel is by automobile and the terrain is flat. Let (a, b) denote the headquarters' location and (x_i, y_i) the coordinates of the ith plant. The problem is illustrated graphically for four plants in Figure 5-1.

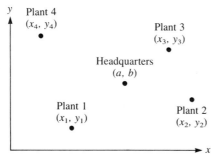

FIGURE 5-1 Locate the headquarters so that the distance traveled to visit the plants is minimized.

The problem is to minimize the sum of the distances from the headquarters to the plants, since each plant is visited with the same frequency. Mathematically, this requirement means to

$$\text{Minimize } D = \sum_{i=1}^{m} \sqrt{(a - x_i)^2 + (b - y_i)^2} \qquad \textbf{(5.5)}$$

Note the similarities and the differences between this problem and the unconstrained optimization problems presented previously. Again distance is being minimized, as in Equation (5.3), but the distance concept has changed because there are two dimensions instead of one. Since the distance D is a function of the two variables a and b, the necessary condition for a relative minimum to exist is that the two partial derivatives $\partial D / \partial a$ and $\partial D / \partial b$ both equal zero (vanish). Solving the system of equations that results from applying this necessary condition presents a difficult task even for a small number of plants. Unfortunately, such is often the case in many optimization problems. Thus, even a conceptually simple problem can lead to computational difficulties.

We conclude this discussion with some brief remarks on approximating solutions to optimization problems—such as those represented by Equations (5.3), (5.4), and (5.5)—which may be difficult to solve analytically. One set

gradient method of numerical techniques is known as the **gradient method.** Such techniques attempt to locate a relative minimum by repeatedly stepping in the direction of the "gradient vector," which is the direction of steepest descent. A critical question is how large a step to take. Moreover, it may be difficult to find **search techniques** and evaluate the gradient function itself. Other methods, known as **search techniques,** avoid this difficulty by employing the principle that $f(x_0 + \Delta x) - f(x_0) > 0$ in the vicinity of a relative minimum: if $f(x_0 + \Delta x) - f(x_0) < 0$, then the point $(x_0 + \Delta x)$ represents a more optimal location than the current x_0. Since the method only finds a relative minimum, knowledge of the general location of the desired global minimum is essential and is usually available for applied problems. Good starting points assist both gradient and search techniques immensely in most situations.

Integer Optimization Programs

The requirements specified for a problem may restrict one or more of its decision variables to integer values. For example, in seeking the right mix of various sized cars, vans, and trucks for a company's transportation fleet to minimize cost under some set of conditions, it would not make sense to determine a fractional part of a vehicle. Integer optimization problems also arise in "coded" problems, where binary (0 and 1) variables represent specific states such as yes/no or on/off. The following is a simple example.

There are various items to be taken on a space shuttle. Unfortunately, there are restrictions on the allowable weight and volume capacities. Suppose there are m different items, each given some numerical value c_j, and having weight w_j and volume v_j. (How might you determine c_j in an actual problem?) Suppose the goal is to maximize the value of the items that are to be taken without exceeding the weight limitation W or volume limitation V. We can formulate a model as follows:

$$\text{Let } y_j = \begin{cases} 1, \text{ if item } j \text{ is taken (yes)} \\ 0, \text{ if item } j \text{ is not taken (no)} \end{cases}$$

Then the problem is to

$$\text{Maximize } \sum_{j=1}^{m} c_j y_j$$

Subject to:

$$\sum_{j=1}^{m} v_j y_j \leqslant V$$

$$\sum_{j=1}^{m} w_j y_j \leqslant W$$

The ability to use binary variables like y_j permits great flexibility in modeling. They can be used to represent yes/no decisions, such as whether or not to finance a given project in a capital budgeting problem, or to restrict variables to being "on" or "off." For example, the variable x can be restricted to the values 0 and a by using the binary variable y as a multiplier:

$$x = ay, \qquad \text{where } y = 0 \text{ or } 1$$

Another illustration restricts x to be either in the interval (a, b) or to be zero by using the binary variable y:

$$ay < x < yb, \qquad \text{where } y = 0 \text{ or } 1$$

The power of using binary variables to represent intervals can be more fully appreciated when it is desired to approximate a nonlinear function by a piecewise linear one. Specifically, suppose the nonlinear function in Figure 5-2a represents a cost function and you want to find its minimum value over the interval $0 \leqslant x \leqslant a_3$. If the function is particularly complicated, it could be approximated by a piecewise linear function like that shown in Figure 5-2b. (The piecewise linear function might occur naturally in a problem, such as where different rates are charged for electrical use based on the amount of consumption.)

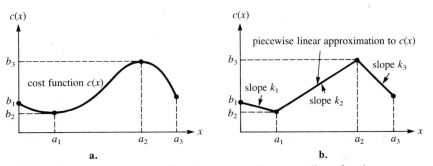

FIGURE 5-2 Using a piecewise linear function to approximate a nonlinear function.

Using the approximation suggested in Figure 5-2, our problem is to find the minimum of the function:

$$c(x) = \begin{cases} b_1 + k_1(x - 0) & \text{if } 0 \leqslant x < a_1 \\ b_2 + k_2(x - a_1) & \text{if } a_1 \leqslant x < a_2 \\ b_3 + k_3(x - a_2) & \text{if } a_2 \leqslant x \leqslant a_3 \end{cases}$$

Define the three new variables $x_1 = (x - 0)$, $x_2 = (x - a_1)$, and $x_3 = (x - a_2)$ for each of the three intervals, and use the binary variables y_1, y_2, and y_3 to restrict the x_i to the appropriate interval:

$$0 \leqslant x_1 \leqslant y_1 a_1$$
$$0 \leqslant x_2 \leqslant y_2(a_2 - a_1)$$
$$0 \leqslant x_3 \leqslant y_3(a_3 - a_2)$$

where y_1, y_2, and y_3 equal 0 or 1

Since we want exactly one x_i to be "active" at any time, we impose the following constraint:

$$y_1 + y_2 + y_3 = 1$$

Now that only one of the x_i is active at any one time, our objective function becomes

$$c(x) = y_1(b_1 + k_1x_1) + y_2(b_2 + k_2x_2) + y_3(b_3 + k_3x_3)$$

Observe that whenever $y_i = 0$, the variable $x_i = 0$ as well. Thus, products of the form x_iy_i are redundant, and the objective function can be simplified to give the model:

$$\text{Minimize } k_1x_1 + k_2x_2 + k_3x_3 + y_1b_1 + y_2b_2 + y_3b_3$$

Subject to:

$$0 \leqslant x_1 \leqslant y_1a_1$$
$$0 \leqslant x_2 \leqslant y_2(a_2 - a_1)$$
$$0 \leqslant x_3 \leqslant y_3(a_3 - a_2)$$
$$y_1 + y_2 + y_3 = 1$$

where y_1, y_2, and y_3 equal 0 or 1

mixed-integer programming The preceding model is classified as a **mixed-integer programming** problem since only some of the decision variables are restricted to integer values. The properties that permit the Simplex Method are lost in integer programming problems, so there is considerable difficulty in solving them. One methodology that has proven successful devises rules to find good feasible solutions quickly, and then uses tests to show which of the remaining solutions can be discarded. Unfortunately, some tests work well for certain classes of problem, but fail miserably for others.

Multiobjective Programming: An Investment Problem

Consider the following problem: An investor has $40,000 to invest. She is considering investments in savings at 7%, municipal bonds at 9%, and stocks that have been consistently averaging 14%. Since there are varying degrees of risk involved in the various investments, the investor has listed the following goals for her portfolio:

1. A yearly return of at least $5,000.
2. An investment of at least $10,000 in stocks.
3. The investment in stocks should not exceed the combined total in bonds and savings.
4. A liquid savings account between $5,000 and $15,000.
5. The total investment must not exceed $40,000.

You can see from the portfolio that the investor has more than one objective. Unfortunately, as is often the case with real-world problems, not all goals can be achieved simultaneously. If the investment returning the lowest yield is set as low as possible (in this example, $5,000 into savings), the best return possible without violating goals 2 through 5 is obtained by investing \$15,000 in bonds and \$20,000 in stocks. However, this portfolio falls short of the desired yearly return of \$5,000. How are problems with more than one objective to be reconciled? Let's begin by formulating each objective mathematically: Let x denote the investment in savings, y the investment in bonds, and z the investment in stocks. Then the goals are as follows:

Goal 1. $.07x + .09y + .14z \geqslant 5000$
Goal 2. $z \geqslant 10000$
Goal 3. $z \leqslant x + y$
Goal 4. $5000 \leqslant x \leqslant 15000$
Goal 5. $x + y + z \leqslant 40000$

We have already seen that the investor will have to compromise on one or more of her goals in order to find a feasible solution. Suppose, then, that the investor feels she must have a return of \$5,000, at least \$10,000 in stocks, and cannot spend more than \$40,000, but she is willing to compromise on Goals 3 and 4. However, she wants to find a solution that minimizes the amounts by which these two goals fail to be achieved. Let's formulate these new requirements mathematically and illustrate a method applicable to similar type problems. Thus, let G_3 denote the amount by which Goal 3, and G_4 the amount by which Goal 4, fails to be satisfied. Then the model is to

$$\text{Minimize } G_3 + G_4$$

Subject to:

$$.07x + .09y + .14z \geqslant 5000$$
$$z \geqslant 10000$$
$$z - G_3 \leqslant x + y$$
$$5000 - G_4 \leqslant x \leqslant 15000$$
$$x + y + z \leqslant 40000$$

where x, y, and z are positive

This last condition is included to ensure that negative investments do not result. This problem is now a linear program that can be solved by the Simplex Method. If the investor feels some goals are more important than others, the objective function can be weighted to emphasize those goals. Furthermore, a sensitivity analysis of the weights in the objective function will identify the "breakpoints" for the range over which various solutions are optimal. This process generates a number of solutions to be carefully considered by the investor before making her investments. Normally, this is the best that can be accomplished where qualitative decisions are to be made.

Calculus of Variations

We now consider several departures from the basic optimization model (5.1). Recall that in that model we seek a vector $\mathbf{X}$ optimizing the value of the

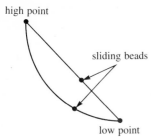

high point

sliding beads

low point

FIGURE 5-3 The path that minimizes the time of descent from the high point to the low point is a cycloid.

cycloid

optimal control problems

dynamic programs

objective function(s) while simultaneously satisfying any side conditions that are present. However, situations arise when an optimizing *function*, rather than a *vector*, is sought. For example, consider a bead sliding under the force of gravity down a wire from a high point to a low point not directly beneath it. The situation is illustrated in Figure 5-3. The problem is to find the path that minimizes the time of descent from the high point to the low point. That is, from the set of all function curves passing through the two points, find the one that minimizes the travel time. At first you might think the solution is the straight line joining the two points, since that curve does minimize the distance between them. However, the straight line proves to be incorrect and the calculus of variations shows that the optimal path is steeper. In fact, the solution curve is a **cycloid,** which is the path traced out by a point on the circumference of a wheel rolling along a horizontal straight line without slipping.

Calculus of variations problems often arise in various types of control problems. For example, it is of interest to find the function that describes the thrust rate (fuel per unit time) for a rocket so that the amount of time it takes the rocket to achieve a certain height is minimized. A related problem is to minimize the amount of fuel it takes to achieve that height while ignoring the time it takes. Problems of this sort are often called **optimal control problems** and, as is typical of most calculus of variations problems, tend to be very difficult to solve.

Dynamic Programming Problems

Often the model being optimized requires that decisions be made at various time intervals rather than all at once. In the 1950s the American mathematician Richard Bellman developed a technique for optimizing such models in *stages* rather than simultaneously, and referred to such problems as **dynamic programs.** Here's a problem scenario that lends itself to solution by dynamic programming techniques:

A rancher enters the cattle business with an initial herd of k cattle. He intends to retire and sell any remaining cattle after N years. Each year the rancher is faced with the decision of how many cattle to sell and how many to keep. If he sells, he estimates his profit per head to be p_i in year i. Also, the number of cattle kept in year i is expected to double by year $i + 1$.

While this scenario neglects many factors an analyst might consider in the real world (name several), you can see that the cattle rancher is faced with a trade-off decision each year: take a profit or build for the future.

The purpose of our discussion has been to provide an overview of the field of optimization. We hope you can appreciate the many applications in which optimization problems arise. The solutions to those problems tend to be difficult, but solution methods are studied in more advanced courses such as linear and nonlinear optimization.

5.2 PROBLEMS

Use the model-building process described in Chapter 2 to analyze the following scenarios. After identifying the problem to be solved using that process, you may find it helpful to answer the following questions in words before formulating the optimization model:

A. *Identify the decision variables:* What decision is to be made?
B. *Formulate the objective function:* How do the decisions to be made affect the objective?
C. *Formulate the constraint set:* What constraints must be satisfied? Be sure to consider whether negative values of the decision variables are allowed by the problem, and ensure that they are so constrained if required.

After constructing a mathematical model, determine whether the assumptions required to satisfy the properties of the linear program seem reasonable.

1. *Resource Allocation*

You have just become the manager of a plant producing plastic products. Though the plant operation involves many products and supplies, you are interested in only three of the products: (1) a vinyl-asbestos floor covering, the output of which is measured in boxed lots, each covering a certain area; (2) a pure vinyl counter top, measured in linear yards; and (3) a vinyl-asbestos wall tile, measured in "squares," each covering 100 sq ft.

Of the many resources needed to produce these plastic products, you have identified four: vinyl, asbestos, labor, and time on a trimming machine. A recent inventory shows that on any given day you have 1500 pounds of vinyl and 200 pounds of asbestos available for use. Additionally, after talking to your shop foreman and to various labor leaders, you realize that you have 3 man-days of labor available for use per day and that your trimming machine is available for 1 machine-day on any given day. The following table indicates the amount of each of the four resources required to produce a unit of the three desired products, where the units are 1 box of floor cover, 1 yard of counter top, and 1 square of wall tiles. Available resources are also tabulated.

	Vinyl (lb)	Asbestos (lb)	Labor (man-days)	Machine (machine-days)	Profit
Floor Cover (*per box*)	30	3	.02	.01	$0.8
Counter Top (*per yard*)	20	0	.1	.05	5
Wall Tile (*per square*)	50	5	.2	.05	5.5
Available (*per day*)	1500	200	3	1	—

Formulate a mathematical model to help determine how to allocate resources in order to maximize profits.

2. Nutritional Requirements

A rancher has determined that the minimum weekly nutritional requirements for an average size horse include 40 lb of protein, 20 lb of carbohydrates, and 45 lb of roughage. These are obtained from the following sources in varying amounts at the prices indicated.

	Protein (lb)	Carbohydrates (lb)	Roughage (lb)	Cost
Hay (per bale)	.5	2.0	5.0	$1.80
Oats (per sack)	1.0	4.0	2.0	3.50
Feeding blocks (per block)	2.0	.5	1.0	.40
High-protein concentrate (per sack)	6.0	1.0	2.5	1.00
Requirements per horse (per week)	40.0	20.0	45.0	

Formulate a mathematical model to determine how to meet the minimum nutritional requirements at minimum cost.

3. Scheduling Production

A manufacturer of an industrial product has to meet the following shipping schedule.

Month	Required shipment (units)
January	10,000
February	40,000
March	20,000

The monthly production capacity is 30,000 units and the production cost per unit is $10. Since the company does not have a warehouse, the service of a storage company is utilized whenever needed. The storage company figures its monthly bill by multiplying the number of units in storage on the last date of the month by $3. On the first day of January the company does not have any beginning inventory, and it doesn't want to have any ending inventory at the end of March. Formulate a mathematical model to assist in minimizing the sum of the production and storage costs for the three-month period.

4. *Mixing Nuts*

A candy store sells three different assortments of mixed nuts, each assortment containing varying amounts of almonds, pecans, cashews, and walnuts. In order to preserve the store's reputation for quality, certain maximum and minimum percentages of the various nuts are required for each type of assortment, as shown in the following table.

Assortment name	Requirements	Selling price per pound
Regular	Not more than 20% cashews Not less than 40% walnuts Not more than 25% pecans No restriction on almonds	$ .89
Deluxe	Not more than 35% cashews Not less than 25% almonds No restriction on walnuts and pecans	1.10
Blue Ribbon	Between 30% and 50% cashews Not less than 30% almonds No restriction on walnuts and pecans	1.80

The following table gives the cost per pound and the maximum quantity of each type of nut available from the store's supplier each week:

Nut type	Cost per pound	Maximum quantity available per week (lb)
Almonds	$.45	2000
Pecans	.55	4000
Cashews	.70	5000
Walnuts	.50	3000

The store would like to determine the exact amounts of almonds, pecans, cashews, and walnuts that should go into each weekly assortment in order to maximize its weekly profit. Formulate a mathematical model that will assist the store management in solving its mixing problem. Hint: How many decisions need to be made? For example, do we need to distinguish between the cashews in the Regular mix and the cashews in the Deluxe?

5. *Producing Electronics Equipment*

An electronics firm is producing three lines of products for sale to the government: transistors, micromodules, and circuit assemblies. The firm has four physical processing areas designated as follows: transistor production, circuit printing and assembly, transistor and module quality control, and circuit assembly test and packing.

The various production requirements are as follows: Production of one transistor requires 0.1 standard hour of transistor production area capacity, 0.5 standard hour of transistor quality control area capacity, and $.70 in direct costs. Production of one micromodule requires 0.4 standard hour of the quality control area capacity, three transistors, and $.50 in direct costs. Production of one circuit assembly requires 0.1 standard hour of the capacity of the circuit printing area, 0.5 standard hour of the test and packing area, one transistor, three micromodules, and $2.00 in direct costs.

Suppose that the three products (transistors, micromodules, and circuit assemblies) may be sold in unlimited quantities at prices of $2, $8, and $25 each, respectively. There are 200 hours of production time open in each of the four process areas in the coming month. Formulate a mathematical model to help determine the production that will produce the highest revenue for the firm.

6. *Purchasing Various Trucks*

A trucking company has allocated $800,000 for the purchase of new vehicles, and it is considering three types. Vehicle A has a 10-ton payload capacity and is expected to average 45 mph; it costs $16,000. Vehicle B has a 20-ton payload capacity and is expected to average 40 mph; it costs $26,000. Vehicle C is a modified form of B and carries sleeping quarters for one driver. This modification reduces its capacity to an 18-ton payload and raises the cost to $30,000, but its operating speed is still expected to average 40 mph.

Vehicle A requires a crew of one man and, if driven on three shifts per day, could be operated for an average of 18 hours per day. Vehicles B and C must have crews of two men each to meet local legal requirements. Vehicle B could be driven an average of 18 hours per day with three shifts, and Vehicle C could average 21 hours per day with three shifts. The company has 150 drivers available each day to make up its crews and will not be able to hire additional trained crews in the near future. The local labor union prohibits any man from working more than one shift per day. Also, maintenance facilities are such that the total number of vehicles must not exceed 30. Formulate a mathematical model to help determine the number of each type of vehicle the company should purchase in order to maximize its shipping capacity in ton-miles per day.

7. *A Farming Problem*

A farm family owns 100 acres of land and has $15,000 in funds available for investment. Its members can produce a total of 3500 man-hours worth of labor during the winter months (mid-September to mid-May) and 4000 man-hours during the summer. If any of these man-hours are not needed, younger members of the family will use them to work on a neighboring farm for $1.80 per hour during the winter months and $2.10 per hour during the summer.

Cash income may be obtained from three crops (soybeans, corn, and oats), and two types of livestock (dairy cows and laying hens). No investment funds are needed for the crops. However, each cow requires an initial investment outlay of $400, and each hen requires $3. Each cow requires 1.5 acres of land and 100 man-hours of work during the winter months, and another 50 man-hours during the summer. Each cow produces a net annual cash income of $400 for the family. The corresponding figures for each hen are: no acreage, 0.6 man-hours during the winter, 0.3 more man-hours during the summer, and an annual net cash income of $2. The chicken house accommodates a maximum of 3000 hens, and the size of the barn limits the cow herd to a maximum of 32 head.

Estimated man-hours and income per acre planted in each of the three crops are as shown in the following table.

Crop	Winter man-hours	Summer man-hours	Net annual cash income (per acre)
Soybean	20	30	$175.00
Corn	35	75	300.00
Oats	10	40	120.00

Formulate a mathematical model to assist in determining how much acreage should be planted in each of the crops, and how many cows and hens should be kept in order to *maximize net cash income*.

5.2 PROJECTS

For Projects 1 through 5, complete the requirements in the referenced UMAP module or monograph.

1. "Unconstrained Optimization," by Joan R. Hundhausen and Robert A. Walsh, UMAP 522. This unit introduces gradient search procedures with examples and applications. Acquaintance with elementary partial differentiation, chain rules, Taylor series, gradients, and vector dot products is required.

2. "Lagrange Multipliers and the Design of Multistage Rockets," by Anthony L. Peressini, UMAP 517. The method of Lagrange multipliers is applied to compute the minimum total mass of an n-stage rocket capable of placing a given payload in an orbit at a given altitude above the earth's surface. Familiarity with elementary minimization techniques for functions of several variables, the method of Lagrange multipliers, and the concepts of linear momentum and conservation of momentum are required.

3. "Lagrange Multipliers: Applications to Economics," by Christopher H. Nevison, UMAP 270. The Lagrange multiplier is interpreted and studied

as the marginal rate of change of a utility function. Differential calculus through Lagrange multipliers is required.

4. "Calculus of Variations with Applications in Mechanics," by Carroll O. Wilde, UMAP 468. This module provides a brief introduction to finding functions that yield the maximum or minimum value of certain definite integral forms, with applications in mechanics. Students learn Euler's equations for some definite integral forms, and learn Hamilton's principle and its application to conservative dynamical systems. The basic physics of kinetic and potential energy, the multivariate chain rules, and ordinary differential equations are required.

5. *The High Cost of Clean Water: Models for Water Quality Management*, by Edward Beltrami, UMAP Expository Monograph. To cope with the severe wastewater disposal problems caused by increases in the nation's population and industrial activity, the U.S. Environmental Protection Agency (EPA) has fostered the development of regional wastewater management plans. This monograph discusses the EPA plan developed for Long Island, and formulates a model that allows for the articulation of the trade-offs between cost and water quality. The mathematics involves partial differential equations and mixed integer linear programming.

MODEL

5.3 AN INVENTORY PROBLEM: MINIMIZING THE COST OF DELIVERY AND STORAGE

Scenario   You have been hired as a consultant by a chain of gasoline stations to determine how often and how much gasoline should be delivered to the various stations. After some questioning, you determine that each time gasoline is delivered the stations incur a charge of d dollars, which is in addition to the cost of the gasoline and is independent of the amount delivered.

Costs are also incurred when the gasoline is stored. One such cost is capital tied up in the inventory—money that is invested in the stored gasoline and that cannot be used elsewhere. This cost is normally computed by multiplying the cost of the gasoline to the company by the current interest rate for the period the gasoline was stored. Other costs include amortization of the tanks and equipment necessary to store the gasoline, insurance, taxes, security measures, and so forth.

The gasoline stations are located near interstate highways, where demand is fairly constant throughout the week. Records indicating the gallons sold daily are available for each station.

Problem identification Assume the firm wishes to maximize its profit and that the demand and price are constant in the short run. Thus, since total revenue is constant, total profit can be maximized by minimizing the total cost. There are many components of total costs such as overhead, employee costs, and the like. If these costs are affected by the amount and timing of the deliveries, they should be considered. Let's assume the costs are not so affected and focus our attention on the following problem: *Minimize the average daily cost of delivering and storing sufficient gasoline at each station to meet consumer demand.* Intuitively we expect such a minimum to exist. For if the delivery charge is very high and the storage cost very low, we would expect very large orders of gasoline delivered infrequently. On the other hand, if the delivery cost is very low and the storage costs very high, we would expect small orders of gasoline delivered very frequently.

Assumptions In the following presentation, we consider some factors important to deciding how large an inventory to maintain. Obvious factors to consider are *delivery costs*, *storage costs*, and *demand rate* for the product. *Perishability* of the product being stored also may be a paramount concern. In the case of gasoline, the effect of condensation may become more and more appreciable as the gasoline level gets lower and lower in the tank. The market stability of the selling price of the product and the cost of raw materials need to be considered. For example, if the *market price* of the product is volatile, the seller would be reluctant to store large quantities of the product. On the other hand, an expected large increase in the *price of raw materials* in the near future argues for large inventories. The *stability of the demand* for the product by the consumer is another factor to be taken into account. There may be seasonal fluctuations in the demand for the product, or a technological breakthrough may cause the product to become obsolete. The *time horizon* being considered also can be extremely important. In the short run, contracts may be signed for warehouse space, some of which will not be needed in the long run. Another consideration is the importance to the owner of an occasional unsatisfied demand (*stock-outs*). Some owners would opt for a more costly inventory strategy to be assured that they will never run out of stock. From this discussion you can see that the inventory decision is not an easy one, and it is not difficult to build scenarios where any one of the preceding factors may dictate a particular strategy. We restrict our initial model here to the following variables:

average daily cost $= f$(storage costs, delivery costs, demand rate)

The Submodels

Storage costs We need to consider how the storage cost per unit varies with the number of units being stored. Are we renting space and receiving a discount when storage exceeds certain levels, as suggested in Figure 5-4a? Or do we rent the cheapest storage first (adding more space as needed), as suggested by Figure 5-4b? Do we need to rent an entire warehouse or floor? If so, the per unit cost is likely to decrease as the quantity stored increases until another warehouse or floor needs to be rented, as suggested in Figure 5-4c. Does the company own its storage facilities? If so, what alternative use can be made of them? In our model we take per unit storage cost as constant.

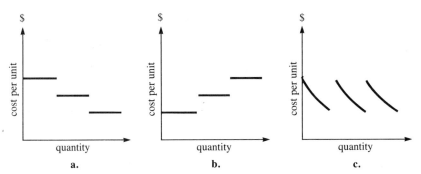

FIGURE 5-4 Three different submodels for storage costs.

Delivery costs In many instances the delivery charge depends on the amount delivered. For example, if a larger truck or an extra flatcar is needed, an additional charge is made. In our model we consider a constant delivery charge independent of the amount delivered.

Demand If we plot the daily demand for gasoline at a particular station, we will very likely get a graph similar to the one shown in Figure 5-5a. If we then plot the frequency of each demand level over a fixed time period (for example, one year) we might get a plot similar to that shown in Figure 5-5b. If the demands are fairly tightly packed about the most frequently occurring demand, then we accept the daily demand as being constant. We

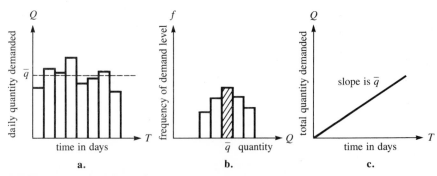

FIGURE 5-5 A constant demand rate.

will assume that such is the case for our model. Finally, even though we realize that the demands occur in discrete time periods, for purposes of simplification we take a continuous submodel for demand. This continuous submodel is depicted in Figure 5-5c, where the slope of the line represents the constant daily demand. Notice the importance of our assumptions in producing the linear submodel. Also, as suggested in Figure 5-5b, about half the time the demand exceeds its average value. We will return to examine the importance of this assumption in the implementation phase when we consider the possibility of unsatisfied demands.

Model formulation   We use the following notation for constructing our model:

   s: storage cost per gallon per day
   d: delivery cost in dollars per delivery
   r: demand rate in gallons per day
   Q: quantity of gasoline in gallons
   T: time in days

Now suppose an amount of gasoline, say $Q = q$, is delivered at time $T = 0$, and that the gasoline is used up after $T = t$ days. The same cycle is then repeated, as illustrated in Figure 5-6. The slope of each line segment in Figure 5-6 is $-r$ (the negative of the demand rate). The problem is to determine an order quantity Q^* and a time between orders T^* that minimizes the delivery and storage costs.

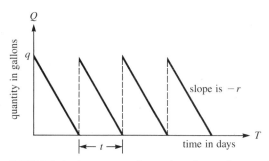

FIGURE 5-6 An inventory cycle consists of an order quantity q that is consumed in t days.

We seek an expression for the average daily cost, so consider the delivery and storage costs for a cycle of length t days. The delivery costs are the constant amount d, since only one delivery is made over the single time period. To compute the storage costs, take the average daily inventory $q/2$, multiply by the number of days in storage t, and multiply that by the storage cost per item per day s. In our notation this gives

$$\text{cost per cycle} = d + s\frac{q}{2}t$$

which upon division by t yields the average daily cost:

$$c = \frac{d}{t} + \frac{sq}{2}$$

Model solution   Apparently the cost function to be minimized has two independent variables, q and t. However, from Figure 5-6 notice that the two variables are related. For a single cyclic period, the amount delivered equals the amount demanded. This translates to $q = rt$. Substitution into the average daily cost equation yields

$$c = \frac{d}{t} + \frac{srt}{2} \tag{5.6}$$

Equation (5.6) is the sum of a hyperbola and a linear function. The situation is depicted in Figure 5-7.

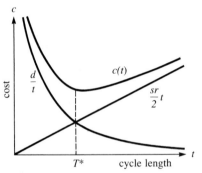

FIGURE 5-7 The average daily cost c is the sum of a hyperbola and a linear function.

Let's find the time between orders T^* that minimizes the average daily cost. Differentiating c with respect to t and setting $c' = 0$ yields

$$c' = -\frac{d}{t^2} + \frac{sr}{2} = 0 \tag{5.7}$$

This last equation gives the (positive) critical point:

$$T^* = \left(\frac{2d}{sr}\right)^{1/2} \tag{5.8}$$

This critical point provides a relative minimum for the cost function since the second derivative:

$$c'' = \frac{2d}{t^3}$$

is always positive for positive values of t. It is clear from Figure 5-7 that

T^* gives a global minimum as well. And note from Equation (5.7) that $d/T^* = (sr/2)T^*$, so that T^* is the point at which the linear function and hyperbola intersect in Figure 5-7.

Interpretation of the model   Given a (constant) demand rate r, Equation (5.8) suggests a proportionality between the optimal time period T^* and $(d/s)^{1/2}$. Intuitively we would expect T^* to increase as the delivery cost d increases and to decrease as storage costs s increase. Thus our model at least makes common sense. However, the nature of the relationship (5.8) is interesting.

For our submodels of demand rate, storage costs, and delivery cost, we assumed very simple relationships. In order to analyze models that have more complex submodels, we need to analyze what we did mathematically. Note that in order to determine the number of item-days of storage, we computed the area under the curve for one cycle. Thus, the storage costs for one cycle could be computed as an integral:

$$s \int_0^t (q - rx)\,dx = s\left(qt - \frac{rt^2}{2} \right) = \frac{sqt}{2}$$

The last equality in this equation follows from the substitution $r = q/t$, and agrees with our previous result for storage costs per cycle. It is important to recognize the underlying mathematical structure in order to facilitate generalization to other assumptions. As you will see, it is also helpful in analyzing the sensitivity of the model to changes in the assumptions.

One of our assumptions was to neglect the cost of the gasoline in the analysis. But does the cost of gasoline actually affect the optimal order quantity and period? Since the amount purchased in each cycle is rt, if the cost per gallon is p dollars, then the constant amount $p(q/t) = pr$ would have to be added to the average daily cost. Because this amount is constant, it cannot affect T^*. (Why?) Thus, we are correct in neglecting the cost of the gasoline. In a more refined model the interest lost in the capital invested in the inventory could be considered.

Implementation of the model   Consider again the graph in Figure 5-6. Now the model assumes the entire inventory is used up in each cyclic period, yet all demands are supposed to be satisfied immediately. Note that this assumption is based on an *average* daily demand of r gallons per day. This assumption means that, over the long run, for roughly half of the time cycles the stations will run out of stock before the end of the period and the next delivery time, and for the other half of the time cycles the stations will still have some gasoline left in the storage tanks when the next delivery arrives. Such a situation probably won't do much for our credibility as consultants! So let's consider recommending a buffer stock to help prevent the stock-outs, as suggested in Figure 5-8. Note in Figure 5-8 that the optimal time period T^* and the optimal quantity $Q^* = rT^*$ are indicated as labels, since we know those values from our model given by Equation (5.8).

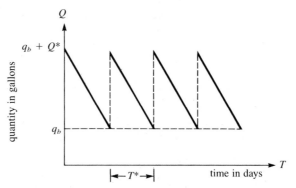

FIGURE 5-8 A buffer stock q_b helps to prevent stock-outs.

Let's examine the effect of the buffer stock on our inventory strategy. We have already observed that the storage costs for one cycle are given by the area under the curve over one time period multiplied by the constant s. Now, the effect of the buffer stock is to add an additional constant area $q_b t$ to the previous area under the curve. Thus, the constant amount $q_b s$ is added to the average daily cost and consequently the value of T^* does not change from what it was without the buffer stock. Thus the stations continue to order $Q^* = rT^*$ as before, although the maximum inventory now becomes $Q^* + q_b$. By determining the daily cost of maintaining the buffer stock, the decision maker can decide how large of a stock to carry. What other information would be useful to the decision maker in determining how large a buffer stock to carry?

Our mathematical analysis has been very straightforward and precise. Yet can we obtain the necessary data to estimate r, s, and d precisely? Probably not. Moreover, how sensitive is the cost to changes in these parameters? These are issues that a consultant would want to consider. Note, too, that we would probably need to round T^* to an integer value. How should it be rounded? Is it better to overestimate or underestimate T^*? Of course, in a given situation, we could substitute various values of T^* for t in Equation (5.6) and see how the cost varies. Let's use the derivative to determine the shape of the average daily cost curve more generally. We know that the average daily cost is minimum at $t = T^*$. The first derivative of the average daily cost represents its rate of change, or the marginal cost, and from Equation (5.7) is $-d/t^2 + sr/2$. The derivative of the marginal cost is $2d/t^3$, which, for positive t, is always positive and decreases as t increases. Note that the derivative of the marginal cost becomes large without bound as t approaches zero and approaches zero as t becomes large. Thus, to the left of T^* the marginal cost is negative and becomes increasingly steeper as t approaches zero. To the right of T^* the marginal cost approaches the constant value $sr/2$. Relate these results to the graph depicted in Figure 5-7 and interpret them economically.

5.3 PROBLEMS

1. Consider an industrial situation where it is necessary to set up an assembly line. Suppose that each time the line is set up a cost c is incurred. Assume c is in addition to the cost of producing any items and is independent of the amount produced. Suggest submodels for the production rate. Now assume a constant production rate k and a constant demand rate r. What assumptions are implied by the model in Figure 5-9? Next assume a storage cost of s (in dollars per unit per day) and compute the optimal length of the production run P^* in order to minimize the costs. List all of your assumptions. How sensitive is the average daily cost to the optimal length of the production run?

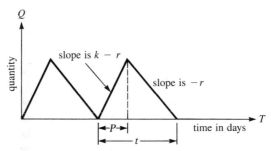

FIGURE 5-9 Determine the optimal length of the production run.

2. Consider a company that allows backordering. That is, the company notifies customers that a temporary stock-out exists and that their order will be filled shortly. What conditions might argue for such a policy? What effect does such a policy have on storage costs? Should costs be assigned to stock-outs? Why? How would you make such an assignment? What assumptions are implied by the model in Figure 5-10? Suppose a "loss of goodwill cost" of w dollars per unit per day is assigned to each stock-out. Compute the optimal order quantity Q^* and interpret your model.

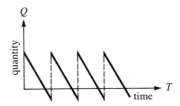

FIGURE 5-10 An inventory strategy that permits stock-outs.

3. In the inventory model discussed in the text we assumed a constant delivery cost that is independent of the amount delivered. Actually, in

many cases, the cost varies in discrete amounts depending on the size of the truck needed, the number of platform cars required, and so forth. How would you modify the model in the text to take into account these changes? We also assumed a constant cost for the raw materials. However, oftentimes bulk-order discounts are given. How would you incorporate these discount effects into the model?

4. Discuss the assumptions implicit in the two graphical models depicted in Figure 5-11. Suggest scenarios in which each model might apply. How would you determine the submodel for demand in Figure 5-11b? Discuss how you would compute the optimal order quantity in each case.

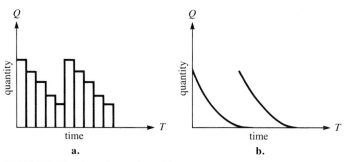

FIGURE 5-11 Two inventory submodels.

5. What is the optimal speed and safe following distance that allows the maximum flow rate (cars per unit time)? The solution to the problem would be of use in controlling traffic in tunnels, for roads under repair, or for other congested areas. In the following schematic, l is the length of a typical car and d is the distance between cars.

Justify that the flow rate is given by

$$f = \frac{\text{velocity}}{\text{distance}}$$

Let's assume a car length of 15 ft. In Section 3.3 the safe stopping distance

$$d = 1.1v + 0.054v^2$$

was determined, where d is measured in feet and v in miles per hour. Find the velocity in miles per hour and the corresponding following distance d that maximizes traffic flow. How sensitive is the solution to changes in v? Can you suggest a practical rule? How would you enforce it?

6. Consider an athlete competing in the shot put. What factors influence the length of his or her throw? Construct a model that predicts the distance

thrown as a function of the initial velocity and angle of release. What is the optimal angle of release? If the athlete cannot maximize the initial velocity at the angle of release you propose, should he or she be more concerned with satisfying the angle of release or generating a high initial velocity? What are the trade-offs?

7. John Smith is responsible for periodically buying new trucks to replace older trucks in his company's fleet of vehicles. He is expected to determine the time a truck should be retained so as to minimize the average cost of owning the truck. Assume the purchase of a new truck is $9000. Also assume the maintenance cost (in dollars) per truck for t years can be expressed analytically by the following empirical model:

$$C(t) = 640 + 180(t + 1)t$$

where t is the time in years that the company owns the truck. (The construction of empirical models is discussed in Chapter 6.)

   a. Determine $E(t)$, the total cost function for a single truck retained for a period of t years.

   b. Determine $E_A(t)$, the *average* annual cost function for a single truck that is kept in the fleet for t years.

   c. Graphically depict $E_A(t)$ as a function of t. Justify the shape of your graph.

   d. Analytically determine t^*, the optimal period that a truck should be retained in the fleet. Remember John Smith's objective is to minimize the average cost of owning a truck.

   e. Suppose we have to round t^* to the nearest whole year. In general, would it be better to round up or round down? Justify your answer.

5.3 PROJECTS

1. "The Human Cough," by Philip M. Tuchinsky, UMAP 211. A model is developed showing how our bodies contract the windpipe during a cough in order to maximize the velocity of the airflow (making the cough maximally effective). Complete the module and prepare a short report for classroom discussion.

2. "An Application of Calculus in Economics: Oligopolistic Competition," by Donald R. Sherbert, UMAP 518. The author analyzes a number of mathematical models that investigate the competitive structure of a market in which a small number of firms compete. Thus a change in price or production level by one firm will cause a reaction in the others. Complete the module and prepare a short report for classroom discussion.

3. "Five Applications of Max-Min Theory from Calculus," UMAP 341 by Thurmon Whitley. In this module several unconstrained optimization problems are solved using the calculus. Scenarios addressed include maximizing profit, minimizing cost, minimizing the travel time of light as it

passes through several mediums (Snell's Law), minimizing the surface area of a bee's cell, and the surgeon's problem of attaching an artery in such a way as to minimize the resistance of blood flow and strain on the heart.

MODEL

5.4 THE ASSEMBLY LINE*

In Section 2.2 we discussed the managerial problem of how to assign employees to the various jobs on an assembly line, taking into account that the jobs are often quite tedious and repetitive and demand various worker skill levels. After discussing several objectives that management might consider in making the assignments, we settled on the following ideas.

Problem identification *Minimize production costs while meeting specified levels of demand, quality control, and job satisfaction.*

Assumptions There are obviously many factors that determine the total cost of production to a firm. For our model we include the following variables: *wages* of the individual workers, individual *productivity* on the various machines, the expected number of *defects* each worker makes on each machine, and the *reduction in both quality and productivity* as a function of the time an individual spends on a particular machine. Here we are assuming that the jobs are assignments to operate a particular machine, and we neglect (among many other factors) the cost per hour of operating the various machines themselves.

Model formulation Assume that the objective is to determine a *daily* production assignment schedule. We define the following symbols:

x_{ij} = the number of hours per day worker i is assigned to machine j

c_{ij} = the hourly dollar wage of worker i when assigned to machine j (an incentive pay may be given for a particularly dangerous or tedious machine)

* Optional section.

a_{ij} = the hourly output of worker i when assigned to machine j, measured in units per hour

d_j = the fixed daily demand for the item produced on machine j, measured in items per day

q_{ij} = the average number of defective items produced per hour by worker i when assigned to machine j, measured in items per hour

Q_j = the maximum number of defective items per day permitted on machine j

For illustrative purposes, assume there are five workers and six machines. Then the objective is to minimize the sum:

$$c_{11}x_{11} + c_{12}x_{12} + c_{13}x_{13} + c_{14}x_{14} + c_{15}x_{15} + c_{16}x_{16} + \cdots$$
$$+ c_{51}x_{51} + c_{52}x_{52} + c_{53}x_{53} + c_{54}x_{54} + c_{55}x_{55} + c_{56}x_{56}$$

or, more succinctly, to

$$\text{Minimize } \sum_{i=1}^{5} \sum_{j=1}^{6} c_{ij}x_{ij}$$

There are a number of constraints that must be considered. Since the schedule is on a daily basis, it is reasonable to assume each worker works no more than a total of 8 hours. Furthermore, assume that the workers can be used productively elsewhere whenever they are not required to work a full 8 hours on the machines in the model. These considerations give the inequality:

$$x_{i1} + x_{i2} + x_{i3} + x_{i4} + x_{i5} + x_{i6} \leqslant 8$$

for each worker i, or

$$\sum_{j=1}^{6} x_{ij} \leqslant 8 \qquad \text{for } i = 1, 2, 3, 4, 5$$

Now consider the issue of satisfying the demand for each machine. We would need to know whether the output of each machine is inspected and whether defective items are removed prior to shipping. (These are questions the modeler would ask of management.) If inspection and removal of defective items are procedures that are followed, then the effective rate of worker i on machine j is $a_{ij} - q_{ij}$ items per hour. However, in many industrial scenarios where the output of each machine is very high, this procedure is not followed because of the high cost involved. Instead, random inspections are made of samples of each worker's output in order to appraise the quality of each worker and machine, and to ensure that quality standards are met. Thus, assuming that defective items are not removed and that excess production is undesirable, the total number of items produced on a single machine is the sum of all the items produced by each worker on that machine, which, in turn, equals the daily demand for the items produced by the machine. This assumption translates into the equation:

$$a_{1j}x_{1j} + a_{2j}x_{2j} + a_{3j}x_{3j} + a_{4j}x_{4j} + a_{5j}x_{5j} = d_j$$

for each machine j, or more succinctly,

$$\sum_{i=1}^{5} a_{ij}x_{ij} = d_j \qquad \text{for } j = 1, 2, 3, 4, 5, 6$$

Next we model quality control. Suppose the firm's quality standards allow no more than a certain percentage for defective items. This assumption means that no more than a certain number, say Q_j, of defective items are permitted for machine j each day. Assume also that there are sufficient historical records of the random inspections made on each worker's performance to determine an average number of defective items q_{ij} produced each hour that worker i is using machine j. These assumptions mean that the sum of all the defective items produced by all workers on a single machine each hour must be less than the number of defective items permitted for that machine. Mathematically, this requirement translates into the inequality:

$$q_{1j}x_{1j} + q_{2j}x_{2j} + q_{3j}x_{3j} + q_{4j}x_{4j} + q_{5j}x_{5j} < Q_j$$

for each machine j, or

$$\sum_{i=1}^{5} q_{ij}x_{ij} < Q_j \qquad \text{for } j = 1, 2, 3, 4, 5, 6$$

Finally, let's consider the requirement of job satisfaction. As indicated in Section 2.2, this requirement is something we would want to discuss with both management and the workers. One criterion might be to limit the total number of hours that any one worker is assigned to a particular machine. For example, it could be decided to limit the number of hours that any worker is assigned to machine 3 to just 2 hours per day. This constraint translates to

$$x_{i3} < 2 \qquad \text{for each worker } i = 1, 2, 3, 4, 5$$

It may be that some workers are not qualified to operate certain machines. For example, if workers 1 and 5 cannot operate machine 3, then $x_{13} = x_{53} = 0$. Note that these last conditions are still compatible with the inequality $x_{i3} < 2$ given before. It is also observed that a worker's productivity and quality diminish as a function of the length of time assigned to a particular machine. These observations are partly accounted for in limiting the number of hours a worker is assigned to certain machines. However, in the preceding formulation, constant values for the parameters a_{ij} and q_{ij} are assumed in order to simplify the formulation of the model (and, later, its solution). Actually, these parameters may vary with time. (We need to be aware of our simplifications, and later refine the model if we are dissatisfied with its ability to reflect accurately the real-world situation.)

Summarizing, our mathematical model for the assembly line problem is to

$$\text{Minimize} \sum_{i=1}^{5} \sum_{j=1}^{6} c_{ij}x_{ij} \qquad \text{(Production costs)}$$

Subject to:

$$\sum_{j=1}^{6} x_{ij} \leqslant 8 \qquad \text{for } i = 1, 2, 3, 4, 5 \qquad \text{(Labor)}$$

$$\sum_{i=1}^{5} a_{ij}x_{ij} = d_j \qquad \text{for } j = 1, 2, 3, 4, 5, 6 \qquad \text{(Demand)}$$

$$\sum_{i=1}^{5} q_{ij}x_{ij} < Q_j \qquad \text{for } j = 1, 2, 3, 4, 5, 6 \qquad \text{(Quality control)}$$

$$x_{i3} < 2 \qquad \text{for } i = 1, 2, 3, 4, 5 \qquad \text{(Satisfaction)}$$

$$x_{13} = x_{53} = 0 \qquad \text{(Skill)}$$

Since the variables are not restricted to integer values, and the objective function and all constraint functions are "linear," the model is an example of a linear program and can be solved by the Simplex Method. Once values for the various parameters in this model are determined, not only can an optimal solution be found (if one exists), but it is also possible to determine how sensitive the solution is to changes in the various parameters a_{ij}, q_{ij}, Q_j, c_{ij}, and d_j. For instance, if you are interested in how the cost varies as the demand for a particular machine is changed, such information is readily available using the Simplex Method. In the assembly line problem this is particularly important since you could not expect to determine the various parameters of the model very precisely, nor have all the important factors been taken into consideration.

5.4 PROBLEMS

1. As director of a cryptoanalysis department you have at your disposal 10 digital computers, 4 analog computers, 10 skilled analysts and 34 semi-skilled analysts. Computers may run 24 hours per day. Skilled analysts may be occupied 8 hours per day, and semiskilled analysts may be occupied 9 hours per day. You determine four possible methods for using the available assets:

    Method 1: A team composed of two skilled and four semiskilled analysts can decode one message in four hours, making use of ten hours of digital computer time.

    Method 2: A team composed of two skilled and six semiskilled analysts can decode one message in two and one half hours making use of four hours of analog computer time.

    Method 3: A team composed of one skilled and five semiskilled analysts can decode one message in three hours, making use of four hours of analog computer time.

    Method 4: A team composed of one skilled and three semiskilled analysts can decode one message in four hours, making use of eight hours of analog computer time and four hours of digital computer time.

Formulate a mathematical model to determine how many messages should be attacked per day by each method in order to decode the maximum number of messages.

2. The army needs to let a contract for C rations in order to increase the amounts on hand at various warehouses throughout the country. The depots located at Philadelphia, Chicago, San Francisco, Fort Lewis, and St. Louis require 20,000, 15,500, 20,000, 20,000, and 25,000 packs, respectively. Bids are made by number of packs, which contain 12 meals or 4 rations. General Foods, General Mills, Best Foods, Armour, and Swift and Co. bid for part of the contract in the respective amounts: 23,500, 20,000, 15,500, 35,000, and 30,000 (packs). The cost of shipping a pack of C rations from a given company to a particular depot is given by the following:

	Phila.	Chicago	San Fran.	Ft. Lewis	St. Louis
General Foods	$7.00	$5.75	$ 9.45	$ 9.00	$6.00
General Mills	5.45	7.00	8.00	10.65	5.60
Best Foods	5.65	7.75	12.00	13.00	7.80
Armour	7.00	5.90	9.95	10.00	5.75
Swift and Co.	7.50	6.30	9.30	9.30	6.25

Formulate a mathematical model that determines how the army should award contracts to provide the lowest cost to the government.

3. How should final exams be scheduled at your school? If a student is assigned to take two different exams at the same time, then one of the instructors must write and monitor a second exam. If a student is scheduled to take two or three exams in succession over the same day rather than a more equitable distribution over the exam week, he or she is disadvantaged, and probably will demonstrate diminished performance on each exam. Discuss appropriate objectives for determining a final exam schedule. What are the constraints? Consider the number of classrooms that will be required.

5.4 PROJECTS

Complete the requirements in the referenced UMAP module.

1. "A Linear Programming Model for Scheduling Prison Guards," UMAP 272 by James M. Maynard. A work-scheduling model for correction officers at state correctional institutions is described. The model is a general one that might be applied to scheduling nurses and doctors at large hospitals, or security guards at industrial plants, and so forth.

2. "The Optimal Assignment Problem," by David Gale, UMAP 317. This unit provides a method of using aptitude scores to assign applicants, who

may have a variety of skills, to jobs in a way that maximizes the benefit to the employer. Familiarity with elementary row and column operations for matrices is required.

3. "A Mathematical Survey of Justice," UMAP 390 by Donald Wittman. This module introduces the concept of a utility function to survey approaches to justice in a systematic way. The module discusses various requirements that a justice function should have and proves that different sets of requirements lead to uniquely different well-specified justice functions.

MODEL

5.5 MANAGING RENEWABLE RESOURCES: THE FISHING INDUSTRY*

Consider the plight of the Antarctic baleen whale, which yielded a peak catch of 2.8 million tons in 1937 but only 50,000 tons in 1978. Or the case of the Peruvian anchoveta, which yielded 12.3 million tons in 1970 but only 500,000 tons just eight years later. The anchoveta fishery was Peru's largest industry and the world's most productive fishery. The seriousness of the economic impact is realized when one considers biologists' estimates that, even without fishing, it will take several years, or even decades in the case of the baleen whale, for these fisheries to reach their former levels of maximum biological productivity.[†]

Resources such as the anchoveta and baleen whale are called *renewable* (as opposed to *exhaustible*) resources. Exhaustible resources can yield only a finite total amount, whereas renewable resources can (theoretically) yield an unlimited total amount and be maintained at some positive level. The management of a renewable resource, such as a fishery, involves several critical considerations. What should the harvest rate be? How sensitive is the survival of the species to population fluctuation caused by harvesting, or to natural disasters such as a temporary alteration in ocean currents (which contributed to the demise of the anchoveta)? The economist Adam Smith (1723–1790) proclaimed that each individual in pursuing only his or her own

---

* Optional section.
[†] Data from Colin W. Clark's article, "The Economics of Over-exploration," *Science 181* (1973a): 630–634.

selfish good was led by an "invisible hand" to achieve the best for all. Will that invisible hand really ensure that market forces work in the best interests of humanity and the renewable resource, or will intervention be required to improve the situation, either for humanity or for the resource? In this example we use some very simple graphical submodels to gain a qualitative understanding of these management issues.

Scenario   Consider the harvesting of a common fish, such as the haddock, in a large competitive fishing industry. *Given the population level of the fish at some time, what happens to future population levels?* Future population levels depend, among other factors, on the harvesting rate of the fish and their natural reproductive rate (births minus deaths per unit time). Let's develop submodels for harvesting and reproduction separately.

A harvesting submodel   The classical theory of the firm was presented in Chapter 1, and the graphical models for total profit and marginal profit are reproduced in Figure 5-12. From Figure 5-12a you can see that a firm breaks even only when it can produce at least q_1 items, where total revenue equals total cost. In order to maximize profits the firm should continue to produce items as long as the marginal revenue exceeds the marginal cost; that is, produce up to the quantity q_2 as depicted in Figure 5-12b.

How does the theory of the firm relate to the fishing industry? For a common fish such as the haddock, and a competitive fishing industry, it is probably reasonable to assume that price is a constant. If a common fish cannot be marketed at that price, consumers will simply switch to another type of fish. Likewise, in a large industry the quantity of a particular fish marketed by an individual firm should not affect the price of the fish. Hence, over a wide range of values, constant price appears to be a reasonable *initial* assumption.

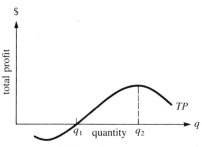

a. Total profit is total revenue minus total cost.

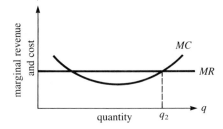

b. Marginal revenue and marginal cost curves.

FIGURE 5-12 A graphical model for the theory of the firm.

Next, let's consider the costs a fisherman encounters. These include such costs as salaries, fuel, capital tied up in equipment, processing, refrigeration, and the like. As usual, once a time horizon is selected, these costs can be divided into two categories: fixed costs (such as capital costs), which are independent of the yield harvested, and variable costs (such as processing), which depend upon the size of the yield. The basic principles underlying the theory of the firm apply reasonably well. For example, you would expect that a typical fishing company must harvest some minimum yield at a given level of effort (including number of boats, man-hours of labor, and so forth) in order to break even.

An interesting condition exists with the fishing industry: the cost of harvesting a given number of fish depends upon *the size of the fish population.* Certainly less effort is required to catch a given number of fish when they are plentiful, so at a given level of effort you would expect to catch more fish when they are plentiful. Thus, we assume the average unit harvesting cost $c(N)$ decreases as the size of the fish population N increases. This assumption is depicted in Figure 5-13, where the harvesting cost per fish is shown as a decreasing function of the size of the fish population. It is important to note that the independent variable N is the size of the fish population, *not* the size of the yield. The market price p of a fish is also shown in the figure.

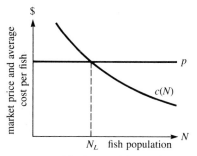

FIGURE 5-13 The (average) harvesting cost per fish decreases as the fish population level increases.

The submodel represented by Figure 5-13 suggests that it would be unprofitable to fish a particular species of fish unless the population level is at least as large as N_L, where the cost of harvesting each fish equals the price paid for it by the consumer. In those instances where it is economically feasible to harvest the species, harvesting is a force driving the population level to N_L. If the population level lies significantly above N_L, then the excess profit potential causes the fishing of the species to be intensified. On the other hand, if the population falls below N_L, the fishing would cease (theoretically), causing the population level to increase.

We have reached an impasse with the harvesting submodel. Although we intended to develop the two submodels independently, we now realize that the average cost of harvesting a fish depends upon the size of the fish population and hence its reproductive capacity. Even if we know the current size of the fish population, we still need to know how the magnitudes of the harvesting and reproduction rates compare in order to determine whether the population will increase or decrease. We turn now to the development of a simple reproduction submodel.

A reproduction submodel   Consider how a species of fish grows in the absence of fishing. Among the many factors affecting its net growth rate are the birthrate, death rate, and environmental conditions. Each of these factors will be developed in detail in Chapter 9, where we construct several population models. For our purposes here, a simple qualitative graphical model will suffice. So, let $N(t)$ denote the size of the fish population at any time t, and let $g(N)$ represent the rate of growth of the function $N(t)$. Assume that $g(N)$ is approximated by the continuous model $g(N) = dN/dt$. Obviously, when $N = 0$, there are neither births nor deaths. Now assume that there is a maximum population level N_u that can be supported by the environment. This level might be imposed by the availability of food, the effect of predators, or some similar inhibitor. At N_u we assume that the births equal the deaths, or $g(N_u) = 0$. Thus, under our assumptions, $g(N)$ has zeros at $N = 0$ and $N = N_u$, and positive values in between. A graphical representation of g is depicted in Figure 5-14.

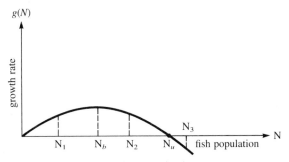

FIGURE 5-14 A submodel for reproduction.

Before proceeding, we offer one analytic submodel for reproduction as an illustration. A simple quadratic function having zeros at $N = 0$ and $N = N_u$ is $g(N) = aN(N_u - N)$, where a is a positive constant. Note that the quadratic is positive for $0 < N < N_u$ and negative for $N > N_u$ as required. Since $g'(N) = a(N_u - 2N)$, $N = N_u/2$ is a critical point of g. Also, $g''(N) = -2a$ implies g is everywhere concave downward, as depicted in Figure 5-14.

Let's see what the submodel for reproduction suggests about the population level. Suppose the population level is currently at level N_1 (see Figure 5-14). Since $g(N_1) > 0$ and $g(N) = dN/dt$, the function N is increasing. As N increases, $g(N) = dN/dt$ increases as well until it reaches a maximum at $N = N_b$. For $N > N_b$ the derivative remains positive, but becomes smaller and approaches zero as N gets closer to N_u. Thus, the population approaches $N = N_u$ as suggested in Figure 5-15. Convince yourself that the curves suggested for starting populations of $N = N_2$ and $N = N_3$ are qualitatively correct in Figure 5-15 as well, where $N_1 < N_b < N_2 < N_u < N_3$. Note that

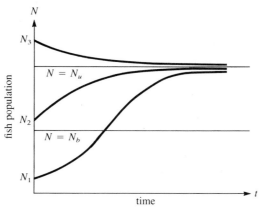

FIGURE 5-15 Regardless of the initial population, the population approaches $N = N_u$ without fishing.

the ordinate in Figure 5-14 is the slope of the curves in Figure 5-15. In Chapter 9 we develop an analytic model using differential equations to represent the population growth in Figure 5-15, which is called "logistic" growth.

The biological optimum population level   We now interpret our submodel for reproduction. In the absence of fishing the population always approaches N_u, which is called the *carrying capacity* of the environment. Moreover, a population level exists where the net biological growth is at a maximum. This population level is called the *biological optimum population* and is denoted by N_b. In our particular submodel, $N_b = N_u/2$, as established earlier and depicted in Figure 5-14. (We establish this result more precisely in Chapter 9.)

The social optimum yield   Now that we have developed a submodel for reproduction, let's return to our harvesting submodel. Assume that the fishing industry seeks to maximize total profit, which is the product of the yield and the average profit per fish. The average profit per fish is the difference between the price p and the average cost $c(N)$ to harvest a fish. Thus, the expression for total profit TP is

$$TP = (\text{yield}) \times [p - c(N)] \qquad (5.9)$$

What shall we assume about the yield? From the fisherman's point of view, the ideal situation is to have a constant yield from year to year. This situation would allow planning for efficient staffing and capitalization of the resources allocated to fishing that particular species. Otherwise, in lean years there would be resources, such as fishing vessels and staff, that would be underutilized. It is also reasonable to assume that consumer demand for a certain kind of fish is pretty much constant from year to year, say for a three- or four-year period. Let's see where the assumption of constant yield leads us.

First, how is a constant yield obtained from year to year? One way is to harvest the difference between the births and deaths over time. Since the function $g(N)$ approximates this difference for a given population N of fish, this idea suggests that $g(N)$ equals the yield. To illustrate the idea, consider Figure 5-16. Suppose the fish population is presently at N_2. If precisely the amount of fish $g(N_2)$ is harvested, the population will remain at N_2 (zero net growth) and the fishing operation could be repeated annually indefinitely. Convince yourself that any population level N with $0 < N < N_u$ can be maintained by harvesting the amount $g(N)$ per year.

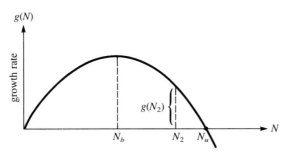

FIGURE 5-16 Harvesting $g(N)$ annually permits a sustainable yield.

Note that N_b, the population level for which the biological growth is a maximum, is of particular importance under the interpretation of a sustainable yield. This is true because N_b is the population level that allows the largest sustainable yield, $g(N_b)$. Since this population level has important social implications, we refer to N_b as the *social optimum population level*. Note that the social and biological optimum population levels are coincident in this example.

The economic optimum population level   Assuming the yield to be $g(N)$, Equation (5.9) for total profit becomes

$$TP = g(N)[p - c(N)] \tag{5.10}$$

The fishing industry wants to maximize TP. What level of fish population permits the greatest profit? Since $g(N)$ reaches a maximum at $N = N_b$, you might be tempted to conclude that profit is maximized there also. After all, the greatest yield occurs there. However, the average profit continues to increase as N increases, based upon our submodel depicted in Figure 5-13. Thus, by choosing $N > N_b$, we may increase the profit while simultaneously catching fewer fish since $g(N) < g(N_b)$. Let's see if this possibility is indeed the case.

What does the graph of the total profit function TP look like? The factor $g(N)$ has zeros at $N = 0$ and $N = N_u$, and the factor $[p - c(N)]$ has a zero at $N = N_L$. One representation of a continuous function meeting these requirements is shown in Figure 5-17 and suggests that a population level does exist at which profits can be maximized. We call that population level the *economic optimum population level* and denote it by N_p.

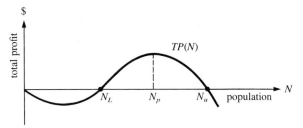

FIGURE 5-17 A continuous total profit function.

Relating the biological, social, and economic optimum population levels   Up to this point several population levels of interest to us have been introduced. They are summarized as follows:

N_L = the minimum population level for economic feasibility
N_b = the social and biological optimum population levels
N_p = the economic optimum population level
N_u = the maximum population sustainable by the environment
      (the environmental carrying capacity)

Next we relate these population levels. In order that it be worthwhile to fish a species on a prolonged basis, it must be the case that $N_L < N_u$, with N_b and N_p somewhere in between. But how are N_b and N_p related? If we believe that the fishing industry will eventually find the population level that maximizes profit, we would hope for social and biological considerations that $N_p = N_b$. Let's see if Adam Smith's invisible hand is at work.

Consider Figure 5-18. For the case $N_L < N_p < N_u$, three possible locations for N_b are shown on the graph for the total profit function TP. Study the graph and convince yourself of the following situations:

1. $N_b < N_p$ if and only if $TP'(N_b) > 0$.
2. $N_b = N_p$ if and only if $TP'(N_b) = 0$.
3. $N_b > N_p$ if and only if $TP'(N_b) < 0$.

Now, taking the derivative of Equation (5.10), we have

$$TP' = g'(N)[p - c(N)] - g(N)c'(N) \qquad \textbf{(5.11)}$$

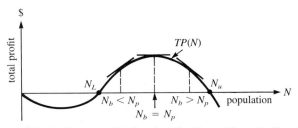

FIGURE 5-18 Three possible locations for N_b are $N_b < N_p$, $N_b = N_p$, and $N_b > N_p$.

From the definition of N_b, $g'(N_b) = 0$ and $g(N_b) > 0$, and from the submodel for average cost depicted in Figure 5-13, $c'(N) < 0$ for all N. Substitution of N_b into Equation (5.11) yields $TP'(N_b) > 0$, which implies that $N_b < N_p$ from Figure 5-18. Thus, under our assumptions, by allowing the fish population to exceed N_b, the fisherman can catch fewer fish yet reap greater profits than is possible at N_b! Let's interpret our model to assist management in determining the alternatives for controlling the location of the various population levels N_L, N_b, and N_p.

Model interpretation   As you will see in Chapter 9, the assumptions underlying the reproduction submodel are extremely simplistic, so we cannot expect to draw precise conclusions from the graphical models developed here. However, our purpose in constructing the model was to identify and analyze qualitatively some of the key issues in managing a renewable resource. Consider again the baleen whale and Peruvian anchoveta in terms of the models we have developed.

Several considerations must be taken into account with whales. First, the whale population displays a low natural growth rate (typically only 5–10% a year), and even a small harvest can result in a negative net growth rate when the population is low. Second, many conservationists argue that there is a minimum population N_s for the whale below which the species will not survive. If harvesting or natural disasters cause the whale population to fall below N_s, the species will be driven to extinction. These ideas are incorporated in the graphical representations depicted in Figure 5-19, showing both the reproduction and population submodels.

Note from Figure 5-19 how sensitive the future whale population is to levels in the vicinity of N_s. The relatively high market price for the whale causes the minimum population N_L for economic feasibility to be quite low (see Figure 5-13) so that N_L can be quite near the level N_s. Thus, the location of N_L becomes critical. (Argue that in the case of the whale, N_L could be smaller than N_s. Management of the fishery in such a case requires sepa-

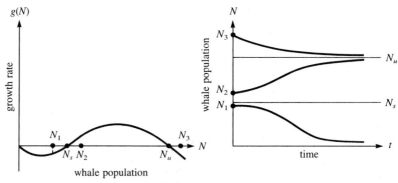

FIGURE 5-19 Reproduction and population models incorporating a minimum survival level.

rating N_s and N_L.) Ideally, one would like N_L and N_b to coincide. Taxation and the establishment of quotas are two alternatives that are discussed in the following problem set.

In the case of the Peruvian anchoveta, it is estimated that the maximum sustainable yield is 10 million tons annually. Thus any additional catch (such as the 12.3 million tons caught in 1970) would result in a population decrease, even if the population were at the biological optimum level N_b. From Figure 5-16 we suspect that as the population decreases from the biological optimum population, the sustainable yield decreases, suggesting that the harvest should be reduced. However, from a practical point of view, once the fishing fleet and staff are established to catch 12.3 million tons, there is strong economic motivation to continue harvesting at that level by fishing more intensively. This critical situation existed during the 1970s in Peru, where there were few economic alternatives, and government intervention to enforce strict regulation of the fishery was politically volatile. You can well imagine how these practical considerations become substantially more complex when more than one country is involved in managing the resource.

5.5 PROBLEMS

1. Assume that the environmental carrying capacity N_u is determined principally by the availability of food. Argue that under such an assumption, as N approaches N_u, the physical condition of the average fish deteriorates as competition for the food supply becomes more severe. What does this suggest about the survival of the species when natural disasters such as storms, severe winters, and similar circumstances, further restrict the food supply? Where should conservationists attempt to maintain the population level?

2. In 1981 and 1982 the deer population in the Florida Everglades was very high. Although the deer were plentiful, they were very thin and on the brink of starvation. Hunting permits were issued to "thin out" the herd. This action caused much furor on the part of environmentalists and conservationists. Explain the poor health of the deer and the purpose of the special hunting permits in terms of population growth and population submodels.

3. Argue that for many species a minimum population level is required for survival and give several examples. Call this minimum survival level N_s. Suggest a simple cubic growth submodel meeting these requirements, as depicted in Figure 5-19. Answer the questions posed in Problem 1 using your graphical submodel.

4. Suppose $N_u < N_L$. What does this inequality suggest about the economic feasibility of fishing that species? Give several examples.

Problems 5 and 6 relate to *fishing regulation*:

5. One of the key assumptions underlying the models developed in this section is that the harvest rate equals the growth rate for a sustainable yield. The reproduction submodels in Figures 5-16 and 5-19 suggest that if the current population levels are known, it is possible to estimate the growth rate. The implication of this knowledge is that if a quota for the season is established based upon the estimated growth rate, then the fish population can be maintained, increased, or decreased as desired. This quota system might be implemented by requiring all commercial fishermen to register their catch daily, and then closing the season when the quota is reached. Discuss the difficulties in determining reproduction models precise enough to be used in this manner. How would you estimate the population level? What are the disadvantages of having a quota that varies from year to year? Discuss the practical and political difficulties of implementing such a procedure.

6. One of the difficulties in managing a fishery in a free enterprise system is that excess capacity may be created through overcapitalization. This happened in 1970 when the capacity of the Peruvian anchoveta fishermen was sufficient to catch and process the maximum annual growth rate in less than three months. A disadvantage of restricting access to the fishery by closing the season after a quota is reached is that this excess capacity is idle during much of the season, which creates a politically and economically unsatisfying situation. An alternative is to control the capacity in some manner. Suggest several procedures for controlling the capacity that is developed. What difficulties would be involved in implementing a procedure like restricting the number of commercial fishing permits issued?

Problems 7–9 relate to *taxation*:

7. Figure 5-13 suggests that market forces tend to drive the population to N_L. Use that figure to show how taxation or subsidization may be used to control the location of N_L. What forms might the taxation and subsidization take? (Hint: One of the costs to the fisherman are the various taxes he pays.) Apply your ideas to the whale fishery.

8. Taxation is appealing from a theoretical perspective since with a properly designed tax the desired goals can be achieved through normal market forces rather than by some artificial method (such as restricting the number of commercial permits). Assume a fish population is currently at N_b and you desire to maintain it at that level by harvesting $g(N_b)$. How can a tax be determined for each fish caught to cause N_L, N_b, and N_p to coincide? (Hint: Consider Equation (5.11) and the condition required for $N_b = N_p$.)

9. A constant price has been assumed in all the models developed in this section. Suggest some fisheries for which that assumption is not realistic.

How might you alter the assumption? How would you determine the appropriate tax to be applied?

5.5 FURTHER READING

Clark, Colin W. *Mathematical Bioeconomics: The Optimal Management of Renewable Resources.* New York: Wiley, 1976.

May, Robert M., John R. Beddington, Colin W. Clark, Sidney J. Holt, and Richard M. Laws. "Management of Multispecies Fisheries." *Science 205* (July 1979): 267–277.

PART THREE

EMPIRICAL MODEL CONSTRUCTION

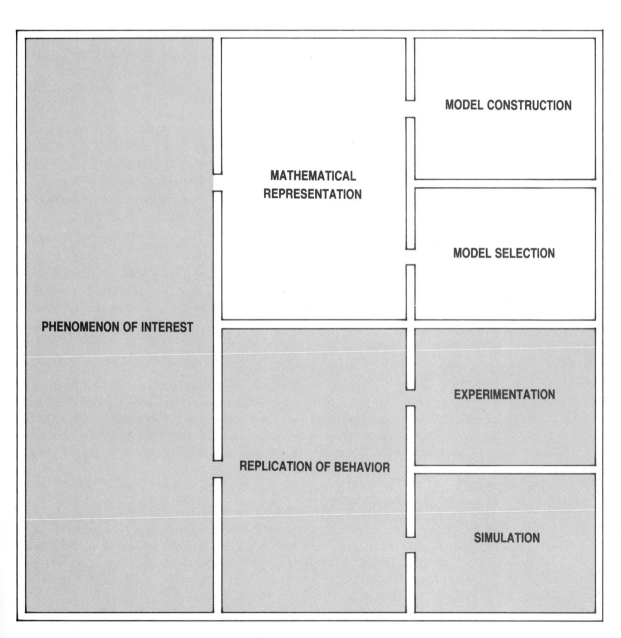

PHENOMENON OF INTEREST

MATHEMATICAL
REPRESENTATION

MODEL CONSTRUCTION

MODEL SELECTION

REPLICATION OF BEHAVIOR

EXPERIMENTATION

SIMULATION

EXPERIMENTAL MODELING

INTRODUCTION

In Chapter 4 we discussed the philosophical differences between curve fitting and interpolation. When fitting a curve, the modeler is using some assumptions that select a particular type of model that "explains" the behavior being observed. If collected data then corroborate the reasonableness of those assumptions, the modeler's task is to choose the parameters of the selected curve that "best fits" the data according to some criterion (like least squares). In this situation the modeler expects, and willingly accepts, some deviations between the fitted model and the collected data in order to obtain a model *explaining* the behavior. The problem with this approach is that in many cases the modeler is unable to construct a tractable model form that satisfactorily explains the behavior. Thus the modeler doesn't know what kind of a curve actually describes the behavior. If it is necessary to predict the behavior nevertheless, the modeler may conduct experiments (or otherwise gather data) to investigate the behavior of the dependent variable(s) for selected values of the independent variable(s) within some range. In essence, **empirical model** the modeler desires to construct *an **empirical model** based on the collected data* rather than select a model based on certain assumptions. In such cases the modeler is strongly influenced by the data that have been carefully collected and analyzed, so he or she seeks a curve that captures the trend of the data in order to *predict* in between the data points. For example, consider the data shown in Figure 6-1a. If the modeler's assumptions lead to the expectation of a quadratic model, a parabola would be fit to the data points, as illustrated in Figure 6-1b. However, if the modeler has no reason to expect a model of a particular type, a smooth curve may be passed through the data points instead, as illustrated in Figure 6-1c.

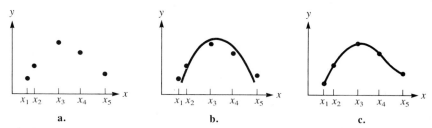

FIGURE 6-1 If the modeler expects a quadratic relationship, a parabola may be fit to the data, as in b. Otherwise a smooth curve may be passed through the points, as in c.

In this chapter we address the construction of empirical models. In Section 6.1 we study the selection process for simple one-term models that capture the trend of the data. A scenario is addressed for which few modelers would attempt to construct an explicative model—namely, predicting the size of a population as a function of the "pace of life." In Section 6.2, we discuss the construction of higher-order polynomials that pass through the collected data points. In Section 6.3, we investigate the "smoothing" of data using lower-order polynomials. Finally, in Section 6.4, we present the technique of cubic spline interpolation, where a distinct cubic polynomial is used across successive pairs of data points.

MODEL

6.1 THE "PACE OF LIFE" AND OTHER ONE-TERM MODELS*

Let's consider a situation in which the modeler has collected some data, but is unable to construct an "explicative" model. In early 1976, Marc and Helen Bornstein reported the results of some simple observations they had made at 15 locations around the world (see 6.1 "Further Reading"). Taking precautions to ensure equivalent conditions, they systematically observed the mean time required for pedestrians to walk 50 ft on the main streets of their cities and towns. That this was considered to be part of a larger problem

* This section is adapted from the UMAP Unit 551, *The Pace of Life: An Introduction to Empirical Model Fitting* by Bruce King. The adaptation is presented with the permission of COMAP, Inc./UMAP, 271 Lincoln St., Lexington, Mass. 02173.

(population effects on the "quality of life") is obvious from the interpretive paragraph at the end of the Bornsteins' paper. Some of their data are shown in Table 6-1. A scatterplot of population P versus time T is shown in Figure 6-2. Consider the difficulties one would have in attempting to develop an "explicative" model for this scenario.

TABLE 6-1 Population and mean time to walk 50 ft, for 15 locations

Location	Population P	Mean time T (sec)
(1) Brno, Czechoslovakia	341,948	10.4
(2) Prague, Czechoslovakia	1,092,759	8.5
(3) Corte, Corsica	5,491	15.1
(4) Bastia, France	49,375	10.2
(5) Munich, West Germany	1,340,000	8.9
(6) Psychro, Crete	365	18.1
(7) Itea, Greece	2,500	22.0
(8) Iraklion, Greece	78,200	13.0
(9) Athens, Greece	867,023	9.6
(10) Safed, Israel	14,000	13.5
(11) Dimona, Israel	23,700	15.3
(12) Netanya, Israel	70,700	11.6
(13) Jerusalem, Israel	304,500	11.3
(14) New Haven, U.S.A.	138,000	11.4
(15) Brooklyn, U.S.A.	2,602,000	9.9

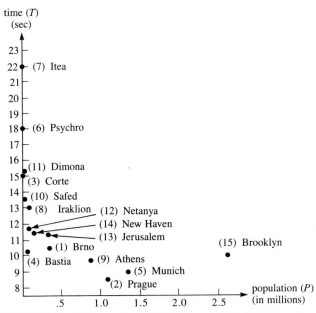

FIGURE 6-2 Population and mean time to walk 50 ft, for 15 locations.

Figure 6-2 clearly shows a tendency for slow times to be associated with small populations, and fast times with large populations. But a more precise description of the relationship is not so obvious. That is, what curve fits the data? The Bornsteins chose, first, to transform the mean time T into a variable of greater interest—the "pace of life," as they called it—by letting $V = 50/T$, so that velocity V is measured in feet per second. The transformed data are shown in Table 6-2, and Figure 6-3 displays the associated scatterplot.

TABLE 6-2 Population and mean velocity over a 50-foot course, for 15 locations

Location*	Population P	Mean velocity V (ft/sec)
(1)	341,948	4.81
(2)	1,092,759	5.88
(3)	5,491	3.31
(4)	49,375	4.90
(5)	1,340,000	5.62
(6)	365	2.76
(7)	2,500	2.27
(8)	78,200	3.85
(9)	867,023	5.21
(10)	14,000	3.70
(11)	23,700	3.27
(12)	70,700	4.31
(13)	304,500	4.42
(14)	138,000	4.39
(15)	2,602,000	5.05

* For location names, see Table 6-1.

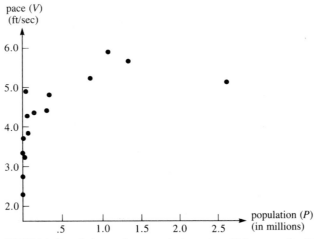

FIGURE 6-3 Population and mean velocity over a 50-ft course, for 15 locations.

It is tempting to superimpose on Figure 6-3 a curve that seems to approximate the relationship between P and V. Figure 6-4 shows one such curve, which was drawn freehand just by inspecting the scatterplot of Figure 6-3.

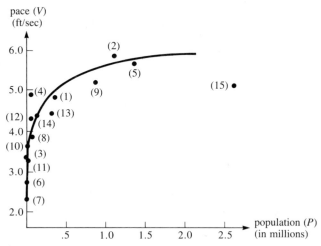

FIGURE 6-4 A curve that approximates the relation between population P and mean pedestrian velocity V.

Figure 6-4 suggests some mathematical questions:

1. Is it possible to find an equation of a summarizing function like the one in Figure 6-4?
2. Is there one function that "best fits" the data shown in Figure 6-3 according to some criterion (like least squares)?

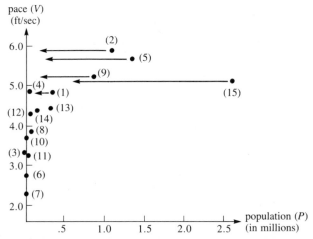

FIGURE 6-5 A strategy for linearizing the pace versus population data.

In the rest of this section, we will suggest how one might begin to answer such questions. Our strategy will be to transform the data of Table 6-2 in such a way that a summarizing function may be taken as a straight line—for then the equation of the line can be derived from its graph.

A transformation that might have the desired effect would make large P-values much smaller, and smaller P-values only a little smaller. (See Figure 6-5.) Two such transformations that might accomplish this effect are $\sqrt{P}$ or log P in place of P. Experience with such matters suggests that a "linearizing transformation" often can be found among those in the "Ladder of Powers" of a variable z.*

Ladder of Powers

$$\vdots$$

$$z^2$$

$$z$$

$$\sqrt{z}$$

$$\log z$$

$$\frac{1}{\sqrt{z}}$$

$$\frac{1}{z}$$

$$\frac{1}{z^2}$$

$$\vdots$$

(Note, however, the modification of this ladder contained in Table 6-3.) Indeed, in many cases a linearizing transformation can be found among just these four "rungs" of the ladder: $\sqrt{z}$, log z, $1/\sqrt{z}$, $1/z$. Before returning to the Bornsteins' data, let's consider this Ladder of Powers a bit more thoroughly.

Figure 6-6 shows a set of five data points (x, y) along the line $y = x$, for $x > 1$. Suppose we change the y value of each point with the $\sqrt{y}$. This procedure yields a new relation $y = \sqrt{x}$ whose y values are closer together over the domain in question. Note that all the y values are reduced, but the larger values are reduced more than the smaller ones.

Changing the y value of each point to log y has a similar, but more pronounced effect; and each additional step down the ladder produces a stronger version of the same effect.

We started in Figure 6-6, in the simplest way—with a linear function. But that was only a convenience. Given any positive-valued function $y = f(x)$,

---

* See Paul F. Velleman and David C. Hoaglin, *Applications, Basics, and Computing of Exploratory Data Analysis* (Boston, Mass.: Duxbury Press, 1981), p. 49.

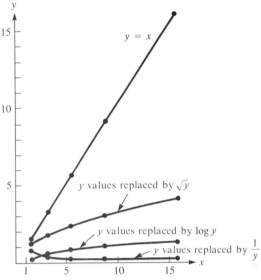

FIGURE 6-6 The relative effects of five transformations.

$x > 1$, that is concave up, like this:

then some transformation in the ladder below y—changing the y values to $\sqrt{y}$, or log y, or a more drastic change—squeezes the right-hand tail downward and has a chance of generating a new function more nearly linear than the original. Which transformation should be used is a matter of trial and error, and experience. Another possibility is to stretch the right-hand tail to the right (so, try changing the x values to x^2, x^3 values, and so forth).

Given a positive-valued function $y = f(x)$, $x > 1$, that is increasing and concave down, like this:

one can hope to (more nearly) linearize it by stretching the right-hand tail upward (so, try changing the y values to y^2, y^3 values, and so forth). Another possibility is to squeeze the right-hand tail to the left (so, try changing the x values to $\sqrt{x}$, or log x, or by a more drastic choice from the ladder).

A final comment on this: note that, although replacing z by $1/\sqrt{z}$, $1/z$, or $1/z^2$, and so on may sometimes have a desirable effect, such replacements also have an undesirable one—an increasing function is converted into a decreasing one. As a result, when using the transformations in the ladder below

TABLE 6-3 Ladder of transformations

$$\vdots$$
$$z^3$$
$$z^2$$
$$z \text{ (no change)}$$

$$*\begin{cases} \sqrt{z} \\ \log z \\ \dfrac{-1}{\sqrt{z}} \\ \dfrac{-1}{z} \end{cases}$$

$$\dfrac{-1}{z^2}$$
$$\vdots$$

* Most often used transformations

log z, data analysts usually use a negative sign to keep the transformed data in the same linear order as the original data. Thus, the Ladder of Transformations, as usually used, is shown in Table 6-3.

Now to the Bornsteins' data. Because of the shape of Figure 6-3, the Bornsteins replaced P by log P (to squeeze the right tail to the left). These data appear in Table 6-4, and Figure 6-7 shows the corresponding scatterplot.

TABLE 6-4 Log (population) and mean velocity over a 50-ft course for 15 locations

Location*	Log (population)	Mean velocity V (ft/sec)
(1)	5.53	4.81
(2)	6.04	5.88
(3)	3.74	3.31
(4)	4.69	4.90
(5)	6.13	5.62
(6)	2.56	2.76
(7)	3.40	2.27
(8)	4.89	3.85
(9)	5.94	5.21
(10)	4.15	3.70
(11)	4.37	3.27
(12)	4.85	4.31
(13)	5.48	4.42
(14)	5.14	4.39
(15)	6.42	5.05

* For location names, see Table 6-1.

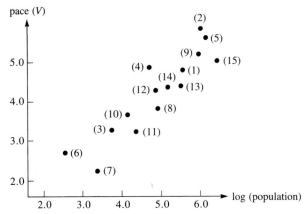

FIGURE 6-7 Log(population) versus mean pedestrian velocity over a 50-ft course for 15 locations.

And now observe that, as we had hoped, the data seem to lie roughly along some line *l*. We could use the least-squares technique from Chapter 4 to find the best-fitting straight line, but it is just as easy here to "eyeball" a suitable line on the scatterplot of Figure 6-7. The result might appear as shown in Figure 6-8.

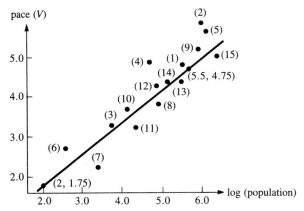

FIGURE 6-8 A summarizing line for the log(population) versus pace data.

Now, if we use *V* for velocity and *P* for population, the line *l* can be expressed in the form:

$$V = m(\log P) + b$$

By selecting a convenient pair of points on *l*, we can compute, approximately, its slope *m*: using (2, 1.75) and (5.5, 4.75), for instance, we find the slope:

$$m \approx \frac{4.75 - 1.75}{5.5 - 2} = \frac{3}{3.5} = 0.86$$

Then, using the fact that (2, 1.75) is on l, we can see that its V-intercept b satisfies the equation:

$$1.75 = \frac{3}{3.5}(2) + b$$

so that $b \approx 0.04$. Consequently, an equation for l is

$$\hat{V} = 0.86(\log P) + 0.04 \tag{6.1}$$

(It is customary to discriminate between the observed V-values in Figure 6-8 and the V-values predicted by the summarizing line; hence, the "hat" over the V in Equation (6.1).) Note that the equation cannot be interpreted as a deterministic model: it describes only an *approximate* relation between P and V.

Using the property that $y = \log n$ if and only if $10^y = n$, we can rewrite Equation (6.1) (with the aid of a calculator) as

$$10^{1.16V - 0.0465} = P$$

or,

$$\hat{P} = 0.9(14.5)^V \tag{6.2}$$

The coefficients in this calculation are necessarily only approximate. (The "hat" over the P in Equation (6.2) discriminates the predicted P-values from the observed P-values.)

Figure 6-9 shows the graph of (6.2), superimposed on the scatterplot. Note that the curve seems to be reasonable since it fits the data about as well as you would expect for a simple curve. (Compare Figure 6-9 with Figure 6-4.)

Verifying the Model

How good will the predictions based on Models (6.1) or (6.2) be? Part of the answer to this question lies in comparing observed values of V with those predicted by (6.1). For example, (6.1) predicts that people in the town of New

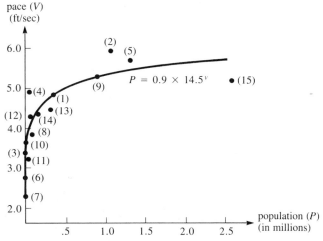

FIGURE 6-9 A summarizing curve for the pace versus population data.

TABLE 6-5 Populations (P) and populations predicted ($\hat{P}$), for 15 locations

Location*	P	$\hat{P}$
(1)	341,948	347,074
(2)	1,092,759	6,068,570
(3)	5,491	6,286
(4)	49,375	441,516
(5)	1,340,000	3,027,830
(6)	365	1,444
(7)	2,500	390
(8)	78,200	26,639
(9)	867,023	1,011,510
(10)	14,000	17,836
(11)	23,700	5,648
(12)	70,700	91,146
(13)	304,500	122,318
(14)	138,000	112,888
(15)	2,602,000	659,404

* For location names, see Table 6-1.

Milford, Connecticut—a town of about 16,900 people*—should walk, on average, about 3.7 ft/sec. Observations of 30 people walking on Main Street, New Milford on 19 January 1980 yielded an average pace of 4.8 ft/sec.

In the same way, we could use (6.2) to predict the population of a region, given a set of observations of the "pace of life" in that region. For example, observations of 30 people walking on East Main Street, Rochester, New York on 2 November 1979 yielded a mean pace of 4.4 ft/sec. If we use (6.2), this suggests that Rochester's population should be about 120,000. But Rochester's actual population is about 242,000.[†]

Let's see how well (6.2) predicts the populations of the 15 sites used by the Bornsteins. With P representing the predictions, the results (obtained with the help of a calculator) are shown in Table 6-5.

These results suggest that (6.2) can be expected to predict a region's population from "pace of life" data only to within an order of magnitude. That is, if a region's population is about 50,000, then (6.2) can be expected to yield a prediction from $.1 \times (50,000) = 5,000$ to $10 \times (50,000) = 500,000$ people. Thus (6.2) could not be used reliably for serious regional planning—for anticipating future needs for transportation, schools, police, or fire-protection services, for example. Nevertheless, (6.2) approximates an actual relationship and confirms our intuition relating large populations to a fast pace of life.

Finally, it is important to recognize that, even if a model is considered to be valid over a certain range of values of its variables, it may not be valid outside that range. For example, (6.1) was generated from data for which $365 \leqslant P \leqslant 2,602,000$. If we ask the model to predict pace V when popu-

* State of Connecticut, *Register and Manual*, 1979.
† U.S. Bureau of the Census, 1980.

lation $P = 1$, we find that $V = 0.9$ ft/sec—an absurd result. Similarly, the Bornsteins observed average paces V satisfying $2.27 < V < 5.88$. An observed mean pace of 7 ft/sec does not seem inconceivable—perhaps between 8 A.M. and 9 A.M. in Tokyo, Japan. Yet, for such a pace, (6.2) would predict a regional population in excess of 120 million! Thus it is important you be extremely wary of extrapolations, even with a reasonable model.

It must be understood that, even if models like (6.1) and (6.2) are regarded to be "good" models, they are not to be taken literally. One simply cannot expect, as a rule, that a simple equation like (6.1) will completely describe a complex relationship. Modelers usually adopt a pragmatic point of view; they use a model if it "works" better than its competitors, and they do not demand that it be "true." That is, most modelers would prefer to construct an explicative model whenever possible. However, if they cannot do so, then they choose the model that does the best job of predicting.

Our analysis of the "pace of life" data has led to an exponential model of the form $P = kb^V$, where the constants $k = 0.9$ and $b = 14.5$ were estimated from the data. On the other hand, an examination of Figure 6-4 reveals that the data in the scatterplot lie, roughly, along the upper branch of some parabola $V = c\sqrt{P} = cP^{0.5}$. This observation suggests that we try fitting a general model of the form $V = cP^a$. Using the data and one of the curve-fitting techniques discussed in Chapter 4, we could then determine the "best-fit" parameters c and a (see Problem 2).

Let's summarize the ideas of this section. When we are constructing an empirical model, we always begin with a careful analysis of the collected data. Do the data suggest the existence of a trend? Are there data points that obviously lie outside the trend? If such "outliers" do exist, it may be desirable to discard them or, if obtained experimentally, to repeat the experiment as a check for a data collection error. Once it is clear that a trend exists, we next attempt to find a function that will transform the data into a straight line (approximately). In addition to trying the functions listed in the Ladder of Transformations presented in this section, we can also attempt the transformations discussed in Chapter 4. Thus if the model $y = ax^b$ is selected, then we would plot $\ln y$ versus $\ln x$ to see if a straight line results. Likewise, when investigating the appropriateness of the model $y = ae^{bx}$, we would plot $\ln y$ versus x to see if a straight line results. Keep in mind our discussion in Chapter 4 about how the use of a transformation may be deceiving, especially if the data points get squeezed together. Our judgment here is strictly qualitative; the idea is to determine if a particular model type appears promising. Once we are satisfied that a certain model type does seem to capture the trend of the data, then we can estimate the parameters of the model graphically or using the analytic techniques discussed in Chapter 4. Eventually we must analyze the goodness of fit using the indicators discussed in Chapter 4. Remember to graph the proposed model against the *original* data points, not the transformed data. If you are dissatisfied with the fit, other one-term models can be investigated. However, because of their inherent simplicity, one-term models cannot fit all data sets. In such situations other techniques can be used, and we will discuss these methods in the next several sections.

6.1 PROBLEMS

1. The Bornsteins wrote, in part:

   The locations ... were ... 'downtown' or commercial areas .... passers-by were timed unobtrusively .... the data were collected on dry, sunny days of moderate temperatures (mean 24.1 degrees C). Subjects ... were selected from those who walked the distance alone and unencumbered. They were selected unsystematically within these constraints and ... equal numbers were chosen from people walking in each direction.

   a. Can you think of any factors not mentioned by the Bornsteins that might have a significant effect on the pace at which people walk?

   b. Can you think of reasons why the predictions of pace in New Milford and of population in Rochester weren't very accurate?

   c. Use a stopwatch to time a number of people—say, about 25—walking on a main street in your town. You will have to set a few ground rules for the observation process.

   i. Calculate the mean time T and mean velocity V for your sample.

   ii. Use (6.2) to estimate the population P of your region. (Note that it sometimes is difficult to decide what your region is. For example, is the "region" for Flatbush Avenue in the Bornsteins' data Flatbush? Or is it Brooklyn? Or New York City? Or something else?)

   iii. Compare your result from Part (ii) with an official estimate of the population of your region. Are the results reasonably close?

2. In fitting a model of the form $V = cP^a$ to the "pace of life" data, use the representation

$$\log V = a \log P + \log c$$

   giving a straight-line representation of $\log V$ versus $\log P$.

   a. From Table 6-2, make a table of values showing $\log V$ versus $\log P$ for each location.

   b. From your table in Part a, construct a scatterplot diagram of $\log V$ versus $\log P$.

   c. "Eyeball" a line l onto your scatterplot.

   d. Using any convenient pair of points of l (not too close together and reasonably close to the line), calculate the slope of l.

   e. Using the result of Part d, find a linear equation that relates $\log V$ and $\log P$.

   f. Using the result of Part e, find an equation of the form $V = cP^a$ that expresses V in terms of P. (Your answer should be close to $\tilde{V} = 1.397P^{0.096}$.)

3. Draw a graph of the equation you found in Part f of Problem 2.
4. Using Table 6-2, a calculator, and the equations of $\hat{V}$ and $\tilde{V}$ (Problem 2f), complete Table 6-6.

TABLE 6-6 Observed mean velocity and two predicted mean velocities, for 15 locations

Location*	Observed velocity V	Predicted Velocities	
		$\hat{V}$	$\tilde{V}$
(1)	4.81	4.80	4.75
(2)	5.88		
(3)	3.31		
(4)	4.90		
(5)	5.62		
(6)	2.76		
(7)	2.27		
(8)	3.85		
(9)	5.21		
(10)	3.70		
(11)	3.27		
(12)	4.31		
(13)	4.42		
(14)	4.39		
(15)	5.05		

* For location names, see Table 6-1.

5. From the data of Table 6-6, calculate the mean (that is, the average) of the Bornstein errors $|V - \hat{V}|$. Do the same for the power curve errors $|V - \tilde{V}|$. What do the results of these calculations suggest about the relative merits of the two models $\hat{V}$ and $\tilde{V}$?
6. In Table 6-7, X is the Fahrenheit temperature, and Y is the number of times a cricket chirps in one minute. Fit a model to these data.

TABLE 6-7 Temperature and chirps per minute for 20 crickets

Observation number	X	Y	Observation number	X	Y
(1)	46	40	(11)	61	96
(2)	49	50	(12)	62	88
(3)	51	55	(13)	63	99
(4)	52	63	(14)	64	110
(5)	54	72	(15)	66	113
(6)	56	70	(16)	67	120
(7)	57	77	(17)	68	127
(8)	58	73	(18)	71	137
(9)	59	90	(19)	72	132
(10)	60	93	(20)	71	137

Data inferred from a scatterplot in Frederick E. Croxton, Dudley J. Cowden, and Sidney Klein, *Applied General Statistics*, 3rd. ed. (Englewood Cliffs, N.J.: Prentice-Hall, 1967), p. 390.

7. Fit a model to Table 6-8. Do you recognize these data? What relationship can be inferred from them?

TABLE 6-8

Observation number	X	Y
(1)	35.97	0.241
(2)	67.21	0.615
(3)	92.96	1.000
(4)	141.70	1.881
(5)	483.70	11.860
(6)	886.70	29.460
(7)	1783.00	84.020
(8)	2794.00	164.800
(9)	3666.00	248.400

8. **a.** The data of Table 6-9 are measures of two characteristics of ponderosa pine trees. The variable X is the diameter of the tree, in inches, measured at breast height; Y is a measure of volume—number of board feet divided by 10. Fit a model to these data. Then express Y in terms of X.

TABLE 6-9 Diameter and volume for 20 ponderosa pine trees

Observation number	X	Y	Observation number	X	Y
(1)	36	192	(11)	31	141
(2)	28	113	(12)	20	32
(3)	28	88	(13)	25	86
(4)	41	294	(14)	19	21
(5)	19	28	(15)	39	231
(6)	32	123	(16)	33	187
(7)	22	51	(17)	17	22
(8)	38	252	(18)	37	205
(9)	25	56	(19)	23	57
(10)	17	16	(20)	39	265

Data reported in Croxton, Cowden, and Klein, *Applied General Statistics*, p. 421.

b. This is an example in which a little "theory" can play a role in generating an alternative to the model generated in Part a. We may think of a tree as (approximately) a cylinder, whose volume V (for tree heights not too different from one another) is proportional to the square of the radius r of the "cylinder": $V = \pi r^2 h$. Since Y is proportional to V, and X to r, it may be hoped that Y is approximately proportional to X^2.

Plot $\sqrt{Y}$ against X and see if the plot is approximately a straight line. If so, generate a model of the form $\sqrt{Y} = mX + b$. Then discuss whether or not the result is an improvement over the model generated in Part a.

Finally, allow the height of the trees to vary. Suggest a simple proportionality argument for the height of the tree versus its radius. Test your model.

For Problems 9 and 10, construct scatterplots of the data given. Is there a trend in the data? Do you think the trend can be captured with a simple one-term model? If so, fit your chosen model using both graphical and analytical techniques. Analyze the goodness of fit using the indicators presented in Chapter 4. Graph the model and the original data points.

9. The following data represent the length and weight of a set of fish (bass). Predict the weight as a function of the length of the fish.

length (in.)	12.5	12.625	14.125	14.5	17.25	17.75
weight (oz)	17	16.5	23	26.5	41	49

10. The following data give the population of the United States from 1800 to 1980. Predict the population as a function of the year.

year	1800	1820	1840	1860	1880	1900	1920	1940	1960	1980
population (thousands)	5308	9638	17,069	31,443	50,156	75,995	105,711	131,669	179,323	226,505

6.1 FURTHER READING

Bornstein, Marc H., and Helen G. Bornstein. "The Pace of Life." *Nature 259* (19 February 1976): 557–559.

Croxton, Frederick E., Dudley J. Cowden, and Sidney Klein. *Applied General Statistics*, 3rd. ed. Englewood Cliffs, N.J.: Prentice-Hall, 1967.

Neter, John, and William Wasserman. *Applied Linear Statistical Models*. Homewood, Ill.: Irwin, 1974.

Velleman, Paul F., and David C. Hoaglin. *Applications, Basics, and Computing of Exploratory Data Analysis*. Boston, Mass.: Duxbury Press, 1981.

Yule, G. Udny. "Why Do We Sometimes Get Nonsense-Correlations between Time Series?—A Study in Sampling and the Nature of Time Series." *Journal of the Royal Statistical Society 89* (1926): 1–69.

6.2 INTERPOLATION USING HIGHER-ORDER POLYNOMIALS

In the previous section we investigated the possibility of finding a simple one-term model that captures the trend of the collected data. Because of their inherent simplicity, one-term models facilitate model analysis including sensitivity analysis, optimization, estimation of rates of change and area under the curve, and so forth. However, due to their inherent mathematical simplicity, one-term models are limited in their ability to capture the trend of any collection of data. In some cases models with more than one term must be considered. The remainder of this chapter considers one type of multiterm model—namely, the polynomial. Since polynomials are easy to integrate and to differentiate, they are especially popular to use. Yet polynomials also have their disadvantages as well. For example, it is far more appropriate to approximate a data set having a vertical asymptote using a quotient of polynomials $p(x)/q(x)$ rather than a single polynomial. Let's begin by studying polynomials that pass through each point in a data set that includes only one observation for each value of the independent variable.

Consider the data in Figure 6-10a. Through the two given data points a unique line $y = a_0 + a_1 x$ can be passed. The constants a_0 and a_1 are determined by the conditions that the line pass through the points (x_1, y_1) and (x_2, y_2). Thus,

$$y_1 = a_0 + a_1 x_1$$

and

$$y_2 = a_0 + a_1 x_2$$

In a similar manner a *unique* polynomial function of (at most) degree two $y = a_0 + a_1 x + a_2 x^2$ can be passed through three distinct data points, as shown in Figure 6-10b. The constants a_0, a_1 and a_2 are determined by solving the following system of linear equations:

$$y_1 = a_0 + a_1 x_1 + a_2 x_1{}^2$$
$$y_2 = a_0 + a_1 x_2 + a_2 x_2{}^2$$
$$y_3 = a_0 + a_1 x_3 + a_2 x_3{}^2$$

Let's explain why the qualifer "at most" is needed in the next-to-last sentence of the preceding paragraph. Note that if the three points in Figure 6-10b just happen to lie along a straight line, then the unique polynomial function of "at most" degree two passing through the points would necessarily be a straight line (a polynomial of degree one) rather than a qua-

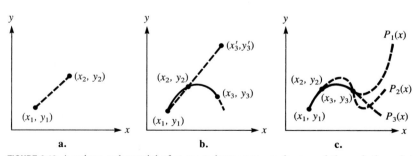

FIGURE 6-10 A unique polynomial of at most degree two can be passed through three data points (Figures 6-10a and 6-10b). However, an infinite number of polynomials of degree greater than two can be passed through three data points (Figure 6-10c).

dratic function, as generally would be expected. The descriptor "unique" is also very important. There are an infinite number of polynomials of degree greater than two and passing through the three points depicted in Figure 6-10b. (Convince yourself of this fact before proceeding by using Figure 6-10c.) However, there is only one polynomial of degree two or less. While that fact may not be obvious, we later state a theorem in its support. For now remember from high school geometry that a unique circle, which is also represented by an algebraic equation of degree two, is determined by three points in the plane. Next, we illustrate these ideas in an applied problem, and then discuss the advantages and disadvantages of the procedure.

Example: The Elapsed Time of a Tape Recorder

In Section 3.5 we presented the problem relating the counter on a particular cassette deck or tape recorder with the elapsed playing time. From certain assumptions we developed a proportionality argument to construct a model apparently explaining the tape recorder's behavior, and then tested the reasonableness of the assumptions. In Chapter 4 we even fit the proposed model to a set of collected data points. But suppose we are unable first to construct a model "explaining" the behavior, and yet are still interested in "predicting" what may occur. How can this difficulty be resolved? As an example, let's construct an empirical model to predict the amount of elapsed time of a tape recorder as a function of its counter reading.

Thus, let c_i represent the counter reading and t_i the corresponding amount of elapsed time. Consider again the data presented in Section 4.4:

c_i	100	200	300	400	500	600	700	800
t_i(sec)	205	430	677	945	1233	1542	1872	2224

One empirical model is a polynomial that passes through each of the data points. Since we have eight data points, a unique polynomial of at most degree seven is expected. Denote that polynomial symbolically by

$$P_7(c) = a_0 + a_1 c + a_2 c^2 + a_3 c^3 + a_4 c^4 + a_5 c^5 + a_6 c^6 + a_7 c^7$$

The eight data points require that the constants a_i satisfy the following system of linear algebraic equations:

$$205 = a_0 + 1a_1 + 1^2 a_2 + 1^3 a_3 + 1^4 a_4 + 1^5 a_5 + 1^6 a_6 + 1^7 a_7$$
$$430 = a_0 + 2a_1 + 2^2 a_2 + 2^3 a_3 + 2^4 a_4 + 2^5 a_5 + 2^6 a_6 + 2^7 a_7$$
$$\cdot \qquad \cdot \qquad \cdot \qquad \cdot \qquad \cdot \qquad \cdot \qquad \cdot$$
$$2224 = a_0 + 8a_1 + 8^2 a_2 + 8^3 a_3 + 8^4 a_4 + 8^5 a_5 + 8^6 a_6 + 8^7 a_7$$

Large systems of linear equations can be very difficult to solve with great numerical precision. In the preceding illustration we divided each counter reading by 100 to lessen numerical difficulties. Since the counter data values are being raised to the seventh power, it is easy to generate numbers differing by several orders of magnitude. For that reason it is important to have as much accuracy as possible in the coefficients a_i, because each is being multiplied by a number raised to a power as high as seven. For instance, a small a_7 may become significant as c becomes large. This observation suggests why there may be dangers using even good polynomial functions that capture the trend of the data when we are beyond the range of the observations. The following solution to this system was obtained with the aid of a handheld calculator program:

$$a_0 = -13.99999923 \qquad a_4 = -5.354166491$$
$$a_1 = 232.9119031 \qquad a_5 = 0.8013888621$$
$$a_2 = -29.08333188 \qquad a_6 = -0.0624999978$$
$$a_3 = 19.78472156 \qquad a_7 = 0.0019841269$$

Let's see how well the empirical model fits the data. Denoting the polynomial predictions by $P_7(c_i)$, we find:

c_i	100	200	300	400	500	600	700	800
t_i	205	430	677	945	1233	1542	1872	2224
$P_7(c_i)$	205	430	677	945	1233	1542	1872	2224

Rounding the predictions for $P_7(c_i)$ to four decimal places gives complete agreement with the observed data (as would be expected) and results in zero absolute deviations. Now you see the folly of applying any of the criteria of "best fit" studied in Chapter 4 as the *sole judge* for the best model. Can we really consider this model to be *better* than that proposed in Chapter 4?

Let's see how well this new model $P_7(c)$ captures the trend of the data. The model is graphed in Figure 6-11.

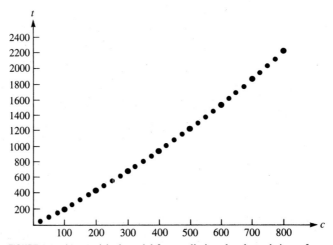

FIGURE 6-11 An empirical model for predicting the elapsed time of a tape recorder.

The Lagrangian Form of a Polynomial

From the preceding discussion you might expect that given $(n + 1)$ distinct data points, there is a unique polynomial of "at most" degree n that passes through all the data points. Because there are the same number of coefficients in the polynomial as there are data points, intuitively we would think that only one such polynomial exists. This hypothesis is indeed the case, although we will not prove that fact here. Rather, we next present the Lagrangian form of a cubic polynomial followed by a brief discussion of how the coefficients may be found for higher order polynomials.

Suppose the following data have been collected:

x	x_1	x_2	x_3	x_4
y	y_1	y_2	y_3	y_4

Consider the following cubic polynomial:

$$P_3(x) = \frac{(x - x_2)(x - x_3)(x - x_4)}{(x_1 - x_2)(x_1 - x_3)(x_1 - x_4)} y_1$$

$$+ \frac{(x - x_1)(x - x_3)(x - x_4)}{(x_2 - x_1)(x_2 - x_3)(x_2 - x_4)} y_2$$

$$+ \frac{(x - x_1)(x - x_2)(x - x_4)}{(x_3 - x_1)(x_3 - x_2)(x_3 - x_4)} y_3$$

$$+ \frac{(x - x_1)(x - x_2)(x - x_3)}{(x_4 - x_1)(x_4 - x_2)(x_4 - x_3)} y_4$$

Convince yourself that the polynomial is indeed cubic and agrees with the value y_i when $x = x_i$. Notice that the x_i values must all be different to avoid division by zero. Observe the pattern for forming the numerator and denominator for the coefficient of each y_i. This same pattern is followed when forming polynomials of any desired degree. The procedure is justified by the following result.

Theorem If $x_0, x_1, \ldots, x_n$ are $(n + 1)$ distinct points and $y_0, y_1, \ldots, y_n$ are corresponding observations at these points, then there exists a unique polynomial $P(x)$, of at most degree n, with the property that

$$y_k = P(x_k) \qquad \text{for each } k = 0, 1, \ldots, n$$

This polynomial is given by

$$P(x) = y_0 L_0(x) + \cdots + y_n L_n(x) \tag{6.3}$$

where

$$L_k(x) = \frac{(x - x_0)(x - x_1) \cdots (x - x_{k-1})(x - x_{k+1}) \cdots (x - x_n)}{(x_k - x_0)(x_k - x_1) \cdots (x_k - x_{k-1})(x_k - x_{k+1}) \cdots (x_k - x_n)}$$

Since the polynomial (6.3) passes through each of the data points, the resultant sum of absolute deviations is zero. Considering the various criteria of "best" presented in Chapter 4, we are tempted to use higher-order polynomials to fit larger sets of data. After all, the fit is precise! Let's examine both the advantages and disadvantages of using interpolating polynomials.

Advantages and Disadvantages of Interpolating Polynomials

As you have seen on several occasions in previous chapters, it may be of interest to determine the area under the curve representing our model or its rate of change at a particular point. Polynomial functions have the distinct advantage of being easily integrated and differentiated. If a polynomial can be found that reasonably represents the underlying behavior, it would then be easy to approximate the integral and derivative of the "unknown true model" as well. Now consider some of the disadvantages of higher-order polynomials. For the 17 data points presented in Table 6-10, it is clear that the "trend" of the data is $y = 0$ for all x over the interval $-8 \leqslant x \leqslant 8$.

TABLE 6-10

x_i	-8	-7	-6	-5	-4	-3	-2	-1	0	1	2	3	4	5	6	7	8
y_i	0	0	0	0	0	0	0	0	0	0	0	0	0	0	0	0	0

Suppose Equation (6.3) is used to determine a polynomial that passes through the points. Since there are 17 distinct data points, it is possible to pass a unique polynomial of at most 16th degree through the given points. The graph of a polynomial passing through the data points is depicted in Figure 6-12.

Note that while the polynomial does pass through the data points (within tolerances of computer round-off error), there is severe oscillation of the poly-

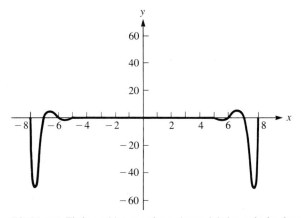

FIGURE 6-12 Fitting a higher-order polynomial through the data points given in Table 6-10.

nomial near each end of the interval. Thus there would be gross errors in estimating y between the data points near $+8$ or -8. Likewise, consider the error in using the derivative of the polynomial to estimate the rate of change of the data or the area under the polynomial to estimate the area trapped by the data! This tendency of higher-order polynomials to oscillate severely near the ends of the interval is a serious disadvantage to using them.

Let's illustrate another disadvantage of higher-order polynomials. Consider the three sets of data presented in Table 6-11.*

TABLE 6-11

x_i	0.2	0.3	0.4	0.6	0.9
Case 1: y_i	2.7536	3.2411	3.8016	5.1536	7.8671
Case 2: y_i	2.754	3.241	3.802	5.154	7.867
Case 3: y_i	2.7536	3.2411	3.8916	5.1536	7.8671

Let's discuss the three cases. Case 2 is merely Case 1 with less precise data: there is one less significant digit for each of the observed data points. Case 3 is Case 1 with an error introduced in the observation corresponding to $x = 0.4$. Note that the error occurs in the third significant digit (3.8916 versus 3.8016). Intuitively, we would think that the three interpolating polynomials should be quite similar because the trends of the data are all similar. However, let's determine the interpolating polynomials and see if that is really the situation.

Since there are five distinct data points in each case, a unique polynomial of at most fourth degree can be passed through each set of data. Denote the fourth degree polynomial symbolically as follows:

$$P_4(x) = a_0 + a_1 x + a_2 x^2 + a_3 x^3 + a_4 x^4$$

Table 6-12 tabulates the coefficients a_0, a_1, a_2, a_3, a_4 (to four decimal places) determined by fitting the data points in each of the three cases. Note how sensitive the values of the coefficients are to the data. Nevertheless *the graphs* of the polynomials are nearly the same over the interval of observations (0.2, 0.9).

TABLE 6-12

	a_0	a_1	a_2	a_3	a_4
Case 1	2	3	4	-1	1
Case 2	2.0123	2.8781	4.4159	-1.5714	1.2698
Case 3	3.4580	-13.2000	64.7500	-91.0000	46.0000

* This example was furnished by Professor George Morris, Naval Postgraduate School, Monterey, California.

The graphs of the three fourth-degree polynomials representing each case are presented in Figure 6-13. This example illustrates the sensitivity of the coefficients of higher-order polynomials to small changes in the data. Since we do *expect* measurement errors to occur, the tendency of higher-order polynomials to oscillate, as well as the sensitivity of their coefficients to small changes in the data, is a disadvantage that restricts their usefulness in modeling. In the next two sections we consider techniques that address the deficiencies noted in this section.

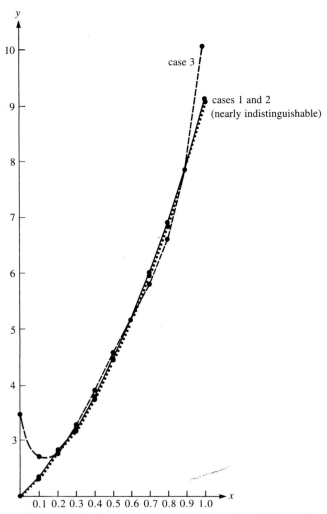

FIGURE 6-13 Small measurement errors can cause huge differences in the coefficients of the higher-order polynomials that result. Note that the polynomials diverge outside the range of observations.

6.2 PROBLEMS

1. For the data you collected for your tape recorder (see 3.5 Problems), give a system of equations determining the coefficients of a polynomial that passes through each of the data points. If a computer is available, determine and sketch the polynomial. Does it represent the trend of the data?

2. Consider again the "pace of life" data presented in Table 6-2 and plotted in Figure 6-3. Consider passing a 14th-degree polynomial through the following data, where P is the population size and V is the mean velocity. Discuss the disadvantages of using the polynomial to make predictions. If you have a computer available, fit a 14th-degree polynomial to the data and graph the results.

P	365	2500	5491	14000	23700	49375	70700	78200
V	2.76	2.27	3.31	3.70	3.27	4.90	4.31	3.85

P	138000	304500	341948	867023	1092759	1340000	2602000
V	4.39	4.42	4.81	5.21	5.88	5.62	5.05

3. In the following data, X is the Fahrenheit temperature and Y is the number of times a cricket chirps in one minute (see Problem 6 in 6.1 Problems). Make a scatterplot of the data and discuss the appropriateness of using an 18th-degree polynomial that passes through the data points as an empirical model. If you have a computer available, fit a polynomial to the data and graph the results.

X	46	49	51	52	54	56	57	58	59	60
Y	40	50	55	63	72	70	77	73	90	93

X	61	62	63	64	66	67	68	71	72
Y	96	88	99	110	113	120	127	137	132

4. In the following data, X represents the diameter of a ponderosa pine measured at breast height and Y is a measure of volume—number of board feet divided by 10. (See Problem 8 of 6.1 Problems. Multiple observations at the same x_i have been averaged.) Make a scatterplot of the data. Discuss the appropriateness of using a 13th-degree polynomial that passes through the data points as an empirical model. If you have a computer available, fit a polynomial to the data and graph the results.

X	17	19	20	22	23	25	31	32	33	36	37	38	39	41
Y	19	25	32	51	57	71	141	123	187	192	205	252	248	294

6.3 SMOOTHING USING POLYNOMIALS

We seek methods that retain many of the conveniences found in higher-order polynomials without incorporating their disadvantages. One popular technique is to choose a low-order polynomial regardless of the number of data points. This choice normally results in a situation where the number of data points exceeds the number of constants necessary to determine the polynomial. Since there are fewer constants to determine than there are data points, the lower-order polynomial generally will not pass through all the points. For example, suppose it is decided to fit a quadratic polynomial to a set of ten data points. Since it is generally impossible to force a quadratic to pass through ten data points, it must be decided which quadratic "best" fits the data (according to some criterion, as discussed in Chapter 4). This **smoothing** process, which is called **smoothing,** is illustrated in Figure 6-14. The combination of using a low-order polynomial while not requiring that it pass through each data point reduces both the tendency of the polynomial to oscillate and its sensitivity to small changes in the data.

The process of smoothing requires two decisions. First, the order of the interpolating polynomial must be selected. Second, the coefficients of the polynomial must be determined according to some criterion for the "best-fitting" polynomial. The problem that results is an optimization problem of the form addressed in Chapter 4. For example, it may be decided to fit a quadratic model to ten data points using the least-squares "best-fitting" criterion. We will review the process of fitting a polynomial to a set of data points using the least-squares criterion and later return to the more difficult question of how best to choose the order of the interpolating polynomial.

Example: The Elapsed Time of a Tape Recorder Revisited

Consider again the tape recorder problem identified in Section 3.5: for a particular cassette deck or tape recorder equipped with a counter, relate the counter to the amount of playing time that has elapsed. If we are interested

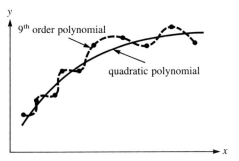

FIGURE 6-14 The quadratic function "smooths" the data since it is not required to pass through all the data points.

in predicting the elapsed time, but are unable to construct an "explicative" model, it may be possible to construct an empirical model instead. Let's fit a second-order polynomial of the following form to the data:

$$P_2(c) = a + bc + dc^2$$

where c is the counter reading, $P_2(c)$ is the elapsed time, and a, b, and d are constants to be determined. Using the data in Table 6-13 (which was first presented in Chapter 3), our problem is to determine the constants a, b, and d so that the resultant quadratic model "best" fits the data. While other criteria might be used, we will find the quadratic that minimizes the sum of the squared deviations. Mathematically, the problem is to

$$\text{Minimize } S = \sum_{i=1}^{m} [t_i - (a + bc_i + dc_i^2)]^2$$

TABLE 6-13 Data collected for the tape recorder problem

c_i (counts)	100	200	300	400	500	600	700	800
t_i (sec)	205	430	677	945	1233	1542	1872	2224

The necessary conditions for a minimum to exist ($\partial S/\partial a = \partial S/\partial b = \partial S/\partial d = 0$) yield the equations:

$$ma + \left(\sum c_i\right)b + \left(\sum c_i^2\right)d = \sum t_i$$
$$\left(\sum c_i\right)a + \left(\sum c_i^2\right)b + \left(\sum c_i^3\right)d = \sum c_i t_i$$
$$\left(\sum c_i^2\right)a + \left(\sum c_i^3\right)b + \left(\sum c_i^4\right)d = \sum c_i^2 t_i$$

For the data given in Table 6-13, the preceding system of equations becomes

$$8a + \qquad 3{,}600b + \qquad 2{,}040{,}000d = \qquad 9{,}128$$
$$3{,}600a + \qquad 2{,}040{,}000b + 1{,}296{,}000{,}000d = \qquad 5{,}318{,}900$$
$$2{,}040{,}000a + 1{,}296{,}000{,}000b + \quad 8.772 \times 10^{11}d = 3{,}435{,}390{,}000$$

Solution of the preceding system yields the values $a = 0.14286$, $b = 1.94226$, and $d = 0.00105$, giving the quadratic:

$$P_2(c) = 0.14286 + 1.94226c + 0.00105c^2$$

We can compute the deviations between the observations and the predictions made by the model $P_2(c)$:

c_i	100	200	300	400	500	600	700	800
t_i	205	430	677	945	1233	1542	1872	2224
$t_i - P_2(c_i)$	.167	−.452	.000	.524	.119	−.214	−.476	.333

Note that the deviations are very small compared to the order of magnitude of the times. You may remember that in Chapter 3, using a proportionality argument, we constructed a model of the form:

$$t = k_2 c + k_3 c^2$$

Comparison of the fit of $P_2(c)$ with this proportionality model (see Section 4.4) shows that $P_2(c)$ fits even better. You might expect this improvement since the extra parameter (the constant a in this case) absorbs some of the error expected in the modeling process. However, consider what can happen when an empirical model is used to predict outside of the range of observations. When $c = 0$, P_2 predicts $P_2(0) = 0.1429$, whereas $t = k_2 c + k_3 c^2$ predicts (correctly) $t = 0$. This observation again underscores the foolishness of judging a model based solely on how well it fits the observed data.

At this point you are probably thinking that the choice of a quadratic is somewhat contrived since our work in Chapter 3 suggested a quadratic relationship. However, suppose we had completely "struck out" in determining the underlying relationship between the variables in Chapter 3. How could the collected data be used? When we are considering the use of a low-order polynomial for smoothing, two issues come to mind:

1. Should a polynomial be used?
2. If so, what order of polynomial would be appropriate?

The derivative concept can help in answering these two questions.

Divided Differences

Notice that a quadratic function is characterized by the properties that its second derivative is constant and its third derivative is zero. That is, given

$$P(x) = a + bx + cx^2$$

we have

$$P'(x) = b + 2cx$$
$$P''(x) = 2c$$
$$P'''(x) = 0$$

However, the only information available is a set of discrete data points. How can these points be used to estimate the various derivatives? Refer to Figure 6-15 and recall the definition of the derivative:

$$\frac{dy}{dx} = \lim_{\Delta x \to 0} \frac{\Delta y}{\Delta x}$$

Since dy/dx at $x = x_1$ can be interpreted geometrically as the slope of the line tangent to the curve there, you see from Figure 6-15 that, unless Δx is small, the ratio $\Delta y/\Delta x$ is probably not a good estimate of dy/dx. Nevertheless, if dy/dx is to be zero everywhere, then Δy must go to zero. Thus we can compute the *differences* $y_{i+1} - y_i = \Delta y$ between successive function values in our tabled data to gain insight into what the first derivative is doing.

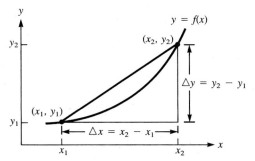

FIGURE 6-15 The derivative of $y = f(x)$ at $x = x_1$ is the limit of the slope of the secant line.

Likewise, since the first derivative is itself a function, the process can be repeated to estimate the second derivative. That is, the differences between successive estimates of the first derivative can be computed to approximate the second derivative. Before describing the entire process, we illustrate this idea with a simple example.

You know that the curve $y = x^2$ passes through the points $(0, 0)$, $(2, 4)$, $(4, 16)$, $(6, 36)$, and $(8, 64)$. Suppose the data displayed in Table 6-14 have been collected.

TABLE 6-14 A hypothetical set of collected data

x_i	0	2	4	6	8
y_i	0	4	16	36	64

Using the data in Table 6-14, we can construct a *difference table*, as shown in Table 6-15.

TABLE 6-15 A difference table for the data of Table 6-14

Data		Differences			
x_i	y_i	Δ	Δ^2	Δ^3	Δ^4
0	0				
		4			
2	4		8		
		12		0	
4	16		8		0
		20		0	
6	36		8		
		28			
8	64				

The first differences, denoted by Δ, are constructed by computing $y_{i+1} - y_i$ for $i = 1, 2, 3, 4$. The second differences, denoted by Δ^2, are computed by finding the difference between successive first differences from the Δ column. The process can be continued, column by column, until Δ^{n-1} is computed for n data points. Note from Table 6-15 that the second differences in our example are constant and the third differences are zero. These results are consistent with the fact that a quadratic function has a constant second derivative and a zero third derivative.

Even if the data are "essentially quadratic" in nature, you would not expect the differences to go to zero precisely due to the various errors present in the modeling and data collection processes. However, you might expect them to become "small." Our judgment of the significance of "small" can be improved by computing **divided differences.** Notice that the differences computed in Table 6-15 are estimates of the numerator for each of the various order derivatives. These estimates can be improved by dividing the numerator by the corresponding estimate of the denominator.

divided differences

Consider the three data points and the corresponding estimates of the first and second derivatives, called the first and second divided differences, respectively, in Table 6-16. The first divided difference follows immediately from the ratio $\Delta y / \Delta x$. Since the second derivative represents the rate of change of the first derivative, we can estimate how much the first derivative changes between x_1 and x_3. That is, we can compute the differences between the adjacent first divided differences and divide by the length of the interval over which that change has taken place ($x_3 - x_1$ in this case). Refer to Figure 6-16 for a geometrical interpretation of the second divided difference.

TABLE 6-16 The first and second divided differences estimate the first and second derivatives respectively

Data		First divided difference	Second divided difference
x_1	y_1		
		$\dfrac{y_2 - y_1}{x_2 - x_1}$	
x_2	y_2		$\dfrac{\dfrac{y_3 - y_2}{x_3 - x_2} - \dfrac{y_2 - y_1}{x_2 - x_1}}{x_3 - x_1}$
		$\dfrac{y_3 - y_2}{x_3 - x_2}$	
x_3	y_3		

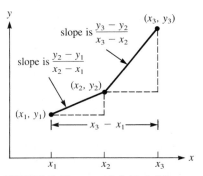

FIGURE 6-16 The second divided difference may be interpreted as the difference between the adjacent slopes (first divided differences) divided by the length of the interval over which the change has taken place.

In practice it is easy to construct a divided difference table. We generate the "next higher" order divided difference by taking the differences between adjacent "current order" divided differences and then dividing by the length of the interval over which the change has taken place. Using $\pmb\Delta^n$ to denote the nth divided difference, a divided difference table for the data of Table 6-14 is displayed in Table 6-17.

TABLE 6-17 A divided difference table for the data of Table 6-14

Data		Divided differences		
x_i	y_i	$\pmb\Delta$	$\pmb\Delta^2$	$\pmb\Delta^3$
0	0			
2	4	$4/2 = 2$		
4	16	$12/2 = 6$	$4/4 = 1$	
6	36	$20/2 = 10$	$4/4 = 1$	$0/6 = 0$
8	64	$28/2 = 14$	$4/4 = 1$	$0/6 = 0$

$\Delta x = 6$

Of course, it is easy to remember what the numerator should be in each divided difference of the table. To remember what the denominator should be for a given divided difference, one can construct diagonal lines back to the y_i of the original data entries and compute the differences in the corresponding x_i. This is illustrated for a third-order divided difference in Table 6-17. This construction becomes more critical when the x_i's are unequally spaced.

Example: The Elapsed Time of a Tape Recorder Revisited Again

Returning now to our construction of an empirical model for the elapsed time of a tape recorder, how might the order of the "smoothing polynomial" be chosen? Let's begin by constructing the divided difference table for the given data from Table 6-13. The divided differences are displayed in Table 6-18.

TABLE 6-18 A divided difference table for the tape recorder data

Data		Divided differences			
x_i	y_i	$\pmb\Delta$	$\pmb\Delta^2$	$\pmb\Delta^3$	$\pmb\Delta^4$
100	205				
200	430	2.2500	0.0011		
300	677	2.4700	0.0011	0.0000	
400	945	2.6800	0.0010	0.0000	0.0000
500	1233	2.8800	0.0011	0.0000	0.0000
600	1542	3.0900	0.0011	0.0000	0.0000
700	1872	3.3000	0.0011	0.0000	
800	2224	3.5200			

Note from Table 6-18 that the second divided differences are essentially constant and that the third divided differences equal zero to four decimal places. The table suggests that the data are essentially quadratic, which supports the use of a quadratic polynomial as an empirical model. The modeler may now want to reinvestigate the assumptions to determine if a quadratic relationship seems reasonable.

Observations on Difference Tables

Several observations about divided difference tables are in order. First, the x_i must be distinct and listed in increasing order. It is important to be sensitive to x_i's that are close together since division by a very small number can cause numerical difficulties. The scales used to measure both the x_i and the y_i must also be considered. For example, suppose the x_i represent distance and are currently measured in miles. If the units of measurement are changed to feet, the denominators become much larger, resulting in divided differences that are much smaller. Thus, judgment on what is "small" is relative and qualitative. Remember, however, that we are trying to decide whether the use of a low-order polynomial is worth further investigation. Before accepting the model, we would want to graph it and analyze the goodness of fit.

When using a divided difference table, we must be sensitive to errors and irregularities that occur in the data. Measurement errors can propagate themselves throughout the table, and even magnify themselves. For example, consider the following difference table:

Δ^{n-1}	Δ^n	Δ^{n+1}
6.01		
6.00	$-.01$	
6.01	.01	.02

Let's suppose that the Δ^{n-1} column is actually constant except for the presence of a relatively small measurement error. Note how the measurement error gives rise to a negative sign in the Δ^n column. Note also that the magnitude of the number in the Δ^{n+1} column is larger than those in the Δ^n column even though the Δ^{n-1} column is essentially constant. The effect of these errors is present not only in subsequent columns but spreads to other rows as well. Since errors and irregularities are normally present in collected data, it is important to be sensitive to the effects they can cause in difference tables.

Historically, divided difference tables were used to determine various forms of interpolating polynomials that passed through a chosen subset of the data points. Today other interpolating techniques, such as smoothing and cubic splines, are more popular. Nevertheless, difference tables are easily constructed on a computer and, like the derivatives they approximate, do provide an inexpensive source of useful information about the data.

Example: Vehicular Stopping Distance

The following problem was presented in Section 2.2: predict a vehicle's total stopping distance as a function of its speed. In previous chapters explicative models were constructed describing the vehicle's behavior. Those models will be reviewed in the next section. For the time being, suppose there is no explicative model but only the data displayed in Table 6-19.

TABLE 6-19 Data relating total stopping distance and speed

Speed, v (mph)	20	25	30	35	40	45	50	55	60	65	70	75	80
Distance, d (ft)	42	56	73.5	91.5	116	142.5	173	209.5	248	292.5	343	401	464

If the modeler is interested in constructing an empirical model by "smoothing" the data, a divided difference table can be constructed as displayed in Table 6-20.

TABLE 6-20 A divided difference table for the data relating total vehicular stopping distance and speed

Data		Divided differences			
v_i	d_i	Δ	Δ^2	Δ^3	Δ^4
20	42				
		2.8000			
25	56		0.0700		
		3.5000		−0.0040	
30	73.5		0.0100		0.0006
		3.6000		0.0080	
35	91.5		0.1300		−0.0007
		4.9000		−0.0060	
40	116		0.0400		0.0004
		5.3000		0.0027	
45	142.5		0.0800		0.0000
		6.1000		0.0027	
50	173		0.1200		−0.0004
		7.3000		−0.0053	
55	209.5		0.0400		0.0005
		7.7000		0.0053	
60	248		0.1200		−0.0003
		8.9000		0.0000	
65	292.5		0.1200		0.0001
		10.1000		0.0020	
70	343		0.1500		−0.0003
		11.6000		−0.0033	
75	401		0.1000		
		12.6000			
80	464				

An examination of the table reveals that the third divided differences are small in magnitude compared to the data and that negative signs have begun to appear. As previously discussed, the negative signs may indicate the presence of measurement error or variations in the data that will not be captured with low-order polynomials. The negative signs will also have a detrimental effect on the differences in the remaining columns. Here we just may decide to use a quadratic model, reasoning that the higher-order terms will not reduce the deviations sufficiently to justify their inclusion, but our judgment is qualitative. The cubic term will probably account for some of the deviations not accounted for by the "best" quadratic model (otherwise the "optimal" value of the coefficient of the cubic term would be zero, causing the quadratic and cubic polynomials to coincide), but the addition of higher-order terms increases the complexity of the model, its susceptibility to oscillation, and its sensitivity to data errors. These considerations are studied in statistics.

If we use the least-squares criterion, the general quadratic model

$$P(v) = a + bv + cv^2$$

can be fit to the data. Therefore the problem is to

$$\text{Minimize } S = \sum [d_i - (a + bv_i + cv_i^2)]^2$$

The necessary conditions for a minimum to exist ($\partial S/\partial a = \partial S/\partial b = \partial S/\partial c = 0$) yield the following system of linear algebraic equations:

$$ma + \left(\sum v_i\right)b + \left(\sum v_i^2\right)c = \sum d_i$$
$$\left(\sum v_i\right)a + \left(\sum v_i^2\right)b + \left(\sum v_i^3\right)c = \sum v_i d_i$$
$$\left(\sum v_i^2\right)a + \left(\sum v_i^3\right)b + \left(\sum v_i^4\right)c = \sum v_i^2 d_i$$

Substitution from the data in Table 6-19 gives the system:

$$13a + \qquad 650b + \qquad 37{,}050c = \qquad 2652.5$$
$$650a + \qquad 37{,}050b + \qquad 2{,}307{,}500c = \qquad 163{,}970$$
$$37{,}050a + 2{,}307{,}500b + 152{,}343{,}750c = 10{,}804{,}975$$

which leads to the solution $a = 50.0594$, $b = -1.9701$, and $c = 0.0886$ (rounded to four decimals). Therefore the empirical quadratic model is given by

$$P(v) = 50.0594 - 1.9701v + 0.0886v^2$$

Finally, the fit of $P(v)$ is analyzed in Table 6-21. As in the tape recorder example, this empirical model fits better than the model

$$d = 1.104v + 0.0542v^2$$

TABLE 6-21 Smoothing the stopping distance data using a quadratic polynomial

v_i	20	25	30	35	40	45	50
d_i	42	56	73.5	91.5	116	142.5	173
$d_i - P(v_i)$	−4.097	−0.182	2.804	1.859	2.985	1.680	−0.054

v_i	55	60	65	70	75	80
d_i	209.5	248	292.5	343	401	464
$d_i - P(v_i)$	−0.719	−2.813	−3.838	−3.292	0.323	4.509

determined in Section 4.4 because there is an extra parameter (the constant a in this case) that absorbs some of the error. However, note that the empirical model predicts a stopping distance of approximately 50 ft when the velocity is zero!

Example: The Growth of Yeast in a Culture

In this example we consider a collection of data points for which a divided difference table can help in deciding whether a low-order polynomial will provide a satisfactory empirical model. The data represent the population of yeast cells in a culture measured over time (in hours). A divided difference table for the population data is given by Table 6-22.

TABLE 6-22 A divided difference table for the growth of yeast in a culture

| Data | | Divided differences | | | |
t_i	P_i	Δ	Δ^2	Δ^3	Δ^4
0	9.60				
		8.70			
1	18.30		1.00		
		10.70		0.92	
2	29.00		3.75		−0.31
		18.20		−0.30	
3	47.20		2.85		0.84
		23.90		3.07	
4	71.10		12.05		−1.46
		48.00		−2.77	
5	119.10		3.75		1.51
		55.50		3.28	
6	174.60		13.60		−1.51
		82.70		−2.75	
7	257.30		5.35		0.11
		93.40		−2.30	
8	350.70		−1.55		−0.05
		90.30		−2.48	
9	441.00		−9.00		0.29
		72.30		−1.32	
10	513.30		−12.95		0.94
		46.40		2.43	
11	559.70		−5.65		−0.16
		35.10		1.80	
12	594.80		−0.25		−1.40
		34.60		−3.78	
13	629.40		−11.60		1.87
		11.40		3.68	
14	640.80		−0.55		−1.10
		10.30		−0.73	
15	651.10		−2.75		0.37
		4.80		0.73	
16	655.90		−0.55		−0.20
		3.70		−0.07	
17	659.60		−0.75		
		2.20			
18	661.80				

Note that the first divided differences Δ are increasing until $t = 8$ hours, when they begin to decrease. This characteristic is reflected in the Δ^2 column with the appearance of a consecutive string of negative signs, indicating a change in concavity. Thus we cannot hope to capture the trend of this data with a quadratic function that has only a single concavity. In the Δ^3 column additional negative signs appear sporadically although the magnitude of the numbers is relatively large. A scatterplot of the data is given in Figure 6-17.

Although the divided difference table suggests that a quadratic function would not be a good model, for illustrative purposes suppose we try to fit a quadratic anyway. Using the least-squares criterion and the equations developed previously for the tape recorder example, we determine the following

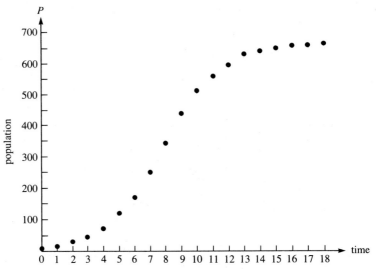

FIGURE 6-17 A scatterplot of the yeast in a culture data.

quadratic for the data of Table 6-22:

$$P = -93.82 + 65.70t - 1.12t^2$$

The model and the data points are plotted in Figure 6-18. The model fits very poorly as we expected and fails to capture the trend of the data. In the next section you will be asked to construct a cubic spline model that fits the data much better. In Section 9.1 we construct an analytic model that explains the behavior of the growth of yeast in a culture.

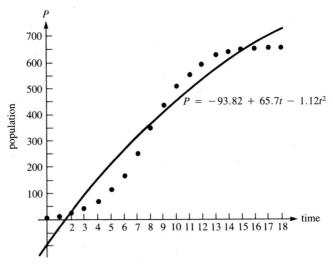

FIGURE 6-18 The best-fitting quadratic model fails to capture the trend of the data. Note the magnitude of the deviations $P_i - P(t_i)$.

6.3 PROBLEMS

For the data sets in Problems 1 through 4, construct a divided difference table. What conclusions can you make about the data? Would you use a low-order polynomial as an empirical model? If so, what order?

1.

x	0	1	2	3	4	5	6	7
y	2	8	24	56	110	192	308	464

2.

x	0	1	2	3	4	5	6	7
y	23	48	73	98	123	148	173	198

3.

x	0	1	2	3	4	5	6	7
y	7	15	33	61	99	147	205	273

4.

x	0	1	2	3	4	5	6	7
y	1	4.5	20	90	403	1808	8103	36316

5. Construct a scatterplot of the data you collected for your tape recorder relating elapsed playing time as a function of the counter (see 3.5 Problems). Do the data seem reasonable? Construct a divided difference table. Does smoothing with a low-order polynomial appear appropriate? If so, choose an appropriate polynomial and fit it using an appropriate criterion. Analyze the fit and compare it with the fit determined in 3.5 Problems. Graph the resulting model, the data points, and the deviations.

In Problems 6 through 12, construct a scatterplot of the given data. Is there a trend in the data? Are any of the data points "outliers"? Construct a divided difference table. Is smoothing with a low-order polynomial appropriate? If so, choose an appropriate polynomial and fit it using the least-squares criterion of "best fit." Analyze the goodness of fit by examining appropriate indicators and graphing the model, the data points, and the deviations.

6. X is the Fahrenheit temperature and Y is the number of times a cricket chirps in one minute (see 6.1 Problems).

X	46	49	51	52	54	56	57	58	59	60	61
Y	40	50	55	63	72	70	77	73	90	93	96

X	62	63	64	66	67	68	71	72
Y	88	99	110	113	120	127	137	132

7. *X* represents the diameter of a ponderosa pine tree measured at breast height and *Y* is a measure of volume—number of board feet divided by 10 (see 6.1 Problems).

X	17	19	20	22	23	25	31	32	33	36	37	38	39	41
Y	19	25	32	51	57	71	141	123	187	192	205	252	248	294

8. The following data represent the population of the United States from 1790 to 1980.

Year	Observed population
1790	3,929,000
1800	5,308,000
1810	7,240,000
1820	9,638,000
1830	12,866,000
1840	17,069,000
1850	23,192,000
1860	31,443,000
1870	38,558,000
1880	50,156,000
1890	62,948,000
1900	75,995,000
1910	91,972,000
1920	105,711,000
1930	122,755,000
1940	131,669,000
1950	150,697,000
1960	179,323,000
1970	203,212,000
1980	226,505,000

9. The following data were obtained for the growth of a sheep population introduced into a new environment on the island of Tasmania. (Adapted from J. Davidson, "On the Growth of the Sheep Population in Tasmania," *Trans. Roy. Soc. S. Australia 62* (1938): 342–346.)

t (year)	1814	1824	1834	1844	1854	1864
P(t)	125	275	830	1200	1750	1650

10. The following data represent the "pace of life" (see Section 6.1). *P* is the population and *V* is the mean velocity in feet per second over a 50-ft course.

P	365	2500	5491	14000	23700	49375	70700	78200	138000
V	2.76	2.27	3.31	3.70	3.27	4.90	4.31	3.85	4.39

P	304500	341948	867023	1092759	1340000	2602000
V	4.42	4.81	5.21	5.88	5.62	5.05

11. The following data represent the length of a fish (bass) and its weight.

length (in.)	12.5	12.625	14.125	14.5	17.25	17.75
weight (oz)	17.0	16.50	23.00	26.5	41.00	49.00

12. The following data represent the weight-lifting results from the 1976 Olympics (see 3.1 Problems).

Bodyweight class (lb)		Total winning lifts (lb)		
	Max. weight	Snatch	Jerk	Total weight
Flyweight	114.5	231.5	303.1	534.6
Bantamweight	123.5	259.0	319.7	578.7
Featherweight	132.5	275.6	352.7	628.3
Lightweight	149.0	297.6	380.3	677.9
Middleweight	165.5	319.7	418.9	738.5
Light-heavyweight	182.0	358.3	446.4	804.7
Middle-heavyweight	198.5	374.8	468.5	843.3
Heavyweight	242.5	385.8	496.0	881.8

13. Fit a cubic to the data of Table 6-22 using the least-squares criterion. (You should be able to write the equations directly by examining closely your results from Problem 5 in Section 4.3.) Graph the resulting model and data points. How well does the cubic capture the trend of the data?

6.4 CUBIC SPLINE INTERPOLATION

The use of polynomials in constructing empirical models that capture the trend of the data is appealing because polynomials are so easy to integrate and differentiate. However, high-order polynomials do tend to oscillate near the endpoints of the data interval, and the coefficients can be quite sensitive to small changes in the data. Smoothing with a low-order polynomial lessens these effects. Yet, unless the data are essentially quadratic or cubic in nature, smoothing with a low-order polynomial may yield a relatively poor fit somewhere over the range of the data. For instance, the quadratic model that was fit to the data collected for the vehicular braking distance problem in Section 6.3 did not fit well at high velocities. In this section a very popular modern technique called **cubic spline interpolation** is introduced. By using different cubic polynomials between successive pairs of data points, we can capture the trend of the data regardless of the nature of the underlying relationship, while simultaneously reducing the tendency toward oscillation and the sensitivity to changes in the data.

cubic spline interpolation

Consider the data displayed in Figure 6-19a. What would a draftsperson do if asked to draw a smooth curve connecting the points? One solution

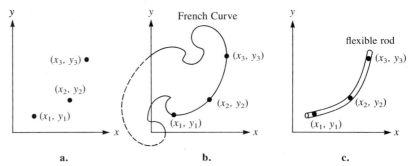

FIGURE 6-19 A draftsperson might attempt to draw a smooth curve through the data points using a French Curve, or a thin flexible rod called a spline.

would be to use a drawing instrument called a French Curve (see Figure 6-19b), which actually contains many different curves. By manipulating the French Curve, one can draw a curve that works reasonably well between two data points and transitions smoothly to another curve for the next pair of data points. Another alternative for the draftsperson is to take a very thin **spline**   flexible rod (called a **spline**) and tack it down at each data point. Cubic spline interpolation is essentially the same idea, except that distinct cubic polynomials are used between the successive pairs of data points in a "smooth" way.

Linear Splines

Probably all of us at one time or another have referred to a table of values (for instance, square root, trigonometric, or logarithmic tables) seeking a value that does not appear. We find two values that bracket the desired value and make a proportionality adjustment. For example, consider Table 6-23. Suppose an estimate of the value of y at $x = 1.67$ is desired. Probably we would compute $y(1.67) \approx 5 + (2/3)(8 - 5) = 7$. That is, we implicitly as-sume that the variation in y, between $x = 5$ and $x = 8$, occurs linearly. Simi-**linear interpolation**   larly, $y(2.33) \approx 13\frac{2}{3}$. This procedure is called **linear interpolation,** and for many applications it yields reasonable results, especially where the data are closely spaced.

Figure 6-20 is helpful in interpreting the process of linear interpolation geometrically in a manner that mimics what is done with cubic spline inter-

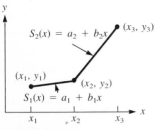

FIGURE 6-20 A linear spline model is a continuous function consisting of line segments.

TABLE 6-23
Linear interpolation

x_i	$y(x_i)$
1	5
2	8
3	25

linear spline

polation. When x is in the interval $x_1 \leqslant x < x_2$ the model that is used is the linear spline $S_1(x)$ passing through the data points (x_1, y_1) and (x_2, y_2):

$$S_1(x) = a_1 + b_1 x \qquad \text{for } x \text{ in } [x_1, x_2)$$

Similarly, when $x_2 \leqslant x \leqslant x_3$, the linear spline $S_2(x)$ passing through (x_2, y_2) and (x_3, y_3) is used:

$$S_2(x) = a_2 + b_2 x \qquad \text{for } x \text{ in } [x_2, x_3]$$

Note that both spline segments meet at the point (x_2, y_2).

Let's determine the constants for the respective splines for the data given in Table 6-23. The spline $S_1(x)$ must pass through the points (1, 5) and (2, 8). Mathematically this implies

$$a_1 + 1b_1 = 5$$
$$a_1 + 2b_1 = 8$$

Similarly, the spline $S_2(x)$ must pass through the points (2, 8) and (3, 25) yielding

$$a_2 + 2b_2 = 8$$
$$a_2 + 3b_2 = 25$$

Solution of these linear systems of equations yields $a_1 = 2, b_1 = 3, a_2 = -26$, and $b_2 = 17$. The linear spline model for the data of Table 6-23 is summarized in Table 6-24.

TABLE 6-24 A linear spline model for the data of Table 6-23

Interval	Spline model
$1 \leqslant x < 2$	$S_1(x) = 2 + 3x$
$2 \leqslant x \leqslant 3$	$S_2(x) = -26 + 17x$

To illustrate how the linear spline model is used, let's predict $y(1.67)$ and $y(2.33)$. Since $1 \leqslant 1.67 < 2$, $S_1(x)$ is selected to compute $S_1(1.67) \approx 7.01$. Likewise, $2 \leqslant 2.33 \leqslant 3$ gives rise to the prediction $S_2(2.33) \approx 13.61$.

While the linear spline method is sufficient for many applications, it fails to capture the trend of the data. Further, if you examine Figure 6-21, you see that the linear spline model does not appear "smooth." That is, in the interval $[1, 2)$ $S_1(x)$ has constant slope 3, whereas in the interval $[2, 3]$ $S_2(x)$ has constant slope 17. Thus at $x = 2$ there is an abrupt change in the slope of the model from 3 to 17 so that the first derivatives $S_1'(x)$ and $S_2'(x)$ fail to agree at $x = 2$. As you will discover, the process of cubic spline interpolation incorporates smoothness into the empirical model by requiring that both the first and second derivatives of adjacent splines agree at each data point.

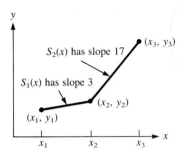

FIGURE 6-21 The linear spline does not appear smooth since the first derivative is not continuous.

Cubic Splines

Consider now Figure 6-22. In a manner analogous to linear splines, we define a separate spline function for the intervals $x_1 \leqslant x < x_2$ and $x_2 \leqslant x \leqslant x_3$ as follows:

$$S_1(x) = a_1 + b_1x + c_1x^2 + d_1x^3 \qquad \text{for } x \text{ in } [x_1, x_2)$$
$$S_2(x) = a_2 + b_2x + c_2x^2 + d_2x^3 \qquad \text{for } x \text{ in } [x_2, x_3]$$

Since we will want to refer to the first and second derivatives, let's define them as well:

$$S_1'(x) = b_1 + 2c_1x + 3d_1x^2 \qquad \text{for } x \text{ in } [x_1, x_2)$$
$$S_1''(x) = 2c_1 + 6d_1x \qquad \text{for } x \text{ in } [x_1, x_2)$$
$$S_2'(x) = b_2 + 2c_2x + 3d_2x^2 \qquad \text{for } x \text{ in } [x_2, x_3]$$
$$S_2''(x) = 2c_2 + 6d_2x \qquad \text{for } x \text{ in } [x_2, x_3]$$

The model is presented geometrically in Figure 6-22.

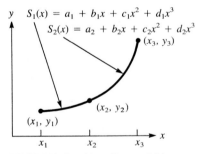

FIGURE 6-22 A cubic spline model is a continuous function with continuous first and second derivatives consisting of cubic polynomial segments.

Cubic splines offer the possibility of matching up not only slopes, but also the curvatures at each interior data point. To determine the constants defining each cubic spline segment, we appeal to the requirement that each spline pass through the two data points specified by the interval over which the spline is defined. For the spline model depicted in Figure 6-22, this requirement yields the equations:

$$y_1 = S_1(x_1) = a_1 + b_1 x_1 + c_1 x_1{}^2 + d_1 x_1{}^3$$
$$y_2 = S_1(x_2) = a_1 + b_1 x_2 + c_1 x_2{}^2 + d_1 x_2{}^3$$
$$y_2 = S_2(x_2) = a_2 + b_2 x_2 + c_2 x_2{}^2 + d_2 x_2{}^3$$
$$y_3 = S_2(x_3) = a_2 + b_2 x_3 + c_2 x_3{}^2 + d_2 x_3{}^3$$

Note that there are eight unknowns $(a_1, b_1, c_1, d_1, a_2, b_2, c_2, d_2)$ and only four equations in the preceding system. An additional four independent equations are needed in order to determine the constants uniquely. However, since smoothness of the spline system is also required, adjacent first derivatives must match at each interior data point (in this case when $x = x_2$). This requirement yields the equation:

$$S_1'(x_2) = b_1 + 2c_1 x_2 + 3d_1 x_2{}^2 = b_2 + 2c_2 x_2 + 3d_2 x_2{}^2 = S_2'(x_2)$$

It can also be required that adjacent second derivatives match at each interior data point as well:

$$S_1''(x_2) = 2c_1 + 6d_1 x_2 = 2c_2 + 6d_2 x_2 = S_2''(x_2)$$

To determine unique constants, we still require two additional independent equations. While conditions on the derivatives at interior data points have been applied, nothing has been said about the derivative at the exterior endpoints (x_1 and x_3 in Figure 6-22). Two popular conditions may be specified. One is to require that there be no change in the first derivative at the exterior endpoints. Mathematically, since the first derivative is constant, the second derivative must then be zero. Application of this condition at x_1 and x_3 yields

$$S_1''(x_1) = 2c_1 + 6d_1 x_1 = 0$$
$$S_2''(x_3) = 2c_2 + 6d_2 x_3 = 0$$

natural spline   A cubic spline formed in this manner is called a **natural spline.** If you think again of our analogue with the thin flexible rod tacked down at the data points, a natural spline allows the rod to be free at the endpoints to assume whatever direction there that the data points dictate. The natural spline is interpreted geometrically in Figure 6-23a.

Alternatively, if the values of the first derivatives at the exterior endpoints are known, the first derivatives of the exterior splines can be required to match the known values. Suppose the derivatives at the exterior endpoints are known and given by $f'(x_1)$ and $f'(x_3)$. Mathematically, this matching requirement yields the equations:

$$S_1'(x_1) = b_1 + 2c_1 x_1 + 3d_1 x_1{}^2 = f'(x_1)$$
$$S_2'(x_3) = b_2 + 2c_2 x_3 + 3d_2 x_3{}^2 = f'(x_3)$$

clamped spline   A cubic spline formed in this manner is called a **clamped spline.** Again referring to our flexible rod analogue, this situation corresponds to clamping

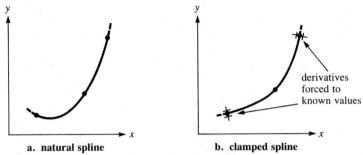

a. natural spline   **b. clamped spline**

FIGURE 6-23 The conditions for the natural and clamped cubic splines result in first derivatives at the two exterior endpoints that are constant. That constant value for the first derivative is specified in the clamped spline whereas it is free to assume a natural value in the natural spline.

the flexible rod in a vise at the exterior endpoints to ensure that the flexible rod has the proper angle there. The clamped cubic spline is interpreted geometrically in Figure 6-23b. Unless *precise* information about the first derivatives at the endpoints is known, the natural spline is generally used.

Let's illustrate the construction of a natural cubic spline model using the data displayed in Table 6-23. We illustrate the technique with this simple example since the procedure readily extends to problems with more data points.

Requiring the spline segment $S_1(x)$ to pass through the two endpoints $(1, 5)$ and $(2, 8)$ of its interval requires that $S_1(1) = 5$ and $S_1(2) = 8$, or

$$a_1 + 1b_1 + 1c_1 + 1d_1 = 5$$
$$a_1 + 2b_1 + 2^2c_1 + 2^3d_1 = 8$$

Similarly, $S_2(x)$ must pass through the endpoints of the second interval so that $S_2(2) = 8$ and $S_2(3) = 25$, or

$$a_2 + 2b_2 + 2^2c_2 + 2^3d_2 = 8$$
$$a_2 + 3b_2 + 3^2c_2 + 3^3d_2 = 25$$

Next the first derivatives of $S_1(x)$ and $S_2(x)$ are forced to match at the interior data point $x_2 = 2$: $S_1'(2) = S_2'(2)$, or

$$b_1 + 2c_1(2) + 3d_1(2)^2 = b_2 + 2c_2(2) + 3d_2(2)^2$$

Forcing the second derivatives of $S_1(x)$ and $S_2(x)$ to match at $x_2 = 2$ requires $S_1''(2) = S_2''(2)$, or

$$2c_1 + 6d_1(2) = 2c_2 + 6d_2(2)$$

Finally, a natural spline is built by requiring that the second derivatives at the two endpoints be zero: $S_1''(1) = S_2''(3) = 0$, or

$$2c_1 + 6d_1(1) = 0$$
$$2c_2 + 6d_2(3) = 0$$

Thus the procedure has yielded a linear algebraic system of eight equations in eight unknowns that can be solved uniquely. The resulting model is summarized in Table 6-25 and graphed in Figure 6-24.

TABLE 6-25 A natural cubic spline model for the data of Table 6-23

Interval	Model
$1 \leqslant x < 2$	$S_1(x) = 2 + 10x - 10.5x^2 + 3.5x^3$
$2 \leqslant x \leqslant 3$	$S_2(x) = 58 - 74x + 31.5x^2 - 3.5x^3$

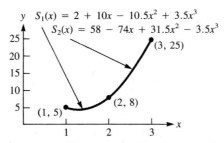

FIGURE 6-24 The natural cubic spline model for the data in Table 6-23 is a smooth curve that is easily integrated and differentiated.

Let's illustrate the use of the model by again predicting $y(1.67)$ and $y(2.33)$:

$$S_1(1.67) \approx 5.72$$
$$S_2(2.33) \approx 12.32$$

Compare these values with the values predicted by the linear spline. In which prediction values do you have the most confidence?

The construction of cubic splines for more data points proceeds in the same manner. That is, each spline is forced to pass through the endpoints of the interval over which it is defined, the first and second derivatives of adjacent splines are forced to match at the interior data points, and either the clamped or natural conditions are applied at the two exterior data points. Of course, for computational reasons it would be necessary to implement the procedure on a computer. The procedure we described here does not give rise to a computationally or numerically efficient computer algorithm. Our approach was designed to facilitate your understanding of the basic concepts underlying cubic spline interpolation.*

It is revealing to view how the graphs of the different cubic splines fit together to form a single composite interpolating curve between the data

---

* For a computationally efficient algorithm, see R. L. Burden, J. D. Faires, and A. C. Reynolds, *Numerical Analysis*, 2nd (Boston, Mass.: Prindle, Weber, & Schmidt, 1978), pp. 112–115.

points. Consider the following data (from Problem 4, Section 4.3):

x	7	14	21	28	35	42
y	8	41	133	250	280	297

Since there are 6 data points, 5 distinct cubic polynomials S_1 through S_5, are calculated to form the composite natural cubic spline. Each of these cubics is graphed and overlaid on the same graph to obtain Figure 6-25. Between any two consecutive data points only one of the five cubic poly-

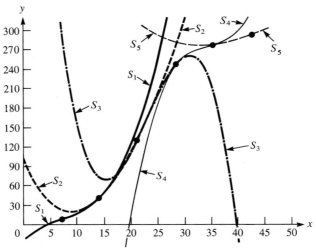

FIGURE 6-25 Between any two consecutive data points only one cubic spline polynomial is active. (*Graphics by Jim McNulty and Bob Hatton.*)

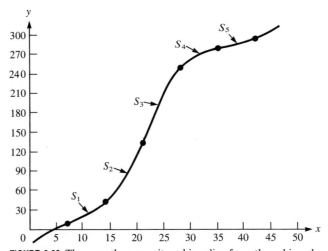

FIGURE 6-26 The smooth composite cubic spline from the cubic polynomials in Figure 6-25.

nomials is "active," giving the smooth composite cubic spline shown in Figure 6-26.

You should be concerned with whether the procedure just described results in a unique solution. Also you may be wondering why we jumped from a linear spline to a cubic spline without discussing quadratic splines. Intuitively, we would think that first derivatives could be matched with a quadratic spline. These and related issues are discussed in most numerical analysis texts (for example, see Burden, Faires, and Reynolds cited earlier).

Example: Vehicular Stopping Distance Revisited

Consider again the problem posed in Section 2.2: predict a vehicle's total stopping distance as a function of its speed. In Section 3.3 we reasoned that the model should have the form:

$$d = k_1 v + k_2 v^2$$

where d is the total stopping distance, v the velocity, and k_1 and k_2 are constants of proportionality resulting from the submodels for reaction distance and mechanical braking distance, respectively. We found reasonable agreement with the data furnished for the submodels and graphically estimated k_1 and k_2 to obtain the model:

$$d = 1.1v + 0.054v^2$$

In Section 4.4 we estimated k_1 and k_2 using the least-squares criterion and obtained the model:

$$d = 1.104v + 0.0542v^2$$

The fit of the preceding two models is analyzed in Table 4-7. Note in particular that both models break down at high speeds where they increasingly underestimate the stopping distances. In Section 6.3 we constructed an empirical model by smoothing the data with a quadratic polynomial, and we analyzed the fit.

Now suppose that we are not satisfied with the predictions made by our analytic models or that we are unable to construct an analytic model, yet find it necessary to make predictions. If we are reasonably satisfied with the collected data, we might consider constructing a cubic spline model for the data presented in Table 6-26.

Using a computer code, we obtained the cubic spline model summarized in Table 6-27. The first three spline segments are plotted in Figure 6-27. Note how each segment passes through the data points at either end of its interval, and note the smoothness of the transition across adjacent segments.

TABLE 6-26 Data relating total stopping distance and speed

Speed, v (mph)	20	25	30	35	40	45	50	55	60	65	70	75	80
Distance, d (ft)	42	56	73.5	91.5	116	142.5	173	209.5	248	292.5	343	401	464

TABLE 6-27 A cubic spline model for vehicular braking distance

Interval	Model
$20 \leqslant v < 25$	$S_1(v) = 42 + 2.596(v - 20) + 0.008(v - 20)^3$
$25 \leqslant v < 30$	$S_2(v) = 56 + 3.208(v - 25) + 0.122(v - 25)^2 - 0.013(v - 25)^3$
$30 \leqslant v < 35$	$S_3(v) = 73.5 + 3.472(v - 30) - 0.070(v - 30)^2 + 0.019(v - 30)^3$
$35 \leqslant v < 40$	$S_4(v) = 91.5 + 4.204(v - 35) + 0.216(v - 35)^2 - 0.015(v - 35)^3$
$40 \leqslant v < 45$	$S_5(v) = 116 + 5.211(v - 40) - 0.015(v - 40)^2 + 0.006(v - 40)^3$
$45 \leqslant v < 50$	$S_6(v) = 142.5 + 5.550(v - 45) + 0.082(v - 45)^2 + 0.005(v - 45)^3$
$50 \leqslant v < 55$	$S_7(v) = 173 + 6.787(v - 50) + 0.165(v - 50)^2 - 0.012(v - 50)^3$
$55 \leqslant v < 60$	$S_8(v) = 209.5 + 7.503(v - 55) - 0.022(v - 55)^2 + 0.012(v - 55)^3$
$60 \leqslant v < 65$	$S_9(v) = 248 + 8.202(v - 60) + 0.161(v - 60)^2 - 0.004(v - 60)^3$
$65 \leqslant v < 70$	$S_{10}(v) = 292.5 + 9.489(v - 65) + 0.096(v - 65)^2 + 0.005(v - 65)^3$
$70 \leqslant v < 75$	$S_{11}(v) = 343 + 10.841(v - 70) + 0.174(v - 70)^2 - 0.005(v - 70)^3$
$75 \leqslant v < 80$	$S_{12}(v) = 401 + 12.245(v - 75) + 0.106(v - 75)^2 - 0.007(v - 75)^3$

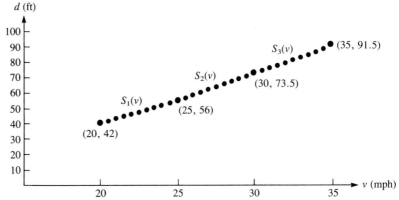

FIGURE 6-27 A plot of the cubic spline model for vehicular braking distance for $20 \leqslant v \leqslant 35$.

Summary: Constructing Empirical Models

We conclude this chapter by presenting a summary and suggesting a procedure for constructing empirical models using the techniques presented in this chapter. The procedure begins by examining the data in search of "suspect" data points that need to be examined more closely and perhaps discarded or obtained anew. Simultaneously, we look to see if a trend exists in the data. Normally, these questions are best considered by constructing a scatterplot, possibly with the aid of a computer routine if a large amount of data is being considered. If a trend does appear to exist, simple models are investigated first to see if one adequately captures the trend of the data. A one-term model can be identified using a transformation that converts the data into a straight line (approximately). The transformation may be found among the Ladder of Transformations or among the transformations discussed in Chapter 4. A graphical plot of the transformed data is often useful to determine whether the one-term model "linearizes" the data. If a model that appears adequate is found, the chosen model can be fit graphically or analytically using one of the criteria discussed in Chapter 4. A careful analysis determining how well the model fits the data points is obtained next by examining indicators such as the sum of the absolute deviations, the largest absolute deviation, the sum of the squared deviations, and so forth. A plot of the deviations as a function of the independent variable(s) may be useful for determining where the model does not fit well. If the fit proves to be unsatisfactory, other one-term models may be considered.

If it is determined that one-term models are inadequate, then polynomials can be used. If there are a small number of data points, an $n - 1^{st}$ order polynomial through the n data points can be tried. Be sure to note any oscillations, especially near the endpoints of the interval. A careful plot of the polynomial will help to reveal this feature. If there are a large number of data points, a low-order polynomial to "smooth the data" may be considered. A divided difference table is useful as a qualitative aid for determining whether a low-order polynomial is appropriate and in choosing the order of that polynomial. Once the order of the polynomial is chosen, the polynomial may be fit and analyzed according to the techniques discussed in Chapter 4. If smoothing with a low-order polynomial proves inadequate, a cubic spline that is satisfactory for many data sets can be used.

6.4 PROBLEMS

1. For each of the following data sets, write a system of equations to determine the coefficients of the natural cubic splines passing through the given points. If a computer program is available, solve the system of equations and graph the splines.

   a.

X	2	4	7
Y	2	8	12

   b.

X	3	4	6
Y	10	15	35

   c.

X	0	1	2
Y	0	10	30

   d.

X	0	2	4
Y	5	10	40

For Problems 2 and 3, find the natural cubic splines that pass through the given data points. Use the splines to answer the requirements.

2.

X	3.0	3.1	3.2	3.3	3.4	3.5	3.6	3.7	3.8	3.9
Y	20.08	22.20	24.53	27.12	29.96	33.11	36.60	40.45	44.70	49.40

a. Estimate the derivative evaluated at $X = 3.45$. Compare your estimate with the derivative of e^x evaluated at $X = 3.45$.

b. Estimate the area under the curve from 3.3 to 3.6. Compare with

$$\int_{3.3}^{3.6} e^x \, dx$$

3.

X	0	$\pi/6$	$\pi/3$	$\pi/2$	$2\pi/3$	$5\pi/6$	π
Y	0.00	.50	.87	1.00	.87	.50	0.00

4. For the data you collected for your tape recorder (see 3.5 Problems) relating elapsed time with the counter reading, construct the natural splines that pass through the data points. Compare your results with previous models that you have constructed. Which model makes the best predictions?

5. "The Cost of a Postage Stamp." Consider the following data. Use the procedures described in this chapter to capture the trend of the data if one exists. Would you eliminate any data points? Why? Would you be willing to use your model to predict the price of a stamp on January 1, 2001? What do the various models you construct predict about the price on January 1, 2001? When will the price reach $1? You might enjoy reading the article upon which this problem is based: Donald R. Byrkit and Robert E. Lee, "The Cost of a Postage Stamp, or Up, Up, and Away," *Mathematics and Computer Education 17*, no. 3 (Summer 1983): 184–190.

Date	First-class stamp	
1885–1917	$.02	
1917–1919	.03	(War-time increase)
1919	.02	(Restored by Congress)
July 6, 1932	.03	
August 1, 1958	.04	
January 7, 1963	.05	
January 7, 1968	.06	
May 16, 1971	.08	
March 2, 1974	.10	
December 31, 1975	.13	(Temporary)
July 18, 1976	.13	
May, 1978	.15	
March 22, 1981	.18	
November 1, 1981	.20	

6.4 PROJECTS

1. Construct a computer code for determining the coefficients of the natural splines that pass through a given set of data points. See Burden, Faires, and Reynolds, cited earlier in this chapter, for an efficient algorithm.

For Projects 2 through 7, use the software you developed in Project 1 to find the splines that pass through the given data points. Use graphics software, if available, to sketch the resulting splines.

2. The following data represent the growth of yeast in a culture (data from R. Pearl, "The Growth of Population," *Quart. Rev. Biol. 2* (1927): 532–548.)

t (hours)	0	1	2	3	4	5	6	7	8	9
Observed yeast biomass	9.6	18.3	29.0	47.2	71.1	119.1	174.6	257.3	350.7	441.0

t (hours)	10	11	12	13	14	15	16	17	18
Observed yeast biomass	513.3	559.7	594.8	629.4	640.8	651.1	655.9	659.6	661.8

3. The following data represent the population of the United States from 1790 to 1980.

Year	Observed population
1790	3,929,000
1800	5,308,000
1810	7,240,000
1820	9,638,000
1830	12,866,000
1840	17,069,000
1850	23,192,000
1860	31,443,000
1870	38,558,000
1880	50,156,000
1890	62,948,000
1900	75,995,000
1910	91,972,000
1920	105,711,000
1930	122,755,000
1940	131,669,000
1950	150,697,000
1960	179,323,000
1970	203,212,000
1980	226,505,000

4. The following data were obtained for the growth of a sheep population introduced into a new environment on the island of Tasmania. (Adopted from J. Davidson, "On the Growth of the Sheep Population in Tasmania," *Trans. Roy. Soc. S. Australia 62* (1938): 342–346.)

t (year)	1814	1824	1834	1844	1854	1864
$P(t)$	125	275	830	1200	1750	1650

5. The following data represent the "pace of life" (see Section 6.1). P is the population and V is the mean velocity in feet per second over a 50-ft course.

P	365	2500	5491	14000	23700	49375	70700	78200	138000
V	2.76	2.27	3.31	3.70	3.27	4.90	4.31	3.85	4.39

P	304500	341948	867023	1092759	1340000	2602000
V	4.42	4.81	5.21	5.88	5.62	5.05

6. The following data represent the length of a fish (bass) and its weight.

length (in.)	12.5	12.625	14.125	14.5	17.25	17.75
weight (oz)	17.0	16.50	23.00	26.5	41.00	49.00

7. The following data represent the weight-lifting results from the 1976 Olympics (see 3.1 Problems).

Bodyweight class (lb)		Total winning lifts (lb)		
	Max. weight	Snatch	Jerk	Total weight
Flyweight	114.5	231.5	303.1	534.6
Bantamweight	123.5	259.0	319.7	578.7
Featherweight	132.5	275.6	352.7	628.3
Lightweight	149.0	297.6	380.3	677.9
Middleweight	165.5	319.7	418.9	738.5
Light-heavyweight	182.0	358.3	446.4	804.7
Middle-heavyweight	198.5	374.8	468.5	843.3
Heavyweight	242.5	385.8	496.0	881.8

8. You can use the cubic spline software you developed, coupled with some graphics software, to draw smooth curves to represent a figure you wish to draw on the computer. Overlay a piece of graph paper on a picture or drawing of the figure you wish to produce on the computer. Record enough data points to gain ultimately smooth curves. Take more data points where there are abrupt changes. (See Figure 6-28.)

   Now take the data points and determine the splines that pass through them. Note that if natural discontinuities in the derivative occur in the

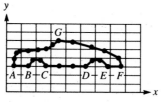

FIGURE 6-28

data such as at Points A, B, C, D, E, F and G in Figure 6-28, you will want to terminate one set of spline functions and begin another. You can then graph the spline functions using graphics software. In essence, we are using the computer to "connect the dots" with smooth curves. Select a figure of interest to you, such as your school mascot, and draw it on the computer.

DIMENSIONAL ANALYSIS AND SIMILITUDE

INTRODUCTION

In the process of constructing a mathematical model, we have seen that the variables influencing the behavior must be identified and classified and then appropriate relationships determined among those retained for consideration. In the case of a single dependent variable this procedure gives rise to some unknown function:

$$y = f(x_1, x_2, \ldots, x_n)$$

where the x_i measure the various factors influencing the phenomenon under investigation. In some situations the discovery of the nature of the function f for the chosen factors comes about by making some reasonable assumption based on a law of nature or previous experience and constructing a mathematical model. We were able to use this methodology in constructing our model on vehicular stopping distance (see Section 3.3). On the other hand, especially for those models designed to predict some physical phenomenon, we may find it difficult or impossible to construct a solvable or tractable "explicative" model due to the inherent complexity of the problem. In certain instances we might conduct a series of experiments to see how the dependent variable y is related to various values of the independent variable(s). In such cases we usually prepare a figure or table and apply an appropriate curve-fitting or interpolation method that can be used to predict the value of y for suitable ranges of the independent variable(s). We employed this technique in modeling the elapsed time of a tape recorder in Sections 6.2 and 6.3.

Dimensional analysis is a method to assist in determining how the selected variables are related and for reducing significantly the amount of experimental data that must be collected. It is based on the premise that physical quantities have dimensions and that physical laws are not altered by changing the units measuring dimensions. Thus the phenomenon under investigation can be described by a dimensionally correct equation among the variables. A dimensional analysis provides qualitative information about the model. It is especially important when it is necessary to conduct experiments in the modeling process, because the method is helpful in testing the validity of including or neglecting a particular factor, in reducing the number of experiments to be conducted in order to make predictions, and in improving the usefulness of the results by providing alternatives for the parameters employed to present them. Dimensional analysis has proven useful in physics and engineering for many years and now even plays a role in the study of the life sciences, economics, and operations research. Let's consider a specific example illustrating how dimensional analysis can be used in the modeling process to increase the efficiency of an experimental design.

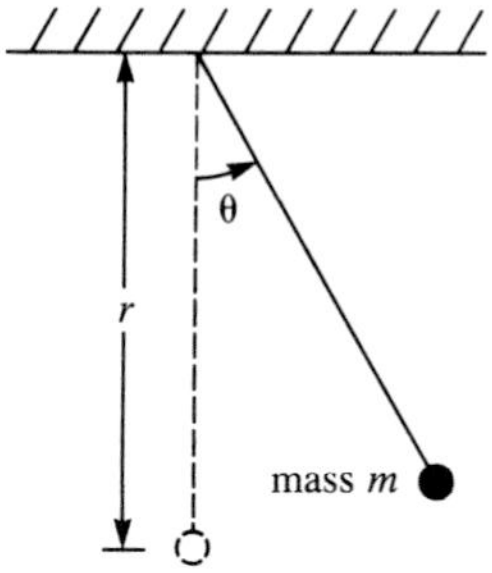

FIGURE 7-1 A simple pendulum.

An Introductory Example: A Simple Pendulum

Consider the situation of a simple pendulum as suggested in Figure 7-1. Let r denote the length of the pendulum, m its mass, and θ the initial angle of displacement from the vertical. One characteristic that is vital in understanding the behavior of the pendulum is its **period,** which is the time required for the pendulum bob to swing through one complete cycle and return to its original position (as at the beginning of the cycle). We represent the period of the pendulum by the dependent variable t.

period

Problem identification *For a given pendulum system, determine its period.*

Assumptions First we list the factors that influence the period. Some of these factors are the length r, the mass m, the initial angle of displacement θ, the acceleration due to gravity g, and frictional forces such as the friction at the hinge and the drag on the pendulum. Assume initially that the hinge is frictionless, that the mass of the pendulum is concentrated at one end of the pendulum, and that the drag force is negligible. Other assumptions about the frictional forces will be examined in Section 7.3. Thus the problem is to determine or approximate the function:

$$t = f(r, m, \theta, g)$$

and test its worthiness as a predictor.

Experimental determination of the model Since gravity is essentially constant under the assumptions, the period t is a function of the three variables length r, mass m, and initial angle of displacement θ. At this point we could systematically conduct experiments to determine how t varies with these three variables. We would want to choose enough values of the independent variables to feel confident in predicting the period t over that range. So how many experiments will be necessary?

For sake of illustration, consider a function of one independent variable $y = f(x)$ and assume that four points have been deemed necessary to predict y over a suitable domain for x. The situation is depicted in Figure 7-2. An appropriate curve-fitting or interpolating method could be used to predict y within the domain for x.

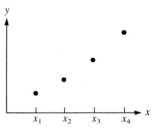

FIGURE 7-2 Four points have been deemed necessary to predict y for this function of one variable x.

Next, consider what happens when a second independent variable affects the situation under investigation. We then have a function

$$y = f(x, z)$$

For each data value of x in Figure 7-2, experiments must be conducted to obtain y for four values of z. Thus, 16 (that is, 4^2) experiments are required altogether. These observations are illustrated in Figure 7-3. Likewise, a function of three variables requires 64 (that is, 4^3) experiments. In general, 4^n experiments are required to predict y when n is the number of arguments of

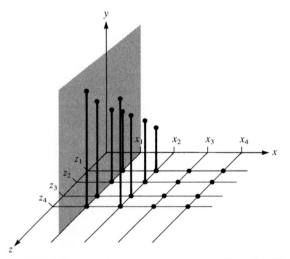

FIGURE 7-3 Sixteen points are necessary to predict y for this function of the two variables x and z.

the function, assuming 4 points for the domain of each argument. Thus, a procedure that reduces the number of arguments of the function f will dramatically reduce the total number of required experiments. Dimensional analysis is one such procedure.

The power of dimensional analysis can also be appreciated when we examine the interpolating curves that would be determined after collecting the data represented in Figures 7-2 and 7-3. Let's assume that it is decided to pass a cubic polynomial through the four points shown in Figure 7-2. That is, the four points are used to determine the four constants C_1, C_2, C_3, C_4 in the interpolating curve:

$$y = C_1 x^3 + C_2 x^2 + C_3 x + C_4$$

Now consider interpolating from Figure 7-3. If for a fixed value of x, say $x = x_1$, it is decided to connect our points using a cubic polynomial in z, the equation of the interpolating surface is

$$\begin{aligned}
y = \ &D_1 x^3 + D_2 x^2 + D_3 x + D_4 \\
&+ (D_5 x^3 + D_6 x^2 + D_7 x + D_8)z \\
&+ (D_9 x^3 + D_{10} x^2 + D_{11} x + D_{12})z^2 \\
&+ (D_{13} x^3 + D_{14} x^2 + D_{15} x + D_{16})z^3
\end{aligned}$$

Note from the equation that there are 16 parameters $D_1, D_2, \ldots, D_{16}$ to determine rather than 4 as in the two-dimensional case. This procedure again illustrates the dramatic reduction in effort required when we reduce the number of arguments of the function we will finally investigate.

At this point we make the important observation that the experimental effort required depends more heavily on the number of arguments of the function to be investigated than on the true number of independent variables originally selected by the modeler. For example, consider a function of two arguments, say $y = f(x, z)$. The discussion concerning the number of experiments necessary would not be altered if x were some particular combination of several variables. That is, x could be uv/w, where u, v, and w are the variables originally selected in the model.

Consider now the following preview of dimensional analysis, which describes how it reduces our experimental effort. Beginning with a function of n variables (hence n arguments), the number of arguments is reduced (ordinarily by 3) by combining the original variables into products. These resulting $(n - 3)$ products are called "dimensionless products" of the original variables. After applying dimensional analysis, we still need to conduct experiments to make our predictions, but the amount of experimental effort that is required will have been reduced exponentially.

In Chapter 2 we discussed the trade-offs of considering additional variables for increased precision versus neglecting variables for simplification. In constructing models based on experimental data, the preceding discussion suggests that the "cost" of each additional variable is an exponential increase in the number of experimental trials that must be conducted. In the next two sections we present the main ideas underlying the dimensional analysis process. You may find that some of these ideas are a little more difficult than previous ones we have been investigating, but the methodology is very powerful when modeling physical behavior.

7.1 DIMENSIONS AS PRODUCTS

The study of physics is based on abstract concepts like mass, length, time, velocity, acceleration, force, energy, work, pressure, and so forth. To each such concept there is assigned a unit of measurement. A physical law such as $F = ma$ is true provided the units of measurement are consistent. Thus if mass is measured in kilograms and acceleration in meters per second squared, then the force must be measured in newtons. These units of measurement belong to the MKS (meter-kilogram-second) mass system. It would be inconsistent with the equation $F = ma$ to measure mass in slugs, acceleration in feet per second squared, and force in newtons. In this latter illustration, force must be measured in pounds, giving the American Engineering System of measurement. There are other systems of measurement, but all of them are prescribed by international standards so as to be consistent with the laws of physics.

The three primary physical quantities we consider in this chapter are mass, length, and time. With these quantities we associate the *dimensions M, L,* and *T*, respectively. The dimensions are symbols that reveal how the numerical value of a quantity changes when the units of measurement change in certain ways. The dimensions of other quantities follow from definitions or from physical laws and are expressed in terms of M, L, and T. For example, velocity v is defined as the ratio of distance s (dimension L) traveled to time t (dimension T) of travel—that is, $v = st^{-1}$, so the dimension of velocity is LT^{-1}. Similarly, since area is fundamentally a product of two lengths, its dimension is L^2. These dimension expressions hold true regardless of the particular system of measurement, and they show, for example, that velocity may be expressed in meters per second, feet per second, miles per hour, and so forth. Likewise, area is measured in terms of square meters, square feet, square miles, and so on.

There are still other entities in physics that are more complex in the sense that they are not usually defined directly in terms of mass, length, and time alone, but instead their definitions include other quantities such as velocity. We associate dimensions with these more complex quantities in accordance with the algebraic operations involved in the definitions. For example, since momentum is the product of mass with velocity, its dimension is $M(LT^{-1})$, or simply MLT^{-1}.

The basic definition of a quantity may also involve dimensionless constants; these are ignored in finding dimensions. Thus, the dimension of kinetic energy, which is one half (a dimensionless constant) the product of mass with velocity squared, is $M(LT^{-1})^2$, or simply ML^2T^{-2}. As you will see in Example 2, some constants (dimensional constants) do have an associated dimension, and these must be considered in a dimensional analysis.

These examples illustrate some important concepts regarding dimensions of physical quantities.

1. We have based the concept of dimension on three physical quantities: mass m, length s, and time t. These quantities are measured in some appropriate system of units whose choice does not affect the assignment of dimensions. (However, this underlying system must be "linear." A dimensional analysis will not work if the scale is logarithmic, for example.)

2. There are other physical quantities, such as area and velocity, that are defined as simple products involving only mass, length, or time. Here we use **product** the term **product** to include any quotient, since we may indicate division by negative exponents.

3. There are still other, more complex, physical entities, such as momentum and kinetic energy, whose definitions involve quantities other than mass, length, and time. But since the simpler quantities from (1) and (2) are themselves products, these more complex quantities can also be expressed as products involving mass, length, and time by algebraic simplification. We use the term *product* to refer to any physical quantity from item (1), (2), or (3); a product from (1) is trivial, since it has only one factor.

dimension **4.** To each product, there is assigned a **dimension**—that is, an expression of the form:

$$M^nL^pT^q \tag{7.1}$$

where n, p, and q are real numbers that may be positive, negative, or zero.

When a basic dimension is missing from a product, the corresponding exponent is understood to be zero. Thus, the dimension $M^2L^0T^{-1}$ may also appear as M^2T^{-1}. When n, p, and q are all zero in an expression of the form (7.1), so the dimension reduces to

$$M^0L^0T^0 \tag{7.2}$$

the quantity, or product, is said to be *dimensionless*.

Special care must be taken in forming sums of products because, just as we cannot "add apples and oranges," in an equation we cannot add products that have unlike dimensions. For example, if F denotes force, m mass, and v velocity, we know immediately that the equation

$$F = mv + v^2$$

cannot be correct because mv has dimension MLT^{-1} while v^2 has dimension L^2T^{-2}. These dimensions are unlike, and hence the products mv and v^2 cannot be added. An equation such as this—that is, one that contains among its terms two products having unlike dimensions—is said to be *dimensionally incompatible*. Equations that involve only sums of products having the same dimension are, then, dimensionally compatible.

dimensional homogeneity

The concept of dimensional compatibility is related to another important concept, called **dimensional homogeneity.** In general, an equation that is true regardless of the system of units in which the variables are measured is said to be *dimensionally homogeneous.* For example, $t = \sqrt{2s/g}$ giving the time a body falls a distance s under gravity (neglecting air resistance) is dimensionally homogeneous (true in all systems), whereas the equation $t = \sqrt{s/16.1}$ is not dimensionally homogeneous (because it depends on a particular system). In particular, if an equation involves only sums of dimensionless products (that is, it is a dimensionless equation), then the equation is dimensionally homogeneous. Since the products are dimensionless, the factors used for conversion from one system of units to another would simply cancel.

The application of dimensional analysis to a real-world problem is based on the assumption that the solution to the problem is given by a dimensionally homogeneous equation in terms of the appropriate variables. Thus the task is to determine the form of the desired equation by finding an appropriate dimensionless equation and then solving for the dependent variable. In order to accomplish this task, you must decide which variables enter into the physical problem under investigation, and you must determine all the dimensionless products among them. In general, there may be infinitely many such products, so they will have to be described rather than actually written out. Certain subsets of these dimensionless products are then used to construct dimensionally homogeneous equations. In Section 7.2 we investigate how the dimensionless products are used to find all dimensionally homogeneous equations. The following example illustrates how the dimensionless products may be found.

Example 1 A Simple Pendulum Revisited

Consider again the simple pendulum discussed in the introduction. Analyzing the dimensions of the variables for the pendulum problem, we have

Variable	m	g	t	r	θ
Dimension	M	LT^{-2}	T	L	$M^0L^0T^0$

Next we find all the dimensionless products among the variables. Any product of these variables must be of the form:

$$m^a g^b t^c r^d \theta^e \tag{7.3}$$

and hence must have dimension

$$(M)^a(LT^{-2})^b(T)^c(L)^d(M^0L^0T^0)^e$$

Therefore, a product of the form (7.3) is dimensionless if and only if

$$M^a L^{b+d} T^{c-2b} = M^0 L^0 T^0 \tag{7.4}$$

Equating the exponents on both sides of this last equation leads to the system of linear equations

$$\left. \begin{array}{l} a \qquad\qquad\quad + 0e = 0 \\ \quad b \quad + d + 0e = 0 \\ -2b + c \qquad + 0e = 0 \end{array} \right\} \qquad (7.5)$$

Solution of the system (7.5) gives $a = 0$, $c = 2b$, $d = -b$, where b is arbitrary. Thus there are infinitely many solutions. Notice that the exponent e does not really appear in (7.4) (because it has a zero coefficient in each equation) so that it is arbitrary also. One dimensionless product is obtained by setting $b = 0$ and $e = 1$, yielding $a = c = d = 0$, and a second, independent dimensionless product is obtained when $b = 1$ and $e = 0$, yielding $a = 0$, $c = 2$, and $d = -1$. These solutions give the dimensionless products:

$$\Pi_1 = m^0 g^0 t^0 r^0 \theta^1 = \theta$$

$$\Pi_2 = m^0 g^1 t^2 r^{-1} \theta^0 = \frac{gt^2}{r}$$

You will learn a methodology for relating these products in Section 7.2 in order to carry the modeling process to completion. For now, we will develop a relationship in an intuitive manner.

Assuming that $t = f(r, m, g, \theta)$, to determine more about the function f, we observe that if the units in which we measure mass are made smaller by some factor (say by 10), then the measure of the period t will not change because it is measured in units (T) of time. Since m is the only factor whose dimension contains M, it cannot therefore appear in the model. Similarly, if the scale of the units (L) for measuring length is altered, it cannot change the measure of the period. In order for this to happen, the factors r and g must appear in the model as r/g, g/r, or more generally, $(g/r)^k$. This ensures that any linear change in the way length is measured will be cancelled. Finally, if we make the units (T) that measure time smaller by a factor of, say, 10, the measure of the period will directly increase by this same factor 10. Thus in order to have the dimension of T on the right side of the equation $t = f(r, m, g, \theta)$, g and r must appear as $\sqrt{r/g}$, since T appears to the power -2 in the dimension of g. Note that none of the preceding conditions places any restrictions on the angle θ. Thus the equation of the period should be of the form

$$t = \sqrt{\frac{r}{g}} \, h(\theta)$$

where the function h must be determined or approximated by experimentation.

We note two things in this analysis that are characteristic of a dimensional analysis. In the MLT system, three conditions are placed on the model, so we would generally expect to reduce the number of arguments of the function relating the variables by three. In the example of the pendulum problem we reduced the number of arguments from four to one. Second, all arguments of the function present at the end of a dimensional analysis (in this case, θ) are dimensionless products.

In the problem of the undamped pendulum we assumed that friction and drag were negligible. Before proceeding with experiments (which might be costly), we would like to know if that assumption is reasonable. Consider the model obtained so far:

$$t = \sqrt{\frac{r}{g}}\, h(\theta)$$

Keeping θ constant while allowing r to vary, form the ratio:

$$\frac{t_1}{t_2} = \frac{\sqrt{r_1/g}\; h(\theta_0)}{\sqrt{r_2/g}\; h(\theta_0)} = \sqrt{r_1/r_2}$$

Hence the model predicts that t will vary as $\sqrt{r}$ for constant θ. Thus, if we plot t versus $\sqrt{r}$ with θ fixed for some observations, we would expect to get a straight line (see Figure 7-4). If we do not obtain a reasonably straight line, then we need to reexamine the assumptions. Note that our judgment here is qualitative. The final measure of the adequacy of any model is always how well it predicts or explains the phenomenon under investigation. Nevertheless, this initial test is useful for eliminating obviously bad assumptions and also for choosing among competing sets of assumptions.

Dimensional analysis has helped to construct a model $t = f(r, m, g, \theta)$ for the undamped pendulum as $t = \sqrt{r/g}\, h(\theta)$. If we are interested in predicting the behavior of the pendulum, we could isolate the effect of h by holding r constant and varying θ. This provides the ratio:

$$\frac{t_1}{t_2} = \frac{\sqrt{r_0/g}\; h(\theta_1)}{\sqrt{r_0/g}\; h(\theta_2)} = \frac{h(\theta_1)}{h(\theta_2)}$$

Hence a plot of t versus θ for several observations would reveal the nature of h. This plot is illustrated in Figure 7-5. We may never discover the true function h relating the variables. In such cases an empirical model might be constructed from the experimental data, as discussed in Chapter 6. When we are interested in using our model to predict t, based on experimental results, it is convenient to use the equation $t\sqrt{g/r} = h(\theta)$, and to plot $t\sqrt{g/r}$ versus θ as in Figure 7-6. Then, for a given value of θ, we would determine $t\sqrt{g/r}$, multiply by $\sqrt{r/g}$ for a specific r, and finally determine t.

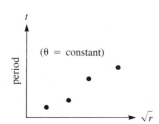

FIGURE 7-4 Testing the assumptions of the simple pendulum model by plotting the period t versus the square root of the length r for constant displacement θ.

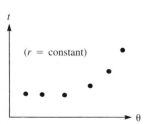

FIGURE 7-5 Determining the unknown function h.

FIGURE 7-6 Presenting the results for the simple pendulum.

Example 2 Wind Force on a Van*

Suppose you are driving a van down a highway with gusty winds. How does the speed of your vehicle affect the wind force you are experiencing?

The force F of the wind on the van is certainly affected by the speed v of the van and the surface area A of the van directly exposed to the wind's direction. Thus we might hypothesize that the force is proportional to some power of the speed times some power of the surface area; that is,

$$F = kv^a A^b \qquad (7.6)$$

for some (dimensionless) constant k. Analyzing the dimensions of the variables gives

Variable	F	k	v	A
Dimension	MLT^{-2}	$M^0 L^0 T^0$	LT^{-1}	L^2

Hence, dimensionally, Equation (7.6) becomes

$$MLT^{-2} = (M^0 L^0 T^0)(LT^{-1})^a (L^2)^b$$

However, this last equation cannot be correct because the dimension M for mass does not enter into the right-hand side with nonzero exponent.

So consider again Equation (7.6). What is missing in our assumption concerning the wind force? Wouldn't the strength of the wind be affected by its density? After some reflection you would probably agree that density does have an effect. If we include the density ρ as a factor, then our refined model becomes

$$F = kv^a A^b \rho^c \qquad (7.7)$$

Since density is mass per unit volume, the dimension of density is ML^{-3}. Therefore, dimensionally, Equation (7.7) becomes

$$MLT^{-2} = (M^0 L^0 T^0)(LT^{-1})^a (L^2)^b (ML^{-3})^c$$

Equating the exponents on both sides of this last equation leads to the system of linear equations:

$$\left. \begin{array}{r} c = 1 \\ a + 2b - 3c = 1 \\ -a \qquad\qquad = -2 \end{array} \right\} \qquad (7.8)$$

Solution of the system (7.8) gives $a = 2$, $b = 1$, and $c = 1$. When substituted into (7.7), these values give the model:

$$F = kv^2 A\rho$$

---

* This example was suggested from a preprint titled "Dimensional Analysis," by Ralph Mansfield.

At this point we make an important observation. When it was assumed that $F = kv^a A^b$, the constant k was assumed to be dimensionless. Subsequently, our analysis revealed that for a particular medium (so ρ is constant)

$$F \propto Av^2$$

giving $F = k_1 Av^2$. However, k_1 does have a dimension associated with it and is called a dimensional constant. In particular, the dimension of k_1 is

$$\frac{MLT^{-2}}{L^2(L^2 T^{-2})} = ML^{-3}$$

Dimensional constants contain important information and must be considered when performing a dimensional analysis. We consider dimensional constants again in Section 7.3 when we investigate a damped pendulum.

If we assume the density ρ is constant, our model shows that the force of the wind is proportional to the square of the speed of the van times its surface area directly exposed to the wind. We can test the model by collecting some data and plotting the wind force F versus $v^2 A$ to see if the graph approximates a straight line through the origin. This example illustrates one of the ways dimensional analysis can be used to test your assumptions and check whether you have a faulty list of variables identifying the problem.

Table 7-1 gives a summary of the dimensions of some common physical entities.

TABLE 7-1 Dimensions of physical entities in the MLT system

Mass	M	Momentum	MLT^{-1}
Length	L	Work	$ML^2 T^{-2}$
Time	T	Density	ML^{-3}
Velocity	LT^{-1}	Viscosity	$ML^{-1} T^{-1}$
Acceleration	LT^{-2}	Pressure	$ML^{-1} T^{-2}$
Specific Weight	$ML^{-2} T^{-2}$	Surface Tension	MT^{-2}
Force	MLT^{-2}	Power	$ML^2 T^{-3}$
Frequency	T^{-1}	Rotational Inertia	ML^2
Angular Velocity	T^{-1}	Torque	$ML^2 T^{-2}$
Angular Acceleration	T^{-2}	Entropy	$ML^2 T^{-2}$
Angular Momentum	$ML^2 T^{-1}$	Heat	$ML^2 T^{-2}$
Energy	$ML^2 T^{-2}$		

7.1 PROBLEMS

1. Determine whether the equation:

$$s = s_0 + v_0 t - 0.5gt^2$$

is dimensionally compatible, if s is the position (measured vertically from a fixed reference point) of a body at time t, s_0 is the position at $t = 0$, v_0 the initial velocity, and g the acceleration due to gravity.

2. Find a dimensionless product relating the torque τ (ML^2T^{-2}) produced by an automobile engine, the engine's rotation rate ψ (T^{-1}), the volume V of air displaced by the engine, and the air density ρ.
3. The various "constants" of physics often have physical dimensions (dimensional constants) because their values depend on the system in which they are expressed. For example, Newton's law of gravitation asserts that the attractive force between two bodies is proportional to the product of their masses divided by the square of the distance between them, or symbolically,

$$F = \frac{Gm_1m_2}{r^2}$$

where G is the gravitational "constant." Find the dimension of G so that Newton's law is dimensionally compatible.
4. Certain stars, whose light and radial velocities undergo periodic vibrations, are thought to be pulsating. It is hypothesized that the period t of pulsation depends on the star's radius r, its mass m, and the gravitational constant G. (See Problem 3 for the dimension of G.) Express t as a product of m, r, and G so the equation:

$$t = m^a r^b G^c$$

is dimensionally compatible.
5. In checking the dimensions of an equation, you should note that derivatives also possess dimensions. For example, the dimension of ds/dt is LT^{-1}, and the dimension of d^2s/dt^2 is LT^{-2}, where s denotes distance and t denotes time. Determine whether the equation:

$$\frac{dE}{dt} = \left[mr^2 \left(\frac{d^2\theta}{dt^2} \right) + mgr \sin\theta \right] \frac{d\theta}{dt}$$

for the time rate of change of total energy E in a pendulum system with damping force is dimensionally compatible.
6. For a body moving along a straight-line path, if the mass of the body is changing over time, then an equation governing its motion is given by

$$m\frac{dv}{dt} = F + u\frac{dm}{dt}$$

where m is the mass of the body, v is the velocity of the body, F is the total force acting on the body, dm is the mass joining or leaving the body in the time interval dt, and u is the velocity of dm at the moment it joins or leaves the body (relative to an observer stationed on the body). Show that the preceding equation is dimensionally compatible.
7. In the human the hydrostatic pressure of blood contributes to the total blood pressure. The hydrostatic pressure P is a product of blood density ρ, height h of the blood column between the heart and some lower point

in the body, and gravity g. Determine

$$P = k\rho^a h^b g^c$$

where k is a dimensionless constant.

8. Assume the force F opposing the fall of a raindrop through air is a product of viscosity μ, velocity v, and the diameter r of the drop. Assume that density is neglected. Find

$$F = k\mu^a v^b r^c$$

where k is a dimensionless constant.

7.1 PROJECTS

1. Complete the requirements of "Keeping Dimension Straight," by George E. Strecker, UMAP 564. This module is a very basic introduction to the distinction between dimensions and units and provides the student with some practice in using dimensional arguments to properly set up solutions to elementary problems and to recognize errors.

7.2 THE PROCESS OF DIMENSIONAL ANALYSIS

In the preceding section you learned how to determine all dimensionless products among the variables selected in the problem under investigation. Now we investigate how to use the dimensionless products to find all possible dimensionally homogeneous equations among the variables. The key result is Buckingham's theorem,* which summarizes the entire theory of dimensional analysis.

Example 1 in the preceding section shows that in general many dimensionless products may be formed from the variables of a given system. In that example we determined every dimensionless product to be of the form:

$$g^b t^{2b} r^{-b} \theta^e \tag{7.9}$$

where b and e are arbitrary real numbers. Each one of these products corresponds to a solution of the homogeneous system of linear algebraic equations given by (7.5). The two products:

$$\Pi_1 = \theta \quad \text{and} \quad \Pi_2 = \frac{gt^2}{r}$$

obtained when $b = 0$, $e = 1$ and $b = 1$, $e = 0$, respectively, are special in the sense that any of the dimensionless products (7.9) can be given as a product

---

* E. Buckingham, "On Physically Similar Systems; Illustrations of the Use of Dimensional Equations," *Phys. Rev. 4,* no. 4 (1914): 345.

of some power of Π_1 times some power of Π_2. Thus, for instance,

$$g^3 t^6 r^{-3} \theta^{1/2} = \Pi_1^{1/2} \Pi_2^3$$

This observation follows from the fact that $b = 0$, $e = 1$ and $b = 1$, $e = 0$ represent, in some sense, "independent solutions" of the system (7.5). Let's explore these ideas further.

Consider the following system of m linear algebraic equations in the n unknowns $x_1, x_2, \ldots, x_n$:

$$
\begin{aligned}
a_{11}x_1 + a_{12}x_2 + \cdots + a_{1n}x_n &= b_1 \\
a_{21}x_1 + a_{22}x_2 + \cdots + a_{2n}x_n &= b_2 \\
\vdots \qquad \vdots \qquad \vdots \qquad \vdots \qquad \vdots \\
a_{m1}x_1 + a_{m2}x_2 + \cdots + a_{mn}x_n &= b_m
\end{aligned}
\tag{7.10}
$$

coefficients

constants

The symbols a_{ij} and b_i denote real numbers for each $i = 1, 2, \ldots, m$ and $j = 1, 2, \ldots, n$. The numbers a_{ij} are called the **coefficients** of the system and the b_i are referred to as the **constants**. The subscript i in the symbol a_{ij} refers to the ith equation of the system (7.10) and the subscript j refers to the jth unknown x_j to which a_{ij} belongs. Thus the subscripts serve to locate a_{ij}. It is customary to read a_{13} as "a, one, three" and a_{42} as "a, four, two," for example, rather than "a, thirteen" and "a, forty-two."

solution

homogeneous

trivial solution

A **solution** to the system (7.10) is a sequence of numbers $s_1, s_2, \ldots, s_n$ for which $x_1 = s_1$, $x_2 = s_2, \ldots, x_n = s_n$ solves each equation in the system. If $b_1 = b_2 = \cdots = b_m = 0$, the system (7.10) is said to be **homogeneous**. The solution $s_1 = s_2 = \cdots = s_n = 0$ always solves the homogeneous system and is called the **trivial solution**. For a homogeneous system there are two solution possibilities: either the trivial solution is the only solution or there are infinitely many solutions.

sum

scalar multiple

Whenever $s_1, s_2, \ldots, s_n$ and $s_1', s_2', \ldots, s_n'$ are solutions to the *homogeneous* system, then the sequences $s_1 + s_1'$, $s_2 + s_2', \ldots, s_n + s_n'$ and cs_1, $cs_2, \ldots, cs_n$ are also solutions for any constant c. These solutions are called the **sum** and **scalar multiple** of the original solutions, respectively. If S and S' refer to the original solutions, then we use the notation $S + S'$ to refer to their sum and cS to refer to a scalar multiple of the first solution. If S_1, $S_2, \ldots, S_k$ is a collection of k solutions to the homogeneous system, then the solution:

$$c_1 S_1 + c_2 S_2 + \cdots + c_k S_k$$

linear combination

is called a **linear combination** of the k solutions, where $c_1, c_2, \ldots, c_k$ are arbitrary real numbers. It is an easy exercise to show that any linear combination of solutions to the homogeneous system is still another solution to the system.

independent

complete

A set of solutions to a homogeneous system is said to be **independent** if no solution in the set is a linear combination of the remaining solutions in the set. A set of solutions is **complete** if it is independent and every solution is expressible as a linear combination of solutions in the set. For a specific homogeneous system, we seek some complete set of solutions since all other solutions are produced from them using linear combinations. For example, the two solutions corresponding to the two choices $b = 0, e = 1$ and $b = 1, e = 0$ form a complete set of solutions to the homogeneous system (7.5).

It is not our intent to present the theory of linear algebraic equations here. Such a study is appropriate for a course in linear algebra. We do point out that there is an elementary algorithm known as *Gaussian elimination* for producing a complete set of solutions to a given system of linear equations. Moreover, Gaussian elimination is readily implemented on computers and hand-held programmable calculators. The systems of equations you will encounter in this book are simple enough to be solved by the elimination methods you learned in intermediate algebra.

How does our discussion relate to dimensional analysis? Our basic goal thus far has been to find all possible dimensionless products among the variables that influence the physical phenomenon under investigation. We developed a homogeneous system of linear algebraic equations to help us determine these dimensionless products. This system of equations usually has infinitely many solutions. Each solution gives the values of the exponents that result in a dimensionless product among the variables. If you sum two solutions, you produce another solution that yields the same dimensionless product as does multiplication of the dimensionless products corresponding to the original two solutions. For example, the sum of the solutions corresponding to $b = 0, e = 1$ and $b = 1, e = 0$ for (7.5) yields the solution corresponding to $b = 1, e = 1$ with the corresponding dimensionless product from (7.9) given by

$$gt^2 r^{-1} \theta = \Pi_1 \Pi_2$$

The reason for this result is that the unknowns in the system of equations are the exponents in the dimensionless products, and addition of exponents algebraically corresponds to multiplication of numbers having the same base: $x^{m+n} = x^m x^n$. Moreover, multiplication of a solution by a constant produces a solution that yields the same dimensionless product as does raising the product corresponding to the original solution to the power of the constant. For example, -1 times the solution corresponding to $b = 1, e = 0$ yields the solution corresponding to $b = -1, e = 0$ with the corresponding dimensionless product:

$$g^{-1} t^{-2} r = \Pi_2^{-1}$$

The reason for this last result is that algebraic multiplication of an exponent by a constant corresponds to raising a power to a power: $x^{mn} = (x^m)^n$.

In summary, addition of solutions to the homogeneous system of equations results in multiplication of their corresponding dimensionless products, and multiplication of a solution by a constant results in raising the corresponding product to the power given by that constant. Thus if S_1 and S_2 are two solutions corresponding to the dimensionless products Π_1 and Π_2, respectively, then the linear combination $aS_1 + bS_2$ corresponds to the dimensionless product:

$$\Pi_1{}^a\Pi_2{}^b$$

It follows from our preceding discussion that a complete set of solutions to the homogeneous system of equations produces all possible solutions through linear combination. The dimensionless products corresponding to a complete set of solutions is therefore called a *complete set of dimensionless products*. All dimensionless products can be obtained by forming powers and products of the members of a complete set. Next let's investigate how these dimensionless products can be used to produce all possible dimensionally homogeneous equations among the variables.

In Section 7.1 we defined an equation to be dimensionally homogeneous if it remains true regardless of the system of units in which the variables are measured. The fundamental result in dimensional analysis that provides for the construction of all dimensionally homogeneous equations from complete sets of dimensionless products is the following theorem.*

Buckingham's Theorem

An equation is dimensionally homogeneous if and only if it can be put into the form:

$$f(\Pi_1, \Pi_2, \ldots, \Pi_n) = 0 \tag{7.11}$$

where f is some function of n arguments and $\{\Pi_1, \Pi_2, \ldots, \Pi_n\}$ is a complete set of dimensionless products.

Let's apply Buckingham's theorem to the simple pendulum discussed in the preceding sections. The two dimensionless products:

$$\Pi_1 = \theta \quad \text{and} \quad \Pi_2 = \frac{gt^2}{r}$$

form a complete set for the pendulum problem. Thus, according to Buckingham's theorem, there is a function f such that

$$f\left(\theta, \frac{gt^2}{r}\right) = 0$$

Assuming we can solve this last equation for gt^2/r as a function of θ, it follows that

$$t = \sqrt{\frac{r}{g}}\, h(\theta) \tag{7.12}$$

where h is some function of the single variable θ. Notice that this last result agrees with our intuitive formulation for the simple pendulum presented in

* A proof to Buckingham's theorem is given in Henry L. Langhaar, *Dimensional Analysis and Theory of Models* (New York: Wiley, 1951), Chapter 4.

Section 7.1. Observe that Equation (7.12) represents only a general form for the relationship among the variables m, g, t, r, and θ. However, it can be concluded from this expression that t does not depend on the mass m and is related to $r^{1/2}$ and $g^{-1/2}$ by some function of the initial angle of displacement θ. Knowing this much, we can determine the nature of the function h experimentally or approximate it, as discussed in Section 7.1.

Consider Equation (7.11) in Buckingham's theorem. For the case in which a complete set consists of a single dimensionless product, say Π_1, the equation reduces to the form:

$$f(\Pi_1) = 0$$

In this case we assume that the function f has one real root at k (to assume otherwise has little physical meaning). Hence the solution $\Pi_1 = k$ is obtained.

Using Buckingham's theorem, let's reconsider the example from Section 7.1 of the wind force on a van driving down a highway. Since the four variables F, v, A, and ρ were selected and all three equations in (7.8) are independent, a complete set of dimensionless products consists of a single product:

$$\Pi_1 = \frac{F}{v^2 A \rho}$$

Application of Buckingham's theorem gives

$$f(\Pi_1) = 0$$

which implies from the preceding discussion that $\Pi_1 = k$, or

$$F = kv^2 A \rho$$

where k is a dimensionless constant as before. Thus, when a complete set consists of a *single dimensionless product*, as is generally the case when we begin with four variables, the application of Buckingham's theorem yields the desired relationship *up to a constant of proportionality*. Of course, the predicted proportionality must be tested to determine the adequacy of our list of variables. If the list does prove to be adequate, then the constant of proportionality can be determined by experimentation, thereby completely defining the relationship.

For the case $n = 2$, Equation (7.11) in Buckingham's theorem takes the form:

$$f(\Pi_1, \Pi_2) = 0 \tag{7.13}$$

If we choose the products in the complete set $\{\Pi_1, \Pi_2\}$ so that the dependent variable appears in only one of them, say Π_2, we can proceed under the assumption that Equation (7.13) can be solved for that chosen product Π_2 in terms of the remaining product Π_1. Such a solution takes the form:

$$\Pi_2 = H(\Pi_1)$$

and then this latter equation can be solved for the dependent variable. Note that when a complete set consists of more than one dimensionless product, the application of Buckingham's theorem determines the desired relationship *up to an arbitrary function*. After verifying the adequacy of the list of variables, we may be lucky enough to recognize the underlying functional relationship. However, in general you can expect to construct an empirical model, although the task has been eased considerably.

For the general case of n dimensionless products in the complete set for Buckingham's theorem, we again choose the products in the complete set $\{\Pi_1, \Pi_2, \ldots, \Pi_n\}$ so that the dependent variable appears in only one of them, say Π_n for definiteness. Assuming we can solve Equation (7.11) for that product Π_n in terms of the remaining ones, we have the form:

$$\Pi_n = H(\Pi_1, \Pi_2, \ldots, \Pi_{n-1})$$

We then solve this last equation for the dependent variable.

Summary of Dimensional Analysis Methodology

We now summarize the steps in the dimensional analysis process:

Step 1. Decide which variables enter the problem under investigation.

Step 2. Determine a complete set of dimensionless products $\{\Pi_1, \Pi_2, \ldots, \Pi_n\}$ among the variables. Make sure the dependent variable of the problem appears in only one of the dimensionless products.

Step 3. Check to ensure that the products found in the previous step are dimensionless and independent. Otherwise you have an algebra error.

Step 4. Apply Buckingham's theorem to produce all possible dimensionally homogeneous equations among the variables. This procedure yields an equation of the form (7.11).

Step 5. Solve the equation in Step 4 for the dependent variable.

Step 6. Test to ensure that the assumptions made in Step 1 are "reasonable." Otherwise the list of variables is faulty.

Step 7. Conduct the necessary experiments and present the results in a useful format.

Let's illustrate the first five steps of the preceding procedure.

Example 1 Terminal Velocity of a Raindrop

Consider the problem of determining the terminal velocity v of a raindrop falling from a motionless cloud. We looked at this problem from a very simplistic point of view in Chapter 3, but let's take another look using dimensional analysis.

What are the variables influencing the behavior of the raindrop? Certainly the terminal velocity will depend on the size of the raindrop given by, say, its radius r. The density ρ of the air and the viscosity μ of the air will also affect the behavior. (Viscosity measures resistance to motion—a sort of internal molecular friction. In gases this resistance is caused by collisions between fast-moving molecules.) The acceleration due to gravity g is another variable to be considered. Although the surface tension of the raindrop is a factor that does influence the behavior of the fall, we will ignore this factor. If necessary, surface tension can be taken into account in a later, refined model. These considerations give the following table relating the selected variables to their dimensions:

Variable	v	r	g	ρ	μ
Dimension	LT^{-1}	L	LT^{-2}	ML^{-3}	$ML^{-1}T^{-1}$

Next we find all the dimensionless products among the variables. Any such product must be of the form:

$$v^a r^b g^c \rho^d \mu^e \tag{7.14}$$

and hence must have dimension:

$$(LT^{-1})^a (L)^b (LT^{-2})^c (ML^{-3})^d (ML^{-1}T^{-1})^e$$

Therefore, a product of the form (7.14) is dimensionless if and only if the following system of equations in the exponents is satisfied:

$$\left. \begin{array}{r} d + e = 0 \\ a + b + c - 3d - e = 0 \\ -a \quad - 2c \quad - e = 0 \end{array} \right\} \tag{7.15}$$

Solution of the system (7.15) gives $b = (3/2)d - (1/2)a$, $c = (1/2)d - (1/2)a$, and $e = -d$, where a and d are arbitrary. One dimensionless product Π_1 is obtained by setting $a = 1$, $d = 0$; another, independent dimensionless product Π_2 is obtained when $a = 0$, $d = 1$. These solutions give

$$\Pi_1 = vr^{-1/2}g^{-1/2} \quad \text{and} \quad \Pi_2 = r^{3/2}g^{1/2}\rho\mu^{-1}$$

Next we check the results to ensure that the products are indeed dimensionless:

$$\frac{LT^{-1}}{L^{1/2}(LT^{-2})^{1/2}} = M^0 L^0 T^0 \quad \text{and} \quad \frac{L^{3/2}(LT^{-2})^{1/2}(ML^{-3})}{ML^{-1}T^{-1}} = M^0 L^0 T^0$$

Thus, according to Buckingham's theorem, there is a function f such that

$$f\left(vr^{-1/2}g^{-1/2}, \frac{r^{3/2}g^{1/2}\rho}{\mu} \right) = 0$$

Assuming we can solve this last equation for $vr^{-1/2}g^{-1/2}$ as a function of the second product Π_2, it follows that

$$v = \sqrt{rg}\, h\left(\frac{r^{3/2}g^{1/2}\rho}{\mu} \right)$$

where h is some function of the single product Π_2.

The preceding example illustrates a characteristic feature of dimensional analysis. Normally the modeler studying a given physical system has an intuitive idea of the variables involved and has a working knowledge of general principles and laws (such as Newton's second law), but lacks the precise laws governing the interaction of the variables. Of course, the modeler can always experiment with each independent variable separately, holding the others constant and measuring the effect on the system. Often, however, the efficiency of the experimental work can be improved through an application of dimensional analysis. Although we did not illustrate Steps 6 and 7 of the dimensional analysis process for the preceding example, these steps will be illustrated in Section 7.3.

We now make some observations concerning the dimensional analysis process. Suppose n variables have been identified in the physical problem under investigation. When determining a complete set of dimensionless products, we form a system of three linear algebraic equations by equating the exponents for M, L, and T to zero. That is, a system of three equations in n unknowns (the exponents) is obtained. If the three equations are independent, we can solve the system for three of the unknowns in terms of the remaining $n - 3$ unknowns (declared to be arbitrary). In this case, we find $n - 3$ independent dimensionless products that make up the complete set we seek. For instance, in the preceding example there are five unknowns a, b, c, d, e, and we determined three of them (b, c, and e) in terms of the remaining $(5 - 3)$ two arbitrary ones (a and d). Thus we obtained a complete set of two dimensionless products. When choosing the $n - 3$ dimensionless products, we must be sure that the dependent variable appears in only one of them. We can then solve the equation (7.11) guaranteed by Buckingham's theorem for the dependent variable, at least under suitable assumptions on the function f in that equation. (The full story telling when such a solution is possible is the content of an important result in advanced calculus known as the Implicit Function Theorem.)

We acknowledge that we have been rather sketchy in our presentation for solving the system of linear algebraic equations that results in the process of determining all dimensionless products. We have simply assumed that you remember how to solve simple linear systems by the method of elimination of variables. We conclude this section with another example.

Example 2 Automobile Gas Mileage Revisited

Consider again the automobile gasoline mileage problem presented in Section 3.4. One of our submodels in that problem was for the force of propulsion F_p. The variables we identified that affect the propulsion force are C_r, the amount of fuel burned per unit time, the amount K of energy contained in each gallon of gasoline, and the speed v. Let's perform a dimensional analysis. The following table relates the various variables to their dimensions:

Variable	F_p	C_r	K	v
Dimension	MLT^{-2}	L^3T^{-1}	$ML^{-1}T^{-2}$	LT^{-1}

Thus the product:

$$F_p{}^a C_r{}^b K^c v^d \tag{7.16}$$

must have dimension:

$$(MLT^{-2})^a (L^3 T^{-1})^b (ML^{-1}T^{-2})^c (LT^{-1})^d$$

The requirement for a dimensionless product leads to the system:

$$\left. \begin{array}{rcl} a & + c & = 0 \\ a + 3b - & c + d &= 0 \\ -2a - & b - 2c - d &= 0 \end{array} \right\} \tag{7.17}$$

Solution of the system (7.17) gives $b = -a$, $c = -a$, and $d = a$, where a is arbitrary. Choosing $a = 1$, we obtain the dimensionless product:

$$\Pi_1 = F_p C_r{}^{-1} K^{-1} v$$

From Buckingham's theorem there is a function f with $f(\Pi_1) = 0$, so Π_1 equals a constant. Therefore,

$$F_p \propto \frac{C_r K}{v}$$

in agreement with the conclusion reached in Chapter 3.

7.2 PROBLEMS

1. Predict the time of revolution for two bodies of mass m_1 and m_2 in empty space revolving about each other under their mutual gravitational attraction.
2. A projectile is fired with initial velocity v at an angle θ with the horizon. Predict the range R.
3. Consider an object falling under the influence of gravity. Assume that air resistance is negligible. Using dimensional analysis, find the speed v of the object after it has fallen a distance s. Let $v = f(m, g, s)$, where m is the mass of the object and g is the acceleration due to gravity. Does your answer agree with your knowledge of the physical situation?
4. Using dimensional analysis, find a proportionality relationship for the centrifugal force F of a particle in terms of its mass m, velocity v, and radius r of the curvature of its path.
5. One would like to know the nature of the drag forces experienced by a sphere as it passes through a fluid. It is assumed that the sphere has a low speed. Therefore, the drag force is highly dependent on the viscosity of the fluid. The fluid density is to be neglected. Use the dimensional analysis process to develop a model for drag force F as a function of the radius r and velocity v of the sphere, and the viscosity μ of the fluid.

6. The volume flow rate q for laminar flow in a pipe depends on the pipe radius r, the viscosity μ of the fluid, and the pressure drop per unit length dp/dz. Develop a model for the flow rate q as a function of r, μ, and dp/dz.

7. In fluid mechanics, the Reynolds number is a dimensionless number involving the fluid velocity v, density ρ, viscosity μ, and a characteristic length r. Use dimensional analysis to find the Reynolds number.

8. The power P delivered to a pump depends on the specific weight w of the fluid pumped, the height h to which the fluid is pumped, and the fluid flow rate q in cubic feet per second. Use dimensional analysis to determine an equation for power.

9. Find the volume flow rate dV/dt of blood flowing in an artery as a function of the pressure drop per unit length of artery P, the radius r of the artery, the blood density ρ, and the blood viscosity μ.

10. The speed of sound in a gas depends on the pressure and the density. Use dimensional analysis to find the speed of sound in terms of pressure and density.

11. The lift force F on a missile depends upon its length r, velocity v, diameter δ, and initial angle θ with the horizon; it also depends on the density ρ, viscosity μ, gravity g, and speed of sound s of the air. Show that

$$F = \rho v^2 r^2 h\left(\frac{\delta}{r}, \theta, \frac{\mu}{\rho v r}, \frac{s}{v}, \frac{rg}{v^2}\right)$$

12. The height h that a fluid will rise in a capillary tube decreases as the diameter D of the tube increases. Use dimensional analysis to determine how h varies with D and the specific weight w and surface tension σ of the liquid.

MODEL

7.3 A DAMPED PENDULUM

In Section 7.1 we investigated the pendulum problem under the assumptions that the hinge is frictionless, the mass is concentrated at one end of the pendulum, and the drag force is negligible. Suppose we are not satisfied with the results predicted by the constructed model. Then we can refine the model by incorporating drag forces. If F represents the total drag force, the problem now is to determine the function:

$$t = f(r, m, g, \theta, F)$$

Let's consider a submodel for the drag force. As you have seen in previous examples, the modeler is usually faced with a trade-off between simplicity and accuracy. For the pendulum it might seem reasonable to expect the drag force to be proportional to some positive power of the velocity. To keep our model simple, we assume that F is proportional to either v or v^2, as depicted in Figure 7-7.

Now we could experiment to determine directly the nature of the drag force. However, we will first perform a dimensional analysis, since we expect it to reduce our experimental effort. Assume F is proportional to v so that $F = kv$. For convenience we choose to work with the dimensional constant $k = F/v$, which has dimension MLT^{-2}/LT^{-1}, or simply MT^{-1}. Notice that the dimensional constant captures the assumption about the drag force. Thus we apply dimensional analysis to the model:

$$t = f(r, m, g, \theta, k)$$

An analysis of the dimensions of the variables gives

Variable	t	r	m	g	θ	k
Dimension	T	L	M	LT^{-2}	$M^0L^0T^0$	MT^{-1}

Any product of the variables must be of the form:

$$t^a r^b m^c g^d \theta^e k^f \tag{7.18}$$

and hence must have dimension:

$$(T)^a(L)^b(M)^c(LT^{-2})^d(M^0L^0T^0)^e(MT^{-1})^f$$

Therefore, a product of the form (7.18) is dimensionless if and only if

$$\left. \begin{array}{r} c \qquad\qquad +f = 0 \\ b \quad + \ d \qquad\quad = 0 \\ a \qquad - \ 2d \quad -f = 0 \end{array} \right\} \tag{7.19}$$

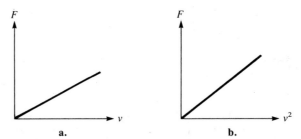

FIGURE 7-7 Possible submodels for the drag force.

The equations in the system (7.19) are independent, so we know we can solve for three of the variables in terms of the remaining (6 − 3) three variables. We would like to choose the solutions in such a way that t appears in only one of the dimensionless products. Thus, we choose a, e, and f as the arbitrary variables with

$$c = -f, \qquad b = -d = \frac{-a}{2} + \frac{f}{2}, \qquad d = \frac{a}{2} - \frac{f}{2}$$

Setting $a = 1$, $e = 0$, and $f = 0$, we obtain $c = 0$, $b = -1/2$, and $d = 1/2$ with the corresponding dimensionless product $t\sqrt{g/r}$. Similarly, choosing $a = 0$, $e = 1$, and $f = 0$, we get $c = 0$, $b = 0$, and $d = 0$, corresponding to the dimensionless product θ. Finally, choosing $a = 0$, $e = 0$, and $f = 1$, we obtain $c = -1$, $b = 1/2$, and $d = -1/2$, corresponding to the dimensionless product $k\sqrt{r}/m\sqrt{g}$. Notice that t appears in only the first of these products. From Buckingham's theorem there is a function h with

$$h\left(t\sqrt{g/r},\, \theta,\, \frac{k\sqrt{r}}{m\sqrt{g}} \right) = 0$$

Assuming we can solve this last equation for $t\sqrt{g/r}$, we obtain

$$t = \sqrt{r/g}\, H\left(\theta,\, \frac{k\sqrt{r}}{m\sqrt{g}} \right)$$

for some function H of two arguments.

Testing the Model (Step 6)

Given $t = \sqrt{r/g}\, H(\theta, k\sqrt{r}/m\sqrt{g})$, our model predicts that $t_1/t_2 = \sqrt{r_1/r_2}$ if the parameters of the function H (namely, θ and $k\sqrt{r}/m\sqrt{g}$) could be held constant. Now there is no difficulty with keeping θ and k constant. However, varying r while simultaneously keeping $k\sqrt{r}/m\sqrt{g}$ constant is more complicated. Since g is constant, we could try to vary r and m in such a manner that $\sqrt{r}/m$ remains constant. This might be done using a pendulum with a hollow mass in order to vary m without altering the drag characteristics. Under these conditions we would expect the plot in Figure 7-8.

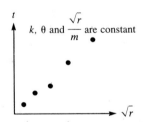

FIGURE 7-8 A plot of t versus $\sqrt{r}$ keeping the variables k, θ, and $\sqrt{r}/m$ constant.

Presenting the Results (Step 7)

As was suggested in predicting the period of the undamped pendulum, we can plot $t\sqrt{g/r} = H(\theta, k\sqrt{r}/m\sqrt{g})$. However, because H is here a function of two arguments, this would yield a three-dimensional figure that is not easy to use. An alternative technique is to plot $t\sqrt{g/r}$ versus $k\sqrt{r}/m\sqrt{g}$ for various values of θ. This is illustrated in Figure 7-9. In order to be safe in predicting t over the range of interest for representative values of θ, it would be necessary to

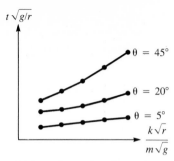

FIGURE 7-9 Presenting the results.

conduct sufficient experiments at various values of $k\sqrt{r}/m\sqrt{g}$. Note that once data are collected, various empirical models could be constructed using an appropriate interpolating scheme for each value of θ.

Choosing Among Competing Models

Because dimensional analysis involves only algebra, one is tempted to develop several models under different assumptions before proceeding with, perhaps quite costly, experimentation. In the case of the pendulum, under different assumptions, we can develop the following three models (see Problem 1):

A: $t = \sqrt{r/g}\, h(\theta)$ ⠀⠀⠀⠀⠀⠀⠀⠀No drag forces

B: $t = \sqrt{r/g}\, h\!\left(\theta, \dfrac{k\sqrt{r}}{m\sqrt{g}}\right)$ ⠀⠀⠀Drag forces proportional to v: $F = kv$

C: $t = \sqrt{r/g}\, h\!\left(\theta, \dfrac{k_1 r}{m}\right)$ ⠀⠀⠀⠀Drag forces proportional to v^2: $F = k_1 v^2$

Since all the preceding models are but approximations, it is reasonable to ask which, if any, is suitable in a particular situation. We now describe the experimentation necessary to distinguish between these models, and we present some experimental results.

Model A predicts that when the angle of displacement θ is held constant, the period t is proportional to $\sqrt{r}$. Model B predicts that when θ and $\sqrt{r}/m$ are both held constant, while maintaining the same drag characteristics k, t is proportional to $\sqrt{r}$. Finally, Model C predicts that if θ, r/m, and k_1 are held constant, then t is proportional to $\sqrt{r}$.

The following discussion describes our experimental results for the pendulum.* Various types of balls were suspended from a string in such a manner

---

* Data collected by Michael Jaye.

as to minimize the friction at the hinge. The kinds of balls included tennis balls as well as various types and sizes of plastic balls. A hole was made in each ball to permit variations in the mass without altering appreciably the aerodynamic characteristics of the ball nor the location of the center of mass. The models were then compared with one another. In the case of the tennis ball, Model A proved to be superior. The period was independent of the mass, and a plot of t versus $\sqrt{r}$ for constant θ is shown in Figure 7-10.

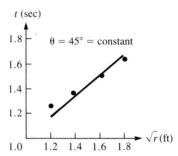

FIGURE 7-10 Model A for a tennis ball.

Having decided that $t = \sqrt{r/g}\, h(\theta)$ is the best of the models for the tennis ball, we isolated the effect of θ by holding r constant in order to gain insight into the nature of the function h. A plot of t versus θ for constant r is shown in Figure 7-11.

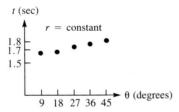

FIGURE 7-11 Isolating the effect of θ.

Note from Figure 7-11 that for small angles of initial displacement θ, the period is virtually independent of θ. However, the displacement effect becomes more noticeable as θ is increased. Thus for small angles we might hypothesize that $t = c\sqrt{r/g}$ for some constant c. If one plots t versus $\sqrt{r}$ for small angles, the slope of the resulting straight line should be constant.

For larger angles, the experiment demonstrates that the effect of θ needs to be considered. In such cases, one may desire to estimate the period for various angles. For example, if $\theta = 45°$ and one knows a particular value of $\sqrt{r}$, he or she can estimate t from Figure 7-10. Although not shown, plots for several different angles can be graphed in the same figure.

Dimensional Analysis in the Model-Building Process

Let's summarize how dimensional analysis provides assistance in the model-building process. In the determination of a model we must first decide which factors to neglect and which to include. A dimensional analysis provides additional information on how the included factors are related. Moreover, in large problems, we often determine one or more submodels before attacking the larger problem. For example, in the pendulum problem we had to develop a submodel for drag forces. A dimensional analysis is helpful in choosing among the various submodels.

A dimensional analysis is also useful for obtaining an initial test of the assumptions in the model. For example, suppose we hypothesize that the dependent variable y is some function of five variables, $y = f(x_1, x_2, x_3, x_4, x_5)$. A dimensional analysis in the MLT system in general yields $\Pi_1 = h(\Pi_2, \Pi_3)$, where each Π_i is a dimensionless product. The model predicts that Π_1 will remain constant if Π_2 and Π_3 are held constant, even though the components of Π_2 and Π_3 may themselves vary. Since there are, in general, an infinite number of ways of choosing the Π_i, one should choose those that can be controlled in laboratory experiments. Having determined that $\Pi_1 = h(\Pi_2, \Pi_3)$, one can isolate the effect of Π_2 by holding Π_3 constant and vice versa. This can help explain the functional relationship among the variables. For instance, we saw in our example that the period of the pendulum did not depend on the initial displacement for small displacements.

Perhaps the greatest contribution of dimensional analysis is that it reduces the number of experiments required to predict the behavior. If we wanted to conduct experiments to predict values of y for the assumed relationship $y = f(x_1, x_2, x_3, x_4, x_5)$ and it was decided that 5 data points would be necessary over the range of each variable, then 5^5 or 3125 experiments would be necessary. Since a two-dimensional chart is required to interpolate conveniently, y might be plotted against x_1 for five values of x_1, holding x_2, x_3, x_4, x_5 constant. Since x_2, x_3, x_4, and x_5 must vary as well, 5^4 or 625 charts would be necessary. However, after a dimensional analysis yields $\Pi_1 = h(\Pi_2, \Pi_3)$, only 25 data points would be required. Moreover, Π_1 can be plotted versus Π_2, for various values of Π_3, on a single chart. Ultimately an empirical model is usually constructed for purposes of interpolation, but the task is far easier after applying a dimensional analysis.

Finally, dimensional analysis is helpful in presenting the results. It is usually best to present experimental results using those Π_i that are "classical" within the field of study. For instance, in the field of fluid mechanics there are eight factors that might be significant in a particular situation. These factors are velocity v, length r, mass density ρ, viscosity μ, acceleration of gravity g, speed of sound c, surface tension σ, and pressure p. Thus a dimensional analysis could require as many as five independent dimensionless products. The five generally used are the Reynolds number, Froude number, Mach number, Weber number, and the pressure coefficient. These numbers are defined as follows (and discussed in Section 7.5):

Reynolds number	$\dfrac{vr\rho}{\mu}$
Froude number	$\dfrac{v^2}{rg}$
Mach number	$\dfrac{v}{c}$
Weber number	$\dfrac{\rho v^2 r}{\sigma}$
Pressure coefficient	$\dfrac{p}{\rho v^2}$

Thus, the application of dimensional analysis becomes quite easy. Depending upon which of the eight variables are considered in a particular problem, the following steps are performed:

1. Choose an appropriate subset from the preceding five dimensionless products.
2. Apply Buckingham's theorem.
3. Test the reasonableness of the choice of variables.
4. Conduct the necessary experiments and present the results in a useful format.

We illustrate an application of these steps to a fluid mechanics problem in Section 7.5.

7.3 PROBLEMS

1. For the damped pendulum,
    a. Assume that F is proportional to v^2 and use dimensional analysis to show that $t = \sqrt{r/g}\, h(\theta, rk_1/m)$.
    b. Assume that F is proportional to v^2 and describe an experiment to test the model $t = \sqrt{r/g}\, h(\theta, rk_1/m)$.
2. Under appropriate conditions, all three models for the pendulum imply that t is proportional to $\sqrt{r}$. Explain how the conditions distinguish between the three models by considering how m must vary in each case.
3.* Use a model employing a differential equation to predict the period of a simple frictionless pendulum for small initial angles of displacement. (Hint: Let $\sin \theta = \theta$.) Under these conditions, what should be the constant of proportionality? Compare your results with those predicted by Model A in the text.

* For students who have had differential equations.

7.4 EXAMPLES ILLUSTRATING DIMENSIONAL ANALYSIS

MODEL

Example 1 Explosion Analysis*

In excavation and mining operations it is important to be able to predict the size of a crater resulting from a given explosive like TNT in some particular soil medium. Direct experimentation is often impossible or too costly. Thus it is desirable to use small laboratory or field tests and then scale these up in some manner in order to predict the results for explosions far greater in magnitude.

You may wonder how the modeler determines which variables to include in the initial list. Experience is necessary in order to determine intelligently which variables can be neglected. Even with experience, however, the task is usually difficult in practice, as this example will illustrate. It also illustrates that the modeler often must change the list of variables in order to get usable results.

Problem identification   *Predict the crater volume V produced by a spherical explosive located at some depth d in a particular soil medium.*

Assumptions and model formulation   Initially, let's assume that the craters are geometrically similar and that the crater size depends on three variables: the radius r of the crater, the density ρ of the soil, and the mass W of the explosive. These three variables are composed of only two primary dimensions, length L and mass M, and a dimensional analysis results in only one dimensionless product (see Problem 1a):

$$\Pi_r = r\left(\frac{\rho}{W}\right)^{1/3}$$

According to Buckingham's theorem, Π_r must equal a constant. Thus the

---

* This example is adapted with permission from R. M. Schmidt, "A Centrifuge Cratering Experiment: Development of a Gravity-scaled Yield Parameter," in *Impact and Explosion Cratering*, ed. D. J. Roddy, R. O. Pepin, and R. B. Merrill (New York: Pergamon Press, 1977), pp. 1261–1276.

crater dimensions of radius or depth vary with the cube root of the mass of the explosive. Since the crater volume is proportional to r^3, it follows that the volume of the crater is proportional to the mass of the explosive for constant soil density. Symbolically, we have

$$V \propto \frac{W}{\rho} \tag{7.20}$$

Experiments have shown that the proportionality (7.20) is satisfactory for small explosions (less than 300 lb of TNT) at zero depth in soils, such as moist alluvium, that have good cohesion. However, the rule proves unsatisfactory for larger explosions. Other experiments suggest that gravity plays a key role in the explosion process, and since we want to consider extraterrestrial craters as well, we need to incorporate gravity as a variable.

If gravity is taken into account, then we assume crater size to be dependent on four variables: crater radius r, density of the soil ρ, gravity g, and charge energy E. Here the charge energy is the mass W of the explosive times its specific energy. Applying a dimensional analysis to these four variables again leads to a single dimensionless product (see Problem 1b):

$$\Pi_{rg} = r \left(\frac{\rho g}{E} \right)^{1/4}$$

Thus Π_{rg} equals a constant and the linear crater dimensions (radius or depth of crater) vary with the one-fourth root of the energy (or mass) of the explosive for constant soil density. This leads to the following proportionality known as quarter-root scaling and is a special case of *gravity scaling*:

$$V \propto \left(\frac{E}{\rho g} \right)^{3/4} \tag{7.21}$$

Experimental evidence indicates that gravity scaling holds for large explosions (greater than 100 tons of TNT) where the stresses in the cratering process are much larger than material strengths of the soil. The proportionality (7.21) predicts that crater volume decreases with increased gravity. The effect of gravity on crater formation is of interest in the study of extraterrestrial craters. Gravitational effects can be tested experimentally using a centrifuge to increase gravitational accelerations.*

A question of interest to explosion analysts is whether the material properties of the soil do become less important with increased charge size as well as with increased gravity. Let's consider the case where the soil medium is characterized only by its density ρ. Thus, the crater volume V depends on the explosive, soil density ρ, gravity g, and the depth of burial d of the charge. In addition, the explicit role of material strength or cohesion has been tested

---

* See the papers by R. M. Schmidt (1977, 1980) and by Schmidt and Holsapple (1980), cited in Further Reading, which discuss the effects when a centrifuge is used to perform explosive cratering tests under the influence of gravitational accelerations up to 480 G, where 1 G is the terrestrial gravity field strength of 981 cm/sec^2.

and the strength–gravity transition is shown to be a function of charge size and soil strength. (See Holsapple and Schmidt, 1979.)

We now describe our explosive in more detail than in the previous models. In order to characterize an explosive, three independent variables are needed: size, energy yield, and explosive density δ. The size can be given as charge mass W, as charge energy E, or as the radius α of the spherical explosive. The energy yield can be given as a measure of the specific energy Q_e or the energy density per unit volume Q_V. The following equations relate these variables:

$$W = \frac{E}{Q_e}$$

$$Q_V = \delta Q_e$$

$$\alpha^3 = \left(\frac{3}{4\pi}\right)\left(\frac{W}{\delta}\right)$$

One choice of these variables leads to the model formulation:

$$V = f(W, Q_e, \delta, \rho, g, d)$$

Since there are seven variables under consideration and the MLT system is being used, a dimensional analysis generally will result in four $(7-3)$ dimensionless products. The dimensions of the variables give the table:

Variable	V	W	Q_e	δ	ρ	g	d
Dimension	L^3	M	L^2T^{-2}	ML^{-3}	ML^{-3}	LT^{-2}	L

Any product of the variables must be of the form:

$$V^a W^b Q_e^{\,c} \delta^e \rho^f g^k d^m \tag{7.22}$$

and hence have dimension:

$$(L^3)^a (M)^b (L^2 T^{-2})^c (ML^{-3})^{e+f} (LT^{-2})^k (L)^m$$

Therefore, a product of the form (7.22) is dimensionless if and only if the exponents satisfy the following homogeneous system of equations:

$$
\begin{aligned}
M: &\quad b \quad\quad\; + e + f \quad\quad\quad\;\; = 0 \\
L: &\quad 3a \;\; + 2c - 3e - 3f + k + m = 0 \\
T: &\quad\quad\quad\; -2c \quad\quad\quad -2k \quad\;\; = 0
\end{aligned}
$$

Solution of this system produces

$$b = \frac{k-m}{3} - a, \quad c = -k, \quad e = a - f + \frac{m-k}{3}$$

where a, f, k, and m are arbitrary. By setting one of these arbitrary exponents equal to 1 and the other three equal to 0, in succession, we obtain the following set of dimensionless products:

$$\frac{V\delta}{W}, \quad \left(\frac{g}{Q_e}\right)\left(\frac{W}{\delta}\right)^{1/3}, \quad d\left(\frac{\delta}{W}\right)^{1/3}, \quad \frac{\rho}{\delta}$$

(Convince yourself that these products are dimensionless.) Since the dimensions of ρ and δ are equal, we can rewrite these dimensionless products as follows:

$$\Pi_1 = \frac{V\rho}{W}$$

$$\Pi_2 = \left(\frac{g}{Q_e}\right)\left(\frac{W}{\delta}\right)^{1/3}$$

$$\Pi_3 = d\left(\frac{\rho}{W}\right)^{1/3}$$

$$\Pi_4 = \frac{\rho}{\delta}$$

so that Π_1 is consistent with the dimensionless product implied by (7.20). Then, applying Buckingham's theorem, we obtain the model:

$$h(\Pi_1, \Pi_2, \Pi_3, \Pi_4) = 0 \tag{7.23}$$

or,

$$V = \frac{W}{\rho} H\left(\frac{gW^{1/3}}{Q_e\delta^{1/3}}, \frac{d\rho^{1/3}}{W^{1/3}}, \frac{\rho}{\delta}\right)$$

Presenting the results   For oil base clay the value of ρ is about 1.53 g/cm³; for wet sand, 1.65; and for desert alluvium, 1.60. For TNT, δ has the value 2.23 g/cm³. Thus, $.69 < \Pi_4 < .74$, so for simplicity we can assume for these soils and TNT that Π_4 is constant. Then (7.23) becomes

$$h(\Pi_1, \Pi_2, \Pi_3) = 0 \tag{7.24}$$

R. M. Schmidt gathered experimental data to plot the surface described by (7.24). A plot of the surface is depicted in Figure 7-12, showing the crater volume parameter Π_1 as a function of the scaled energy charge Π_2 and depth of burial parameter Π_3. Cross-sectional data for the surface parallel to the $\Pi_1\Pi_3$-plane when $\Pi_2 = 1.15 \times 10^{-6}$ are depicted in Figure 7-13.

Experiments have shown that the physical effect of increasing gravity is to reduce crater volume for a given charge yield. This result suggests that increased gravity can be compensated for by increasing the size of the charge in order to maintain the same cratering efficiency. Note also that both Figures 7-12 and 7-13 can be used for prediction once an empirical interpolating model is constructed from the data. Other recommended reading includes the extension of these methods to impact cratering, by Holsapple and Schmidt (1982); and to crater ejecta scaling, by Housen, Schmidt, and Holsapple (1983).

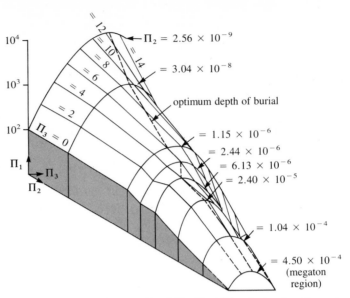

FIGURE 7-12 A plot of the surface $h(\Pi_1, \Pi_2, \Pi_3) = 0$, showing the crater volume parameter Π_1 as a function of gravity-scaled yield Π_2 and depth of burial parameter Π_3. Reprinted by permission of R. M. Schmidt.

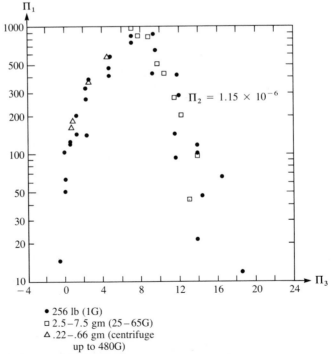

FIGURE 7-13 Data values for a cross section of the surface depicted in Figure 7-12. Data from R. M. Schmidt reprinted by permission.

Example 2 How Long Should You Roast a Turkey?

One rule of thumb for roasting a turkey is the following: set the oven to 400°F and allow 20 minutes per pound for cooking. How good is this rule?

Assumptions Let t denote the cooking time for the turkey. Now, upon what variables does t depend? Certainly the size of the turkey is a factor that must be considered. Let's assume that the turkeys are geometrically similar and use l to denote some characteristic dimension of the meat to be cooked; specifically, we assume that l represents the length of the turkey. Another factor is the difference between the temperatures of the raw meat and the oven, ΔT_m. (We know from experience that it takes longer to cook a bird that is nearly frozen than it does to cook one that is initially at room temperature.) Since the turkey will have to reach a certain interior temperature before it is considered fully cooked, the difference ΔT_c between the temperature of the cooked meat and the oven is a variable determining the cooking time. Finally, we know that different foods require different cooking times quite independent of size: it only takes 10 minutes or so to bake a pan of cookies whereas roast beef or turkey requires several hours. A measure of the factor representing these differences between foods is the *coefficient of heat conduction* for the particular food to be cooked. Let κ denote the coefficient of heat conduction for turkey. Thus, we have the following model formulation for the cooking time:

$$t = f(\Delta T_m, \Delta T_c, \kappa, l)$$

Dimensional analysis Consider now the dimensions of the independent variables. The temperature variables ΔT_m and ΔT_c measure energy per volume and therefore have dimension ML^2T^{-2}/L^3, or simply, $ML^{-1}T^{-2}$. Now what about the heat conduction variable κ? **Thermal conductivity** κ is defined as the amount of energy crossing one unit cross-sectional area per second divided by the temperature gradient perpendicular to the area. That is,

thermal conductivity

$$\kappa = \frac{\text{energy/(area} \times \text{time)}}{\text{temperature/length}}$$

Accordingly, the dimension of κ is $(ML^2T^{-2})(L^{-2}T^{-1})/(ML^{-1}T^{-2})(L^{-1})$, or simply, L^2T^{-1}. Our analysis gives the following table:

Variable	ΔT_m	ΔT_c	κ	l	t
Dimension	$ML^{-1}T^{-2}$	$ML^{-1}T^{-2}$	L^2T^{-1}	L	T

Any product of the variables must be of the form:

$$\Delta T_m{}^a \Delta T_c{}^b \kappa^c l^d t^e \tag{7.25}$$

and hence have dimension:

$$(ML^{-1}T^{-2})^a(ML^{-1}T^{-2})^b(L^2T^{-1})^c(L)^d(T)^e$$

Therefore, a product of the form (7.25) is dimensionless if and only if the exponents satisfy

$$
\begin{aligned}
M: & \quad a + b & = 0 \\
L: & \quad -a - b + 2c + d & = 0 \\
T: & \quad -2a - 2b - c + e & = 0
\end{aligned}
$$

Solution of this system of equations gives

$$a = -b, \quad c = e, \quad d = -2e$$

where b and e are arbitrary. If we set $b = 1$, $e = 0$, we obtain $a = -1$, $c = 0$, and $d = 0$; likewise, $b = 0$, $e = 1$ produces $a = 0$, $c = 1$, and $d = -2$. These independent solutions yield the complete set of dimensionless products:

$$\Pi_1 = \Delta T_m^{-1} \Delta T_c \quad \text{and} \quad \Pi_2 = \kappa l^{-2} t$$

From Buckingham's theorem we obtain

$$h(\Pi_1, \Pi_2) = 0$$

or,

$$t = \left(\frac{l^2}{\kappa}\right) H\left(\frac{\Delta T_c}{\Delta T_m}\right) \tag{7.26}$$

Now the rule of thumb stated in our opening remarks gives the roasting time for the turkey in terms of its weight w. Let's assume the turkeys are geometrically similar, or $V \propto l^3$. If we assume the turkey is of constant density (which is not quite correct because the bones and flesh differ in density), then since weight is density times volume and volume is proportional to l^3, we get $w \propto l^3$. Moreover, if we set the oven to a constant baking temperature and specify that the turkey must initially be near room temperature (65°F), then $\Delta T_c / \Delta T_m$ is a dimensionless constant. Combining these results with (7.26), we get the proportionality

$$t \propto w^{2/3} \tag{7.27}$$

since κ is constant for turkeys. Thus, the required cooking time is proportional to the weight raised to the two-thirds power. Therefore, if t_1 hours are required to cook a turkey weighing w_1 pounds and t_2 is the time for a weight of w_2 pounds

$$\frac{t_1}{t_2} = \left(\frac{w_1}{w_2}\right)^{2/3}$$

It follows that a doubling of the weight of a turkey increases the cooking time by the factor $2^{2/3} \approx 1.59$.

How does our result (7.27) compare to the rule of thumb stated previously? Assume that ΔT_m, ΔT_c, and κ are independent of the length or weight of the turkey, and consider cooking a 23-lb turkey versus an 8-lb bird. According to the rule of thumb, the ratio of cooking times is given by

$$\frac{t_1}{t_2} = \frac{20 \cdot 23}{20 \cdot 8} = 2.875$$

On the other hand, from dimensional analysis and (7.27),

$$\frac{t_1}{t_2} = \left(\frac{23}{8}\right)^{2/3} \approx 2.02$$

Thus, the rule of thumb predicts it will take nearly three times as long to cook a 23-lb bird as it will to cook an 8-lb turkey. Dimensional analysis predicts it will take only about twice as long. Which rule is correct? Why have so many cooks overcooked a turkey?

Testing the results Suppose that various sized turkeys are cooked in an oven preheated to 325°F. The initial temperature of all turkeys is 65°F. All turkeys are removed from the oven when their internal temperature, as measured by a meat thermometer, reaches 195°F. The (hypothetical) cooking times for the various turkeys are recorded as follows:

w (lb)	5	10	15	20
t (hr)	2	3.4	4.5	5.4

A plot of t versus $w^{2/3}$ is shown in Figure 7-14. Since the graph approximates a straight line through the origin, we conclude that $t \propto w^{2/3}$, as predicted by our model.

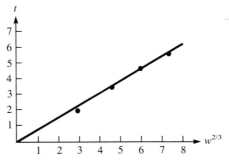

FIGURE 7-14 Plot of cooking times versus weight to the two-thirds power reveals the predicted proportionality.

7.4 PROBLEMS

1. **a.** Use dimensional analysis to establish the cube-root law:

$$r\left(\frac{\rho}{W}\right)^{1/3} = \text{constant}$$

for scaling of explosions, where r is the radius or depth of the crater, ρ the density of the soil medium, and W the mass of the explosive.

   b. Use dimensional analysis to establish the one-fourth root law:

$$r\left(\frac{\rho g}{E}\right)^{1/4} = \text{constant}$$

for scaling of explosions, where r is the radius or depth of the crater, ρ the density of the soil medium, g is gravity, and E is the charge energy of the explosive.

2. **a.** Show that the products $\Pi_1, \Pi_2, \Pi_3, \Pi_4$ for the refined explosion model presented in the text are dimensionless products.

   b. Assume ρ is essentially constant for the soils being used and restrict the explosive to a specific type, say TNT. Under these conditions ρ/δ is essentially constant, yielding

$$\Pi_1 = f(\Pi_2, \Pi_3)$$

You have collected the following data with $\Pi_2 = 1.5 \times 10^{-6}$:

Π_3	0	2	4	6	8	10	12	14
Π_1	15	150	425	750	825	425	250	90

   i. Construct a scatterplot of Π_1 versus Π_3. Does a trend exist?

   ii. How accurate do you think the data are? Find an empirical model that captures the *trend* of the data with accuracy commensurate with your appraisal of the accuracy of the data.

   iii. Use your empirical model to predict the volume of a crater using TNT in desert alluvium with (CGS system) $W = 1500$ g, $\rho = 1.53$ g/cm^3, and $\Pi_3 = 12.5$.

3. Consider a zero-depth burst, spherical explosive, in a soil medium. Assume the value of the crater volume V depends on the explosive size, energy yield, and explosive energy, as well as on the strength Y of the soil (considered as a resistance to pressure with dimensions $ML^{-1}T^{-2}$), soil density ρ, and gravity g. In this problem, assume

$$V = f(W, Q_e, \delta, Y, \rho, g)$$

and use dimensional analysis to produce the following *mass set* of dimensionless products:

$$\Pi_1 = \frac{V\rho}{W} \qquad \Pi_2 = \left(\frac{g}{Q_e}\right)\left(\frac{W}{\delta}\right)^{1/3}$$

$$\Pi_3 = \frac{Y}{\delta Q_e} \qquad \Pi_4 = \frac{\rho}{\delta}$$

4. For the explosion process and material characteristics discussed in Problem 3, consider

$$V = f(E, Q_V, \delta, Y, \rho, g)$$

and use dimensional analysis to produce the following *energy set* of dimensionless products:

$$\bar{\Pi}_1 = \frac{VQ_V}{E} \qquad \bar{\Pi}_2 = \frac{\rho g E^{1/3}}{Q_V^{4/3}}$$

$$\bar{\Pi}_3 = \frac{Y}{Q_V} \qquad \bar{\Pi}_4 = \frac{\rho}{\delta}$$

5. Repeat Problem 4 for

$$V = f(E, Q_e, \delta, Y, \rho, g)$$

to produce the *gravity set* of dimensionless products:

$$\bar{\bar{\Pi}}_1 = V\left(\frac{\rho g}{E}\right)^{3/4} \qquad \bar{\bar{\Pi}}_2 = Q_e^{-1}\left(\frac{g^3 E}{\delta}\right)^{1/4}$$

$$\bar{\bar{\Pi}}_3 = \frac{Y}{\delta Q_e} \qquad \bar{\bar{\Pi}}_4 = \frac{\rho}{\delta}$$

6. An experiment consists of dropping spheres into a tank of heavy oil and measuring the times of descent. It is desired that a relationship for time of descent be determined and verified by experimentation. Assume the time of descent is a function of mass m, gravity g, radius r, viscosity μ, and distance traveled d. Fluid density is to be neglected. That is,

$$t = f(m, g, r, \mu, d)$$

   a. Using dimensional analysis, find a relationship for the time of descent.

   b. How will the spheres be chosen to verify that the time of descent relationship is independent of fluid density? Assuming that you have verified the assumption on fluid density, describe how you would determine the nature of your function experimentally.

   c.* Using differential equations techniques, find the velocity of the sphere as a function of time, radius, mass, viscosity, gravity, and fluid density.

---

* For students who have had differential equations.

Using this result and that found in Part a, predict under what conditions fluid density may be neglected. (Hints: Use the results of Problem 5 of Section 7.2 as a submodel for drag force. Consider the buoyant force.)

7. A windmill is being rotated by air flow to produce power to pump water. It is desired to find the power output P of the windmill. Assume that P is a function of the density of the air ρ, viscosity of the air μ, diameter of the windmill d, wind speed v, and the rotational speed of the windmill ω (measured in radians per second). Thus,

$$P = f(\rho, \mu, d, v, \omega)$$

   a. Using dimensional analysis, find a relationship for P. Be sure to check your products to make certain they are dimensionless.
   b. Do your results make common sense? Explain.
   c. Discuss how you would design an experiment to determine the nature of your function.

8. For a sphere traveling through a liquid, assume that the drag force F_D is a function of the fluid density ρ, fluid viscosity μ, radius of the sphere r, and speed of the sphere v. Use dimensional analysis to find a relationship for the drag force

$$F_D = f(\rho, \mu, r, v)$$

Make sure you provide some justification that the given independent variables influence the drag force.

7.4 PROJECTS

1. Complete the requirements for the module, "Listening to the Earth: Controlled Source Seismology," by Richard G. Montgomery, UMAP 292, 293. This module develops the elementary theory of wave reflection and refraction and applies it to a model of the earth's subsurface. The model shows how information on layer depth and sound velocity may be obtained to provide data on width, density, and composition of the subsurface. This module is a good introduction to controlled seismic methods and requires no previous knowledge of either physics or geology.

7.4 FURTHER READING

Holsapple, K. A., and R. M. Schmidt. "A Material-strength Model for Apparent Crater Volume," *Proc. Lunar Planet Sci. Conf.* no. 10 (1979): 2757–2777.

Holsapple, K. A., and R. M. Schmidt. "On the Scaling of Crater Dimensions-2: Impact Processes," *J. Geophys. Res.* no. 87 (1982): 1849–1870.

Housen, K. R., R. M. Schmidt, and K. A. Holsapple. "Crater Ejecta Scaling Laws 1: Fundamental Forms Based on Dimensional Analysis," *J. Geophys. Res.* no. 88 (1983): 2485–2499.

Schmidt, R. M. "A Centrifuge Cratering Experiment: Development of a Gravity-scaled Yield Parameter," in *Impact and Explosion Cratering*, edited by D. J. Roddy, et al., pp. 1261–1278. New York: Pergamon Press, 1977.

Schmidt, R. M. "Meteor Crater: Energy of Formation—Implications of Centrifuge Scaling," *Proc. Lunar Planet.* Sci. Conf. 11 (1980): 2099–2128.

Schmidt, R. M., and K. A. Holsapple. "Theory and Experiments on Centrifuge Cratering," *J. Geophys. Res.* no. 85 (1980): 235–252.

MODEL

7.5 SIMILITUDE

Suppose we are interested in the effects of wave action on a large ship at sea, or the heat loss of a submarine and the drag force it experiences in its underwater environment, or the wind effects on an aircraft wing. Quite often since it is physically impossible to duplicate the actual phenomenon in the laboratory, we study a scaled-down model in a simulated environment to predict accurately the performance of the physical system. The actual physical system for which the predictions are to be made is called the **prototype.** But how do we scale experiments in the laboratory to ensure that the effects observed for the model will be the same effects experienced by the prototype?

prototype

While extreme care must be exercised in using simulations, the dimensionless products resulting from a dimensional analysis of the problem can provide insight into how the scaling for a model should be done. The idea comes from Buckingham's theorem. If the physical system can be described by a dimensionally homogeneous equation in the variables, then it can be put into the form:

$$f(\Pi_1, \Pi_2, \ldots, \Pi_n) = 0$$

for a complete set of dimensionless products. Assume that the dependent variable of the problem appears only in the product Π_n and that

$$\Pi_n = H(\Pi_1, \Pi_2, \ldots, \Pi_{n-1})$$

In order for the solution to the model and the prototype to be the same, it is sufficient that the value of all the independent dimensionless products Π_1, $\Pi_2, \ldots, \Pi_{n-1}$ be the same for the model and the prototype.

For example, suppose the Reynolds number $vr\rho/\mu$ appears as one of the dimensionless products in a fluid mechanics problem, where v represents fluid velocity, r a characteristic dimension (like the diameter of a sphere or the length of a ship), ρ the fluid density, and μ the fluid viscosity. These values refer to the prototype. Next let v_m, r_m, ρ_m, and μ_m denote the corresponding values for the scaled-down model. In order for the effects on the model and the prototype to be the same, we want their two Reynolds numbers to agree so that

$$\frac{v_m r_m \rho_m}{\mu_m} = \frac{vr\rho}{\mu}$$

This last equation is referred to as a *design condition* to be satisfied by the model. If the length of the prototype is too large for the laboratory experiment so that we have to scale down the length of the model, say $r_m = r/10$, then the same Reynolds number for model and prototype can be achieved by using the same fluid ($\rho_m = \rho$ and $\mu_m = \mu$) and varying the velocity, $v_m = 10v$. If it is impractical to scale the velocity by the factor 10, we can instead scale it by a lesser amount $0 < k < 10$ and use a different fluid so that the equation:

$$\frac{k\rho_m}{10\mu_m} = \frac{\rho}{\mu}$$

is satisfied. We do need to be careful in generalizing the results from the scaled-down model to the prototype. Certain factors (like surface tension, for instance) that may be negligible for the prototype may become significant for the model. Such factors would have to be taken into account before making any predictions for the prototype.

An Example: The Drag Force on a Submarine

We are interested in the drag forces experienced by a submarine to be used for deep-sea oceanographic explorations. We assume that the variables affecting the drag D are fluid velocity v, characteristic dimension r (here, the length of the submarine), fluid density ρ, fluid viscosity μ, and the velocity of sound in the fluid c. We wish to predict the drag force by studying a model of the prototype. But how shall the experiments for the model be scaled?

A major stumbling block in our problem is in describing shape factors related to the physical object being modeled, in this case, the submarine. Let's consider submarines that are ellipsoidal in shape. In two dimensions, if a is the length of the major axis and b is the length of the minor axis of an ellipse, we can define $r_1 = a/b$ and assign a characteristic dimension such as r, the length of the submarine (see Figure 3-9). In three dimensions, define also $r_2 = a/b'$, where a is the original major axis and b' is the second minor axis. Then r, r_1, and r_2 describe the shape of the submarine. In a more irregularly shaped object, additional shape factors would be required. The basic idea is that the object can be described using a characteristic dimension and an appropriate collection of shape factors. In the case of our three-dimensional ellipsoidal submarine, the shape factors r_1 and r_2 are needed. These shape factors are dimensionless constants.

Returning to our list of six fluid mechanics variables D, v, r, ρ, μ, and c, notice that we are neglecting surface tension (because it is small) and that gravity is not being considered. Thus it is expected that a dimensional analysis will produce three $(6 - 3)$ independent dimensionless products. We can choose the following three products for convenience:

$$\text{Reynolds number} \qquad R = \frac{vr\rho}{\mu}$$

$$\text{Mach number} \qquad M = \frac{v}{c}$$

$$\text{Pressure coefficient} \qquad P = \frac{p}{\rho v^2}$$

The added shape factors are dimensionless, so that Buckingham's theorem gives the equation:

$$h(R, M, P, r_1, r_2) = 0$$

Assuming that we can solve for P yields

$$P = H(R, M, r_1, r_2)$$

Substituting $P = p/\rho v^2$ and solving for p gives

$$p = \rho v^2 H(R, M, r_1, r_2)$$

Remembering that the total drag force is the pressure (force per unit area) times the area (which is proportional to r^2 for geometrically similar objects) gives the proportionality $D \propto pr^2$, or

$$D = k\rho v^2 r^2 H(R, M, r_1, r_2) \tag{7.28}$$

Now a similar equation must hold to give the same proportionality for the model:

$$D_m = k\rho_m v_m{}^2 r_m{}^2 H(R_m, M_m, r_{1m}, r_{2m}) \tag{7.29}$$

Since the prototype and model equations refer to the same physical system, both equations are identical in form. Therefore, the design conditions for the model require that

Condition (a)	R_m	$= R$
Condition (b)	M_m	$= M$
Condition (c)	r_{1m}	$= r_1$
Condition (d)	r_{2m}	$= r_2$

Note that if conditions (a), (b), (c), and (d) are satisfied, then Equations (7.28) and (7.29) give

$$\frac{D_m}{D} = \frac{\rho_m v_m^2 r_m^2}{\rho v^2 r^2} \tag{7.30}$$

Thus D can be computed once D_m is measured. Note that design conditions (c) and (d) imply geometric similarity between the model and the prototype submarine:

$$\frac{a_m}{b_m} = \frac{a}{b} \quad \text{and} \quad \frac{a_m}{b_m'} = \frac{a}{b'}$$

If the velocities are small compared to the speed of sound in the fluid, then v/c can be considered constant in accordance with condition (b). If the same fluid is used for both model and prototype, then condition (a) is satisfied if

$$v_m r_m = v r$$

or

$$\frac{v_m}{v} = \frac{r}{r_m}$$

which says that the velocity of the model must increase inversely as the "scaling factor" r_m/r. Under these conditions, Equation (7.30) yields

$$\frac{D_m}{D} = \frac{\rho_m v_m^2 r_m^2}{\rho v^2 r^2} = 1$$

If increasing the velocity of the scaled model proves unsatisfactory in the laboratory, then a different fluid may be considered for the scaled model ($\rho_m \neq \rho$ and $\mu_m \neq \mu$). If the ratio v/c is small enough to neglect, then both v_m and r_m can be varied to ensure that

$$\frac{v_m r_m \rho_m}{\mu_m} = \frac{v r \rho}{\mu}$$

in accordance with condition (a). Having chosen values that satisfy design condition (a), and knowing the drag on the scaled model, we can use Equation (7.30) to compute the drag on the prototype. Consider the additional difficulties if the velocities are sufficiently great that we must satisfy condition (b) as well.

A few comments are in order. One distinction between the Reynolds number and the other four numbers in fluid mechanics is that the Reynolds number contains the viscosity of the fluid. Dimensionally, the Reynolds number is proportional to the ratio of the inertia forces of an element of fluid to the viscous force acting on the fluid. In certain problems the numerical value of the Reynolds number may be significant. For example, the flow of a fluid in a pipe is virtually always parallel to the edges of the pipe (giving *laminar flow*) if the Reynolds number is less than 2000. Reynolds numbers in excess of 3000 almost always indicate turbulent flow. Normally there is a critical Reynolds number between 2000 and 3000 at which the flow becomes turbulent.

The design condition (a), mentioned earlier, requires that the Reynolds number of model and prototype be the same. This requirement precludes the possibility of laminar flow in the prototype being represented by turbulent flow in the model, and vice versa. The equality of the Reynolds number for model and prototype is important in all problems in which viscosity plays a significant role.

The Mach number is the ratio of fluid velocity to the speed of sound in the fluid. It is generally important for problems involving objects moving with a high speed in fluids, such as projectiles, high-speed aircraft, rockets, and submarines. Physically, if the Mach number is the same in model and prototype, the effect of the compressibility force in a fluid relative to the inertia force will be the same for model and prototype. This is the situation that is required by condition (b) in our example on the submarine.

7.5 PROBLEMS

1. A model of an airplane wing is being tested in a wind tunnel. The model wing has an 18-in. chord, and the prototype has a 4-ft chord moving at 250 mph. Assuming the air in the wind tunnel is at atmospheric pressure, at what velocity should the wind tunnel tests be conducted so that the Reynolds number of the model is the same as that of the prototype?

2. Two smooth balls of equal weight but different diameters are dropped from an airplane. The ratio of their diameters is 5. Neglecting compressibility (assume constant Mach number), what is the ratio of the terminal velocities of the balls? Are the flows completely similar?

3. Consider predicting the pressure drop Δp between two points along a smooth horizontal pipe under the condition of steady laminar flow. Assume

$$\Delta p = f(s, d, \rho, \mu, v)$$

where s is the control distance between two points in the pipe, d is the diameter of the pipe, ρ is the fluid density, μ is the fluid viscosity, and v is the velocity of the fluid.

   a. Determine the design conditions for a scaled model of the prototype.
   b. Must the model be geometrically similar to the prototype?
   c. May the same fluid be used for model and prototype?
   d. Show that if the same fluid is used for model and prototype, then the equation is

$$\Delta p = \frac{\Delta p_m}{n^2}$$

   where $n = d/d_m$.

4. It is desired to study the velocity v of a fluid flowing in a smooth open channel. Assume that

$$v = f(r, \rho, \mu, \sigma, g)$$

where r is the characteristic length of the channel cross-sectional area divided by the wetted perimeter, ρ is the fluid density, μ denotes viscosity, σ is the surface tension, and g is the acceleration of gravity.

a. Describe an appropriate pair of shape factors r_1 and r_2.

b. Show that

$$\frac{v^2}{gr} = H\left(\frac{\rho v r}{\mu}, \frac{\rho v^2 r}{\sigma}, r_1, r_2\right)$$

Discuss the design conditions required of the model.

c. Will it be practical to use the same fluid in the model and the prototype?

d. Suppose the surface tension σ is ignored and the design conditions are satisfied. If $r_m = r/n$, what is the equation for the velocity of the prototype? When is this equation compatible with the design conditions?

e. What is the equation for the velocity v if gravity is ignored? What if viscosity is ignored? What fluid would you use if you were to ignore viscosity?

7.5 FURTHER READING

Langhaar, Henry L. *Dimensional Analysis and Theory of Models.* New York: Wiley, 1951.

Massey, Bernard S. *Units, Dimensional Analysis and Physical Similarity,* London: Van Nostrand Reinhold Company, 1971.

Murphy, Glenn. *Similitude In Engineering.* New York: Ronald Press, 1950.

SIMULATION MODELING

INTRODUCTION

In many situations a modeler is unable to construct an analytic (symbolic) model adequately explaining the behavior being observed because of its complexity or the intractability of the proposed explicative models. Yet if it is necessary to make predictions about the behavior, the modeler may conduct experiments (or otherwise gather data) to investigate the relationship between the dependent variable(s) and selected values of the independent variable(s) within some range. We constructed examples of empirical models based on collected data in Chapter 6. To collect the data, the modeler may observe the behavior directly, as in the "pace of life" example of Section 6.1, where the size of a population was predicted as a function of the people's walking speed. In other instances, the behavior might be duplicated (possibly in a scaled-down version) under controlled conditions, as was done when predicting the size of craters in Section 7.4.

In some circumstances, it may *not be feasible* either to observe the behavior directly or to conduct experiments. For instance, consider the service provided by a system of elevators during morning rush hour, as presented in Example 4 of Section 2.2. After identifying an appropriate problem and defining what is meant by "good service," we suggested some alternative delivery schemes, such as assigning elevators to even and odd floors or using express elevators. Theoretically, each alternative could be tested for some period of time to determine which one provides the best service for particular arrival and destination patterns of the customers. However, such a procedure would probably be very disruptive since it would be necessary to harass the customers constantly as the required statistics are collected. Moreover, the customers would become very confused because the elevator delivery system would keep changing. Another problem concerns testing alternative schemes for controlling automobile traffic in a large city. It would be impractical to constantly change directions of the one-way streets and the distribution of traffic signals in order to conduct tests.

In still other situations, the system for which the alternative procedures need to be tested *may not even exist* yet. An example is the situation of several proposed communication networks with the problem of determining which one is best for a given office building. Still another example is the problem of determining the locations of machines in a new industrial plant. The *cost* of conducting the experiments may also be prohibitive. This is the case when trying to predict the effects of various procedures for controlling the spread of a contagious disease, or the various alternatives for protecting and evacuating the population in case of a failure of a nuclear power plant.

In instances where the behavior cannot be explained analytically, or data collected directly, the modeler might *simulate* the behavior indirectly in some manner, and then test the various alternatives being considered to estimate how each affects the behavior. Data can then be collected to determine which alternative is best. For example, in Section 7.5, we wished to determine the drag force on a proposed design for a submarine. Since it was judged infeasible to build a prototype, we decided to build a scaled model to simulate the behavior of an actual submarine. Another example of this type of simulation is using a scaled model of a jet airplane in a wind tunnel to estimate the effects of very high speeds for various designs of the aircraft. There is yet another type of simulation, which you will study in this chapter. This method

Monte Carlo simulation for simulating behavior is called **Monte Carlo simulation** and is typically accomplished with the aid of a computer.

Suppose we are investigating the service provided by a system of elevators at morning rush hour. In Monte Carlo simulation the arrival of customers at the elevators during the hour and the destination floors they select need to be replicated. That is, the distribution of arrival times and the distribution of floors desired on a simulated trial must portray a possible rush hour. Moreover, after we have simulated many trials, the daily distributions of arrivals and destinations that occur must mimic the real-world distributions in the proper proportions. Once we are satisfied that the behavior is adequately duplicated, various alternative strategies for operating the elevators can be investigated. Using a large number of trials, we can gather appropriate statistics, such as the average total delivery time of a customer or the length of the longest queue. These statistics can help determine the best strategy for operating the elevator system.

This chapter provides a brief introduction to Monte Carlo simulation. Additional studies in probability and statistics are required in order to delve into the deeper intricacies of computer simulation and understand its appropriate uses. Nevertheless you will gain here some appreciation of this powerful component of mathematical modeling. There is a danger in placing too much confidence in the predictions resulting from a simulation, especially if the assumptions inherent in the simulation are not clearly stated. Moreover, the appearance of using large amounts of data and huge amounts of computer time, coupled with the fact that laymen can understand a simulation model and computer output with relative ease, often leads to overconfidence in the results.

When any Monte Carlo simulation is performed, random numbers are used. We discuss how to generate random numbers in the Projects in Section 8.1. Loosely speaking, a "sequence of random numbers uniformly distributed in an interval m to n" is a set of numbers with no apparent pattern, where each number between m and n can appear with equal likelihood. For example, if you toss a six-sided die 100 times and write down the number showing on the die each time, you will have written down a sequence of 100 random integers approximately uniformly distributed over the interval 1 to 6. Now, suppose random numbers consisting of six digits can be generated. The tossing of a coin can be duplicated by generating a random number and assigning a head if the random number is even and a tail if it is odd. If this trial is replicated a large number of times, you would expect heads to occur about 50% of the time. However, there is an element of *chance* involved. It *is* possible that a run of 100 trials could produce by chance 51 heads and that the next 10 trials (although not very likely) produce all heads. Thus, the estimate with 110 trials is actually worse than the estimate with **probabilistic** 100 trials. Processes with an element of chance involved are called **prob-** **deterministic** **abilistic,** as opposed to **deterministic,** processes. Monte Carlo simulation is therefore a probabilistic model.

The behavior being modeled may be either deterministic or probabilistic. For instance, the area under a curve is deterministic (even though it may be impossible to find it precisely). On the other hand, the time between arrivals of customers at the elevator on a particular day is probabilistic behavior. Referring to Figure 8-1, you can see that a deterministic model can be used to approximate either a deterministic or a probabilistic behavior and, likewise, a Monte Carlo simulation can be used to approximate a deterministic behavior (as you will see with a Monte Carlo approximation to the area under a curve) or a probabilistic one. But, as you would expect, the real power of Monte Carlo simulation lies in modeling a probabilistic behavior.

A principal advantage of Monte Carlo simulation is the relative ease with which it can sometimes be used to approximate very complex probabilistic systems. Additionally, Monte Carlo simulation provides performance estimation over a wide range of conditions, rather than a very restricted range as often required by an analytic model. Furthermore, because a particular submodel can be changed rather easily in a Monte Carlo simulation (such as the arrival and destination patterns of customers at the elevators), there is the potential of conducting a sensitivity analysis. Still another advantage is that

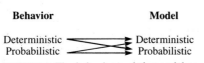

FIGURE 8-1 The behavior and the model can be either deterministic or probabilistic.

the modeler has control over the level of detail in a simulation. For example, a very long time frame can be compressed or a small time frame expanded, giving a great advantage over experimental models. Finally, there are very powerful, high-level simulation languages (such as GPSS, GASP and DYNAMO) that eliminate much of the tedious labor in constructing a simulation model.

On the negative side, simulation models are typically very expensive to develop and operate. They may require many man-hours to construct as well as large amounts of computer time and memory when being run. Another disadvantage is that the probabilistic nature of a simulation model limits the conclusions that can be drawn from a particular run unless a sensitivity analysis is conducted. Such an analysis often requires many more runs just to consider a small number of combinations of conditions that can occur in the various submodels. This limitation then forces the modeler to estimate which combinations might occur for a particular set of conditions.

A simulation model may also be difficult to verify because of its many components and its probabilistic nature. The modeler must ensure that the random number generator is operating correctly and that the various submodels are simulated correctly. Typically, it is difficult to validate a simulation model using real-world data. (This point is particularly important when well-known probability distributions are assumed for various submodels. Each distribution needs to be checked for its appropriateness to the situation.) Nevertheless, the modeler may be forced to develop a simulation model precisely because it is impossible to obtain data (making model validation impossible anyway). Finally, even though a simulation correctly estimates which of the various alternatives seems best, it still cannot provide an optimal solution (as would an optimization model) because all the possible alternatives have not been considered. So considerable judgment is required to determine which alternatives to simulate.

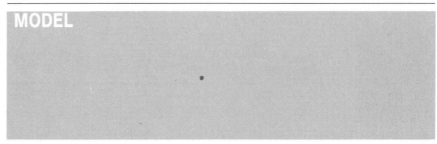

MODEL

8.1 MODELING DETERMINISTIC BEHAVIOR: AREA UNDER A CURVE

In this section we illustrate the use of Monte Carlo simulation to model a deterministic behavior, the area under a curve. We begin by finding an approximate value to the area under a nonnegative curve. Specifically, suppose $y = f(x)$ is some given continuous function satisfying $0 \leqslant f(x) < M$ over the closed interval $a \leqslant x \leqslant b$. Here the number M is simply some constant that *bounds* the function. This situation is depicted in Figure 8-2, below. Notice that the area we seek is wholly contained within the rectangular region of height M and length $b - a$ (the length of the interval over which f is defined).

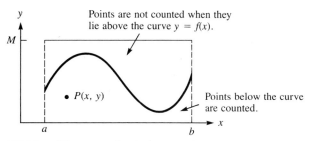

FIGURE 8-2 The area under the nonnegative curve $y = f(x)$ over $a \leqslant x \leqslant b$ is contained within the rectangle of height M and base length $b - a$.

Now select a point $P(x, y)$ at random from within the rectangular region. You do so by generating two random numbers, x and y, satisfying $a \leqslant x \leqslant b$ and $0 \leqslant y \leqslant M$, and interpreting them as a point P with coordinates x and y. Once $P(x, y)$ is selected, ask yourself if it lies within the region below the curve; that is, does the y-coordinate satisfy $0 \leqslant y \leqslant f(x)$? If the answer is yes, then *count* the point P by adding 1 to some counter. Two counters will be necessary: one to count the total number of points generated and a second to count those points that lie below the curve (see Figure 8-2). You can then calculate an approximate value for the area under the curve by the following formula:

$$\frac{\text{area under curve}}{\text{area of rectangle}} \approx \frac{\text{number of points counted below curve}}{\text{total number of random points}}$$

As discussed in the introduction, the Monte Carlo technique is probabilistic and typically requires a large number of trials before the deviation between the predicted and true values becomes small. A discussion of the number of trials needed to ensure a predetermined level of confidence in the final estimate requires a background in statistics. However, as a general rule, in order to double the accuracy of the result (that is, cut the expected error in half) about four times as many experiments are necessary.

The following algorithm gives the sequence of calculations needed for a general computer simulation of this Monte Carlo technique for finding the area under a curve.

Monte Carlo Area Algorithm

Input   Total number n of random points to be generated in the simulation.

Output   AREA = approximate area under the specified curve $y = f(x)$ over the given interval $a \leqslant x \leqslant b$, where $0 \leqslant f(x) < M$.

Step 1   Initialize: COUNTER = 0.

Step 2   For $i = 1, 2, \ldots, n$, do Steps 3–5.

Step 3   Calculate random coordinates x_i and y_i, satisfying $a \leqslant x_i \leqslant b$ and $0 \leqslant y_i < M$.

Step 4   Calculate $f(x_i)$ for the random x_i coordinate.

Step 5   If $y_i \leqslant f(x_i)$, then increment the COUNTER by 1. Otherwise, leave COUNTER as is.

Step 6   Calculate AREA = $M(b - a)$ COUNTER$/n$.

Step 7   OUTPUT (AREA)
           STOP

Table 8-1 gives the results of several different simulations to obtain the area beneath the curve $y = \cos x$ over the interval $-\pi/2 \leqslant x \leqslant \pi/2$, where $0 \leqslant \cos x < 2$.

TABLE 8-1 Monte Carlo approximation to the area under the curve $y = \cos x$ over the interval $-\pi/2 \leqslant x \leqslant \pi/2$

Number of points	Approximation to area	Number of points	Approximation to area
100	2.07345	2000	1.94465
200	2.13628	3000	1.97711
300	2.01064	4000	1.99962
400	2.12058	5000	2.01439
500	2.04832	6000	2.02319
600	2.09440	8000	2.00669
700	2.02857	10000	2.00873
800	1.99491	15000	2.00978
900	1.99666	20000	2.01093
1000	1.96664	30000	2.01186

The actual area under the curve $y = \cos x$ over the given interval is 2 square units. Note that even with the relatively large number of points generated, the error is significant. For functions of one variable, the Monte Carlo technique is generally not competitive with quadrature techniques that you will learn in numerical analysis. The lack of an error bound and the difficulty in finding an upper bound M are disadvantages as well. Nevertheless, the Monte Carlo technique can be extended to functions of several variables (as suggested in the following problem set) and becomes more practical in that situation.

8.1 PROBLEMS

1. Each ticket in a lottery contains a single "hidden" number according to the following scheme: 55% of the tickets contain a "1," 35% contain a "2," and 10% contain a "3." A participant in the lottery wins a prize by obtaining all three numbers 1, 2, and 3. Describe an experiment that could be used to determine how many tickets you would expect to buy in order to win a prize.

2. Two different record companies, A and B, both produce classical music recordings. Label A is a "budget" label, and 5% of A's new records exhibit a significant degree of warpage. Label B is manufactured under tighter quality-control (and consequently is more expensive) than A, so only 2% of its recordings are warped. You purchase one label A and one label B recording at your local store on a regular basis. Describe an experiment that could be used to determine how many times you would expect to make such a purchase before buying two warped records for a given sale.

3. Using Monte Carlo simulation, write an algorithm to calculate an approximation to π by considering the number of random points selected inside the unit quarter-circle:

$$Q: x^2 + y^2 = 1, \qquad x \geqslant 0, \qquad y \geqslant 0,$$

where the quarter-circle is taken to be inside the square:

$$S: 0 \leqslant x \leqslant 1 \quad \text{and} \quad 0 \leqslant y \leqslant 1$$

Use the equation that $\pi/4 = $ area Q/area S.

4. Using Monte Carlo simulation, write an algorithm to calculate that part of the volume of the sphere:

$$x^2 + y^2 + z^2 \leqslant 1$$

that lies in the first octant, $x > 0$, $y > 0$, $z > 0$.

5. Using Monte Carlo simulation, write an algorithm to calculate the volume trapped between the two paraboloids:

$$z = 8 - x^2 - y^2 \quad \text{and} \quad z = x^2 + 3y^2$$

Note that the two paraboloids intersect on the elliptic cylinder:

$$x^2 + 2y^2 = 4$$

8.1 PROJECTS

1. Complete the requirements for the UMAP module 269, "Monte Carlo: The Use of Random Digits to Simulate Experiments," by Dale T. Hoffman. The Monte Carlo technique is presented, explained, and used to find approximate solutions for several realistic problems. Simple experiments are included for student practice.

2. "Random Numbers," by Mark D. Meyerson, UMAP 590. This module discusses methods for generating random numbers, and presents tests for determining the "randomness" of a string of random numbers. Complete the module and prepare a short report for classroom discussion.

Projects 3, 4, and 5 require elementary programming experience. Programs may be written for a programmable calculator or computer.

3. Write a program to generate uniformly distributed random integers in the interval $m < x < n$, where m and n are integers, according to the following algorithm:

Step 1   Let $d = 2^{31}$ and choose N (the number of random numbers to generate).

Step 2   Choose any seed integer Y such that

$$999999 > Y > 100000$$

Step 3   Let $i = 1$.

Step 4   Let $Y = \mathrm{mod}(15625\, Y + 22221, d)$.

Step 5   Let $X_i = m + \mathrm{floor}[(n - m + 1)Y/d]$.

Step 6   Increment i by 1: $i = i + 1$.

Step 7   Go to Step 4 unless $i = N + 1$.

Here the meaning of $\mathrm{mod}(p, q)$ is the remainder obtained on dividing p by q, and $\mathrm{floor}(p)$ means the largest integer not exceeding p.

For most choices of Y the numbers $X_1, X_2, \ldots$ form a sequence of (pseudo) random integers as desired. One possible recommended choice is $Y = 568731$.

To generate random numbers (not just *integers*) in an interval a to b with $a < b$, use the preceding algorithm, replacing the formula in Step 5 by

$$\text{Let } X_i = a + \frac{Y(b - a)}{d - 1}$$

4. Write a program to generate 1000 integers between 1 and 5 in a random fashion so that 1 occurs 22% of the time, 2 occurs 15% of the time, 3 occurs 31% of the time, 4 occurs 26% of the time, and 5 occurs 6% of the time. Over what interval should you generate random integers? How do you then decide which of the integers 1–5 has been generated according to its specified chance of selection?

5. If you have a programmable calculator or computer available, write a program to approximate the areas or volumes designated in Problems 3–5.

8.2 DEVELOPING SUBMODELS FOR PROBABILISTIC PROCESSES

In the previous section a deterministic behavior was modeled using Monte Carlo simulation. In this section you will learn a method for applying Monte Carlo simulation to approximate processes that are probabilistic in nature. Additionally, a check is made to determine how well the simulation duplicates that process. We begin by reconsidering the inventory control problem.

As presented in the scenario for Section 5.3, you have just been hired as a consultant by a chain of gasoline stations to determine how often and how much gasoline should be delivered to the various stations. Each time gasoline is delivered, a cost of *d* dollars is incurred, which is in addition to the cost of the gasoline and is independent of the amount delivered. The gasoline stations are near interstate highways, so demand is fairly constant. Other factors determining the costs include the capital tied up in inventory, the amortization costs of equipment, insurance, taxes, security measures, and so forth. We assumed that, in the short run, the demand and price of gasoline are constant for each station, yielding a constant total revenue as long as the station does not run out of gasoline. Because total profit is total revenue minus total cost, and total revenue is constant by assumption, total profit can be maximized by minimizing total cost. Thus, we identified the following problem: *minimize the average daily cost of delivering and storing sufficient gasoline at each station to meet consumer demand.*

After discussing the relative importance of the various factors determining the average daily cost, we decided to develop the following model:

average daily cost = f(storage costs, delivery cost, demand rate)

Turning our attention to the various submodels, we argued that, while the cost of storage may vary with the amount stored, it is reasonable to assume the cost per unit stored to be constant over the range of values under consideration. Similarly, the delivery cost was assumed constant per delivery, independent of the amount delivered, over the range of values under consideration. Plotting the daily demand for gasoline at a particular station is very likely to give a graph similar to the one shown in Figure 8-3a. If the frequency of each demand level over a fixed time period (say, one year) is plotted, then a plot similar to that shown in Figure 8-3b might be obtained.

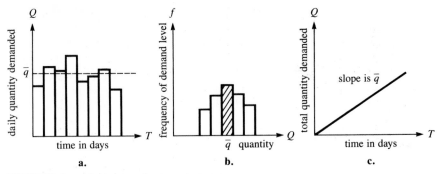

FIGURE 8-3 A constant demand rate.

If the demands are fairly tightly packed about the most frequently occurring demand, then we would accept the daily demand as being constant. We did assume a constant daily demand for the problem at hand. Finally, even though the demands occur in discrete time periods, a continuous submodel for demand is taken for purposes of simplification. This continuous sub-model is depicted in Figure 8-3c, where the slope of the line represents the constant daily demand. Notice the importance of each of the preceding assumptions in producing the linear submodel.

From these assumptions we constructed an analytic model for the average daily cost and used it to compute an optimal time between deliveries as well as an optimal delivery quantity:

$$T^* = \sqrt{2d/sr}$$
$$Q^* = rT^*$$

where

    T^*: optimal time between deliveries in days
    Q^*: optimal delivery quantity of gasoline in gallons
    r: demand rate in gallons per day
    d: delivery cost in dollars per delivery
    s: storage cost per gallon per day

Note how the analytic model depends heavily on a set of conditions that, although reasonable in some cases, would never be met precisely in the real world. It is difficult to develop analytic models that take into account the probabilistic nature of the submodels.

Suppose we decide to check our submodel for constant demand rate by inspecting the sales for the last 1000 days at a particular station. Thus, the data displayed in Table 8-2 are collected.

For each of the ten intervals of demand levels given in Table 8-2, compute the relative frequency of occurrence by dividing the number of occurrences by the total number of days, 1000. This computation results in an estimate of the probability of occurrence for each demand interval. These probabilities are displayed in Table 8-3 and plotted in histogram form in Figure 8-4.

TABLE 8-2 History of demand at a particular gasoline station

Number of gallons demanded	Number of occurrences (in days)
1000–1099	10
1100–1199	20
1200–1299	50
1300–1399	120
1400–1499	200
1500–1599	270
1600–1699	180
1700–1799	80
1800–1899	40
1900–1999	30
	1000

TABLE 8-3 Probability of the occurrence of each demand level

Number of gallons demanded	Probability of occurrence
1000–1099	0.01
1100–1199	0.02
1200–1299	0.05
1300–1399	0.12
1400–1499	0.20
1500–1599	0.27
1600–1699	0.18
1700–1799	0.08
1800–1899	0.04
1900–1999	0.03
	1.00

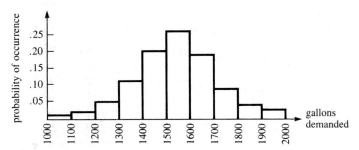

FIGURE 8-4 The histogram displays the relative frequency of each of the demand intervals in Table 8-3.

If we are satisfied with the assumption of a constant demand rate, we might estimate this rate at 1550 gal per day (from Figure 8-4, on previous page). Then the analytic model could be used to compute the optimal time between deliveries and the delivery quantities from the delivery and storage costs.

Suppose, however, that we are not satisfied with the assumption of constant daily demand. How could we then simulate the submodel for the demand suggested by Figure 8-4? First, we could build a cumulative histogram by consecutively adding together the probabilities of each individual demand level, as displayed in Figure 8-5. Note in Figure 8-5 that the difference in height between adjacent columns represents the probability of occurrence of the subsequent demand interval. Thus, we can construct a correspondence between the numbers in the interval $0 \leqslant x \leqslant 1$ and the relative occurrence of the various demand intervals. This correspondence is displayed in Table 8-4.

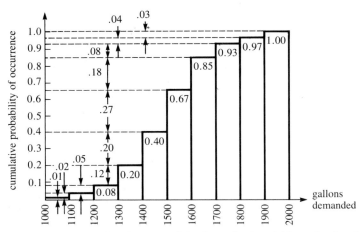

FIGURE 8-5 A cumulative histogram of the demand submodel from the data in Table 8-3.

TABLE 8-4 Using random numbers uniformly distributed over $0 \leqslant x \leqslant 1$ to duplicate the occurrence of the various demand intervals

Random number	Corresponding demand	Percent occurrence
$0 \leqslant x < .01$	1000–1099	.01
$.01 \leqslant x < .03$	1100–1199	.02
$.03 \leqslant x < .08$	1200–1299	.05
$.08 \leqslant x < .20$	1300–1399	.12
$.20 \leqslant x < .40$	1400–1499	.20
$.40 \leqslant x < .67$	1500–1599	.27
$.67 \leqslant x < .85$	1600–1699	.18
$.85 \leqslant x < .93$	1700–1799	.08
$.93 \leqslant x < .97$	1800–1899	.04
$.97 \leqslant x \leqslant 1.00$	1900–1999	.03

Thus, if numbers between 0 and 1 are randomly generated, so that each number has an equal probability of occurring, the histogram of Figure 8-4 can be approximated. Using a random number generator on a hand-held programmable calculator, we generated random numbers between 0 and 1 and then used the assignment procedure suggested by Table 8-4 to determine the demand interval corresponding to each random number. The results for 1,000 and 10,000 trials are presented in Table 8-5.

TABLE 8-5 A Monte Carlo approximation of the demand submodel

Interval	Number of occurrences/expected no. of occurrences	
	1,000 trials	10,000 trials
1000–1099	8/10	91/100
1100–1199	16/20	198/200
1200–1299	46/50	487/500
1300–1399	118/120	1205/1200
1400–1499	194/200	2008/2000
1500–1599	275/270	2681/2700
1600–1699	187/180	1812/1800
1700–1799	83/80	857/800
1800–1899	34/40	377/400
1900–1999	39/30	284/300
	1000/1000	10000/10000

For the inventory problem, we ultimately want to be able to determine a specific demand, rather than a demand interval, for each day simulated. How can this be accomplished? There are several alternatives. Consider the plot of the midpoints of each demand interval as displayed in Figure 8-6. Since we want a continuous model capturing the trend of the plotted data, we can use the methods discussed in Chapter 6.

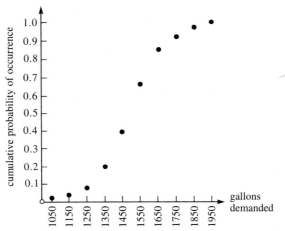

FIGURE 8-6 A cumulative plot of the demand submodel displaying only the center point of each interval.

In many instances, especially where the subintervals are small and the data are fairly approximate anyway, a linear spline model is suitable. A linear spline model for the data displayed in Figure 8-6 is presented in Figure 8-7, and the individual spline functions are given in Table 8-6. The interior spline functions—$S_2(q)$ through $S_9(q)$—were computed by passing a line through the adjacent data points. $S_1(q)$ was computed by passing a line through (1000, 0) and the first data point (1050, 0.01). $S_{10}(q)$ was computed by passing a line through the points (1850, 0.97) and (2000, 1.00). Note that if we use the midpoints of the intervals, we have to make a decision on how to construct the two exterior splines. If the intervals are small, it is usually easy to construct a linear spline function that captures the trend of the data.

Now suppose we wish to simulate a daily demand for a given day. To do this, we generate a random number x between 0 and 1 and compute a corresponding demand q. That is, x is the independent variable from which a unique corresponding q is calculated. This calculation is possible since the function depicted in Figure 8-7 is strictly increasing. (Think about whether

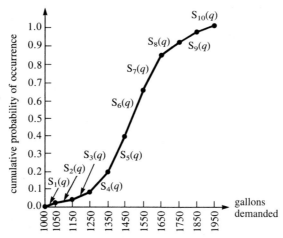

FIGURE 8-7 A linear spline model for the demand submodel.

TABLE 8-6 The linear splines for the empirical demand submodel

Demand interval	Linear spline
$1000 \leqslant q < 1050$	$S_1(q) = 0.0002q - .2$
$1050 \leqslant q < 1150$	$S_2(q) = 0.0002q - .2$
$1150 \leqslant q < 1250$	$S_3(q) = 0.0005q - .545$
$1250 \leqslant q < 1350$	$S_4(q) = 0.0012q - 1.42$
$1350 \leqslant q < 1450$	$S_5(q) = 0.002q - 2.5$
$1450 \leqslant q < 1550$	$S_6(q) = 0.0027q - 3.515$
$1550 \leqslant q < 1650$	$S_7(q) = 0.0018q - 2.12$
$1650 \leqslant q < 1750$	$S_8(q) = 0.0008q - .47$
$1750 \leqslant q < 1850$	$S_9(q) = 0.0004q + .23$
$1850 \leqslant q \leqslant 2000$	$S_{10}(q) = 0.0002q + .6$

TABLE 8-7 The inverse linear splines
provide for the daily demand as a function
of a random number in $[0, 1]$

Random number	Inverse linear spline
$0 \leqslant x < .01$	$q = (x + .2)5000$
$.01 \leqslant x < .03$	$q = (x + .2)5000$
$.03 \leqslant x < .08$	$q = (x + .545)2000$
$.08 \leqslant x < .20$	$q = (x + 1.42)833.33$
$.20 \leqslant x < .40$	$q = (x + 2.5)500$
$.40 \leqslant x < .67$	$q = (x + 3.515)370.37$
$.67 \leqslant x < .85$	$q = (x + 2.12)555.55$
$.85 \leqslant x < .93$	$q = (x + .47)1250$
$.93 \leqslant x < .97$	$q = (x - .23)2500$
$.97 \leqslant x \leqslant 1.00$	$q = (x - .6)5000$

this situation will always be the case.) Thus, the problem is to find the inverse functions for the splines listed in Table 8-6. For instance, given $x = S_1(q) = 0.0002q - .2$, we can solve for $q = (x + .2)5000$. In the case of linear splines, it is very easy to find the inverse functions summarized in Table 8-7.

Let's illustrate how Table 8-7 can be used to represent the daily demand submodel. To simulate a demand for a given day, we generate a random number between 0 and 1, say $x = 0.214$. Since $0.20 \leqslant 0.214 < 0.40$, the spline $q = (x + 2.5)500$ is used to compute $q = 1357$. Thus, 1357 gallons is the simulated demand for that day.

Note that the inverse linear splines presented in Table 8-7 could have been computed directly from the data in Figure 8-6 by choosing x as the independent variable instead of q. We will follow this procedure later when computing the cubic spline demand submodel. (The preceding development was presented to ease your understanding of the process and also because it mimics what you will do after studying probability.) Figure 8-7 is an example of a cumulative distribution function. Many types of behavior approximate well-known probability distributions, which can then be used as the basis for Figure 8-7 rather than experiential data. The inverse function must then be found to use as the demand submodel in the simulation, and this may prove to be difficult. In such cases, the inverse function is approximated with an empirical model, such as a linear spline or cubic spline. For an excellent introduction to some types of behavior that follow well-known probability distributions, see UMAP Module 340, "The Poisson Random Process," by Carroll Wilde, listed in 8.2 Projects.

If we want a smooth continuous submodel for demand, we can construct a cubic spline submodel. We will construct the splines directly as a function of the random number x. That is, using a computer program, we calculate the cubic splines for the data points:

x	0	0.01	0.03	0.08	0.2	0.4	0.67	0.85	0.93	0.97	1.00
q	1000	1050	1150	1250	1350	1450	1550	1650	1750	1850	2000

TABLE 8-8 An empirical cubic spline submodel for demand

Random number	Cubic spline
$0 \leqslant x < .01$	$S_1(x) = 1000 + 4924.92x + 750788.75x^3$
$.01 \leqslant x < .03$	$S_2(x) = 1050 + 5150.18(x-.01) + 22523.66(x-.01)^2 - 1501630.8(x-.01)^3$
$.03 \leqslant x < .08$	$S_3(x) = 1150 + 4249.17(x-.03) - 67574.14(x-.03)^2 + 451815.88(x-.03)^3$
$.08 \leqslant x < .20$	$S_4(x) = 1250 + 880.37(x-.08) + 198.24(x-.08)^2 - 4918.74(x-.08)^3$
$.20 \leqslant x < .40$	$S_5(x) = 1350 + 715.46(x-.20) - 1572.51(x-.20)^2 + 2475.98(x-.20)^3$
$.40 \leqslant x < .67$	$S_6(x) = 1450 + 383.58(x-.40) - 86.92(x-.40)^2 + 140.80(x-.40)^3$
$.67 \leqslant x < .85$	$S_7(x) = 1550 + 367.43(x-.67) + 27.12(x-.67)^2 + 5655.69(x-.67)^3$
$.85 \leqslant x < .93$	$S_8(x) = 1650 + 926.92(x-.85) + 3081.19(x-.85)^2 + 11965.43(x-.85)^3$
$.93 \leqslant x < .97$	$S_9(x) = 1750 + 1649.66(x-.93) + 5952.90(x-.93)^2 + 382645.25(x-.93)^3$
$.97 \leqslant x \leqslant 1.00$	$S_{10}(x) = 1850 + 3962.58(x-.97) + 51870.29(x-.97)^2 - 576334.88(x-.97)^3$

The splines are presented in Table 8-8. If the random number $x = 0.214$ is generated, the empirical cubic spline model yields the demand $q = 1350 + 715.46(.014) - 1572.5(.014)^2 + 2476(.014)^3 = 1359.7$ gal.

An empirical submodel for demand can be constructed in a variety of other ways. For example, rather than using the intervals for gallons demanded as given in Table 8-3, we can use smaller intervals. If the intervals are small enough, the midpoint of an interval could itself be a reasonable approximation to the demand for that entire interval. Thus a cumulative histogram similar to that in Figure 8-5 could serve as a submodel directly. If preferred, a continuous submodel could be constructed readily from the refined data.

The purpose of our discussion has been to demonstrate how a submodel for a probabilistic behavior can be constructed using Monte Carlo simulation and experiential data. Now let's see how the inventory problem can be simulated in general terms. As the problem was presented in Section 5.3, an inventory strategy consists of specifying a delivery quantity Q and a time T between deliveries, given values for storage cost per gallon per day s and a delivery cost per delivery d. If s and d are known, then a specific inventory strategy can be tested using a Monte Carlo simulation algorithm, as follows:

Summary of Monte Carlo Inventory Algorithm Terms

Q delivery quantity of gasoline in gallons

T time between deliveries in days

I current inventory in gallons

d delivery cost in dollars per delivery

s storage cost per gallon per day

C total running cost

c average daily cost

N	number of days to run the simulation
K	days remaining in the simulation
x_i	a random number in the interval $[0, 1]$
q_i	a daily demand
$Flag$	an indicator used to terminate the algorithm

Monte Carlo Inventory Algorithm

Input Q, T, d, s, N

Output c

Step 1   Initialize:
$$K = N$$
$$I = 0$$
$$C = 0$$
$$\text{Flag} = 0$$

Step 2   Begin the next inventory cycle with a delivery:
$$I = I + Q$$
$$C = C + d$$

Step 3   Determine if the simulation will terminate during this cycle:

If $T \geqslant K$, then set $T = K$ and Flag $= 1$

Step 4   Simulate each day in the inventory cycle (or portion remaining):

For $i = 1, 2, \ldots, T$, do Steps 5–9

Step 5   Generate the random number x_i.

Step 6   Compute q_i using the demand submodel.

Step 7   Update the current inventory: $I = I - q_i$.

Step 8   Compute the daily storage cost and total running cost, unless the inventory has been depleted: If $I \leqslant 0$, then set $I = 0$ and GOTO Step 9. Else $C = C + I * s$.

Step 9   Decrement the number of days remaining in the simulation:
$$K = K - 1$$

Step 10   If Flag $= 0$, then GOTO Step 2. Else GOTO Step 11.

Step 11   Compute the average daily cost: $c = C/N$.

Step 12   Output c.

STOP

Various strategies can now be tested with the algorithm to determine their average daily costs. You would probably want to refine the algorithm to keep track of other measures of effectiveness, such as unsatisfied demands, number of days without gas, and so forth, as suggested in the following problem set.

8.2 PROBLEMS

1. Modify the inventory algorithm to keep track of "unfilled demands" and the total number of days that the gasoline station is without gas for at least part of the day.

2. Most gasoline stations have a storage capacity Q_{max} that cannot be exceeded. Refine the inventory algorithm to take this consideration into account. Because of the probabilistic nature of the demand submodel, at the end of the inventory cycle there may still be significant amounts of gasoline remaining. If several such cycles occur in succession, the excess might build up to Q_{max}. Since there is a financial cost in carrying excess inventory, this situation would be undesirable. What alternatives can you suggest? Modify the inventory algorithm to take your alternatives into account.

3. In many situations the time T between deliveries and the order quantity Q is not fixed. Instead, an order is placed for a specific amount of gasoline. Depending on how many orders are placed in a given time interval, the time to fill an order varies. You have no reason to believe that the performance of the delivery operation will change. Therefore, you have examined your records for the last 100 deliveries and found the following "lag times" or "extra days" required to fill your order:

Lag time (in days)	Number of occurrences
2	10
3	25
4	30
5	20
6	13
7	2
	100

Construct a Monte Carlo simulation for the "lag time" submodel. If you have a hand-held calculator or computer available, test your submodel by running 1000 trials and comparing the number of occurrences of the various lag times with the historical data.

4. Problem 3 suggests an alternative inventory strategy. Once the inventory reaches a certain level (an order point), an order can be placed for an

"optimal" amount of gasoline. Construct an algorithm that simulates this process and incorporates probabilistic submodels for demand and lag times. How would you use this algorithm to search for the "optimal" order point and the "optimal" order quantity?

5. In the case that a gasoline station runs out of gas, the customer is simply going to go to another station. However, in many other situations (name a few) a portion of the customers will place a "backorder." If the order is not filled within a time period varying from customer to customer in a probabilistic fashion, the customer will cancel his or her order. Suppose we examine the historical records for 1000 customers and find the data shown in Table 8-9. That is, 200 customers will not even place an order, and an additional 150 customers will cancel if the order is not filled within one day.

TABLE 8-9 Hypothetical data for a backorder submodel

Number of days customer is willing to wait before cancelling	Number of occurrences	Cumulative occurrences
0	200	200
1	150	350
2	200	550
3	200	750
4	150	900
5	50	950
6	50	1000
	1000	

a. Construct a Monte Carlo simulation for the backorder submodel. If you have a calculator or computer available, test your submodel by running 1000 trials and comparing the number of occurrences of the various cancellations with the historical data.

b. Consider the algorithm you modified in Problem 1. Further modify the algorithm to consider backorder. Do you think backorders should be penalized in some fashion? If so, how would you do it?

8.2 PROJECTS

1. Complete the requirements of UMAP module 340, "The Poisson Random Process," by Carroll O. Wilde. Probability distributions are introduced to obtain practical information on random arrival patterns, interarrival times or "gaps" between arrivals, waiting line buildup, and service loss rates. The Poisson distribution, the exponential distribution, and Erlang's formulas are used. The module requires an introductory probability

course, the ability to use summation notation, and basic concepts of the derivative and the integral from calculus. Prepare a 10-minute summary of the module for classroom presentation.

2. For the gasoline inventory problem presented in this section, assume a storage cost of $0.001 per gallon per day and a delivery charge of $500 per delivery. Initially assume a constant demand of 1550 gal per day and use the formulas developed in Section 5.3 to compute the optimal time between deliveries T^* and the optimal delivery quantity Q^*. Now assume the probabilistic demand submodel presented in this section. Construct a computer implementation of the "Monte Carlo Inventory Algorithm" as modified in Problem 1 to keep track of "unfilled demands." Experiment with different inventory strategies "near" the analytic optimum. Compare their average daily costs and unfilled demands. Graph the average daily cost as a function of the time between deliveries and compare your graph with Figure 5-7.

3. Assume a storage cost of $0.001 per gallon per day and a delivery charge of $500 per delivery. Construct a computer implementation of the algorithm you constructed in Problem 4, and compare various "order points" and "order quantity" strategies.

MODEL

8.3 A HARBOR SYSTEM

Consider a small harbor with unloading facilities for ships. Only one ship can be unloaded at any one time. Ships arrive for unloading of cargo at the harbor, and the time between the arrival of successive ships varies from 15 to 145 min. The unloading time required for a ship depends on the type and amount of cargo and varies from 45 to 90 min. We seek answers to the following questions:

1. What is the average and maximum time per ship in the harbor?
2. If the *waiting time* for a ship is the time between its arrival and the start of unloading, what is the average and maximum waiting time per ship?
3. What percent of the time are the unloading facilities idle?

To get some reasonable answers to these questions, we can simulate the activity in the harbor using a computer or programmable calculator. We assume that the arrival times between successive ships and the unloading time per ship are uniformly distributed over their respective time intervals. For instance, the arrival time between successive ships can be any integer between 15 and 145, and any integer within that interval can appear with equal likelihood. Before giving a general algorithm to simulate the harbor system, let's consider a hypothetical situation with five ships.

We have the following data for each ship:

	Ship 1	Ship 2	Ship 3	Ship 4	Ship 5
Time between successive ships	20	30	15	120	25
Unload time	55	45	60	75	80

Since Ship 1 arrives 20 min after the time clock commences at $t = 0$ minutes, the harbor facilities are idle for 20 min at the start. Ship 1 immediately begins to unload. The unloading takes 55 min, and meanwhile Ship 2 arrives on the scene at $t = 20 + 30 = 50$ min after the time clock begins. Ship 2 cannot start to unload until Ship 1 finishes unloading at $t = 20 + 55 = 75$ min. This means that Ship 2 must wait $75 - 50 = 25$ min before unloading begins. The situation is depicted in the following time-line diagram:

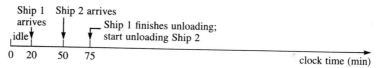

Now before Ship 2 starts to unload, Ship 3 arrives at time $t = 50 + 15 = 65$ min. Since the unloading of Ship 2 starts at $t = 75$ and it takes 45 min to unload, unloading of Ship 3 cannot start until $t = 75 + 45 = 120$ min, when Ship 2 is finished. Thus, Ship 3 must wait $120 - 65 = 55$ min. This situation is depicted in the next time-line diagram.

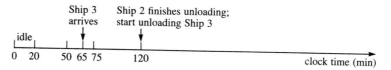

Ship 4 doesn't arrive in the harbor until $t = 65 + 120 = 185$ min. Therefore, Ship 3 has already finished unloading at $t = 120 + 60 = 180$ min, and the harbor unloading facilities are idle for $185 - 180 = 5$ min. Moreover, the unloading of Ship 4 commences immediately upon its arrival, as depicted in the next diagram.

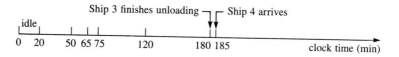

Finally, Ship 5 arrives at $t = 185 + 25 = 210$ min, before Ship 4 finishes unloading at $t = 185 + 75 = 260$ min. Thus Ship 5 must wait $260 - 210 = 50$ min before it starts to unload. The simulation is complete when Ship 5 finishes unloading at $t = 260 + 80 = 340$ min. The final situation is shown in the next diagram.

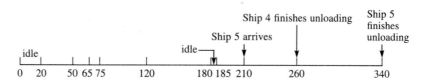

In Figure 8-8 we summarize the waiting and unloading times for each of the five hypothetical ship arrivals. In Table 8-10 we summarize the results of the entire simulation of the five hypothetical ships. Note that the total

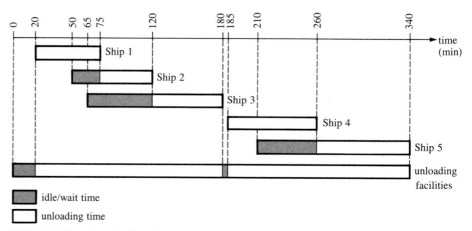

FIGURE 8-8 Idle and unloading times for the ships and docking facilities.

TABLE 8-10 Summary of the harbor system simulation

Ship no.	Random time between ship arrivals	Arrival time	Start service	Queue length at arrival	Wait time	Random unload time	Time in harbor	Dock idle time
1	20	20	20	0	0	55	55	20
2	30	50	75	1	25	45	70	0
3	15	65	120	2	55	60	115	0
4	120	185	185	0	0	75	75	5
5	25	210	260	1	50	80	130	0
Totals (if appropriate):					130			25
Averages (if appropriate):					26	63	89	

Note: All times are given in minutes after the start of the clock at time $t = 0$.

waiting time spent by all five ships before unloading is 130 min. This waiting time represents a cost to the shipowners and is a source of customer dissatisfaction with the docking facilities. On the other hand, the docking facility has only 25 min of total idle time. It is in use 315 of the total 340 min of the simulation, or approximately 93% of the time.

Suppose the owners of the docking facilities are concerned with the quality of service they are providing and want various management alternatives to be evaluated in order to see if improvement in service justifies the added cost. Several statistics can help in evaluating the quality of the service. For example, the maximum time a ship spends in the harbor is 130 min by Ship 5, while the average is 89 min. Generally, customers are very sensitive to the amount of time spent waiting. In this example, the maximum amount of time spent waiting for a facility is 55 min, while the average time spent waiting is 26 min. Some customers are apt to take their business elsewhere if queues are too long. In this case the longest queue length is 2. The following Monte Carlo simulation algorithm will compute such statistics to assess various management alternatives.

Summary of Harbor System Algorithm Terms

$between_i$ — time between successive arrivals of Ships i and $i - 1$ (a random integer varying between 15 and 145 min)

$arrive_i$ — time from start of clock at $t = 0$ when Ship i arrives at the harbor for unloading

$unload_i$ — time required to unload Ship i at the dock (a random integer varying between 45 and 90 min)

$start_i$ — time from start of clock at which Ship i commences its unloading

$idle_i$ — time for which dock facilities are idle immediately *prior to* commencement of unloading Ship i

$wait_i$ — time Ship i waits in the harbor after arrival before unloading commences

$finish_i$ — time from start of clock at which service for Ship i is completed at the unloading facilities

$harbor_i$ — total time Ship i spends in the harbor

$HARTIME$ — average time per ship in the harbor

$MAXHAR$ — maximum time of a ship in the harbor

$WAITIME$ — average waiting time per ship before unloading

$MAXWAIT$ — maximum waiting time of a ship

$IDLETIME$ — percent of total simulation time unloading facilities are idle

Harbor System Simulation Algorithm

Input Total number n of ships for the simulation.

Output HARTIME, MAXHAR, WAITIME, MAXWAIT, and IDLETIME.

Step 1 Randomly generate $between_1$ and $unload_1$. Then set $arrive_1 = between_1$.

Step 2 Initialize all output values:

$$HARTIME = unload_1, \quad MAXHAR = unload_1,$$
$$WAITIME = 0 \quad MAXWAIT = 0, \quad IDLETIME = arrive_1$$

Step 3 Calculate finish time for unloading of $Ship_1$:

$$finish_1 = arrive_1 + unload_1$$

Step 4 For $i = 2, 3, \ldots, n$ do Steps 5–16.

Step 5 Generate the random pair of integers, $between_i$ and $unload_i$ over their respective time intervals.

Step 6 Assuming the time clock begins at $t = 0$ minutes, calculate the time of arrival for $Ship_i$:

$$arrive_i = arrive_{i-1} + between_i$$

Step 7 Calculate the time difference between the arrival time of $Ship_i$ and the finish time for unloading the previous $Ship_{i-1}$:

$$timediff = arrive_i - finish_{i-1}$$

Step 8 For nonnegative timediff, the unloading facilities are idle:

$$idle_i = timediff \quad \text{and} \quad wait_i = 0$$

For negative timediff, $Ship_i$ must wait before it can unload:

$$wait_i = -timediff \quad \text{and} \quad idle_i = 0$$

Step 9 Calculate the start time for unloading $Ship_i$:

$$start_i = arrive_i + wait_i$$

Step 10 Calculate the finish time for unloading $Ship_i$:

$$finish_i = start_i + unload_i$$

Step 11 Calculate the time in harbor for $Ship_i$:

$$harbor_i = wait_i + unload_i$$

Step 12 Sum $harbor_i$ into total harbor time HARTIME for averaging.

Step 13 If $harbor_i > MAXHAR$, then set $MAXHAR = harbor_i$. Otherwise leave MAXHAR as is.

Step 14 Sum $wait_i$ into total waiting time WAITIME for averaging.

Step 15 Sum $idle_i$ into total idle time IDLETIME.

Step 16   If wait$_i$ > MAXWAIT, then set MAXWAIT = wait$_i$. Otherwise leave MAXWAIT as is.

Step 17   Set HARTIME = HARTIME/n, WAITIME = WAITIME/n, and IDLE-TIME = IDLETIME/finish$_n$.

Step 18   OUTPUT (HARTIME, MAXHAR, WAITIME, MAXWAIT, IDLE-TIME)

STOP

Table 8-11 gives the results, according to the preceding algorithm, of six independent simulation runs of 100 ships each.

Now suppose you are a consultant for the owners of the docking facilities. What would be the effect of hiring additional labor or acquiring better equipment for unloading cargo so that the unloading time interval is reduced to between 35 and 75 min per ship? Table 8-12 gives the results based on our simulation algorithm.

TABLE 8-11 Harbor system simulation results for 100 ships

Average time of a ship in the harbor	106	85	101	116	112	94
Maximum time of a ship in the harbor	287	180	233	280	234	264
Average waiting time of a ship	39	20	35	50	44	27
Maximum waiting time of a ship	213	118	172	203	167	184
Percentage of time dock facilities are idle	0.18	0.17	0.15	0.20	0.14	0.21

Note: All times are given in minutes. Time between successive ships is 15 to 145 min. Unloading time per ship varies from 45 to 90 min.

TABLE 8-12 Harbor system simulation results for 100 ships

Average time of a ship in the harbor	74	62	64	67	67	73
Maximum time of a ship in the harbor	161	116	167	178	173	190
Average waiting time of a ship	19	6	10	12	12	16
Maximum waiting time of a ship	102	58	102	110	104	131
Percentage of time dock facilities are idle	0.25	0.33	0.32	0.30	0.31	0.27

Note: All times are given in minutes. Time between successive ships is 15 to 145 min. Unloading time per ship varies from 35 to 75 min.

You can see from Table 8-12 that a reduction of the unloading time per ship by 10 to 15 min decreases the time the ships spend in the harbor, especially the waiting times. However, the percent of the total time during which the dock facilities are idle nearly doubles. This situation is favorable to the shipowners because it increases the availability of each ship for hauling cargo over the long run. Thus the traffic coming into the harbor is likely to increase. If the traffic increases to the extent that the time between successive ships is reduced to between 10 and 120 min, the simulated results are shown in Table 8-13. You can see from the table that the ships again spend more time in the harbor with the increased traffic, but now the harbor facilities are idle much less of the time. Moreover, both the shipowners and the dock owners are benefiting from the increased business.

TABLE 8-13 Harbor system simulation results for 100 ships

Average time of a ship in the harbor	114	79	96	88	126	115
Maximum time of a ship in the harbor	248	224	205	171	371	223
Average waiting time of a ship	57	24	41	35	71	61
Maximum waiting time of a ship	175	152	155	122	309	173
Percentage of time dock facilities are idle	0.15	0.19	0.12	0.14	0.17	0.06

Note: All times are given in minutes. Time between successive ships is 10 to 120 min. Unloading time per ship varies from 35 to 75 min.

Suppose now that we are not satisfied with the assumption that the arrival times between successive ships (that is, their interarrival times) and the unloading time per ship are uniformly distributed over the time intervals $15 \le between_i \le 145$ and $45 \le unload_i \le 90$, respectively. So it is decided to collect experiential data for the harbor system and incorporate the results into our model, as discussed for the demand submodel in Section 8.2. We observe (hypothetically) 1200 ships using the harbor to unload their cargoes, and collect the data displayed in Table 8-14.

TABLE 8-14 Data collected for 1200 ships using the harbor facilities

Time between arrivals	Number of occurrences	Probability of occurrence	Unloading time	Number of occurrences	Probability of occurrence
15–24	11	0.009			
25–34	35	0.029			
35–44	42	0.035	45–49	20	0.017
45–54	61	0.051	50–54	54	0.045
55–64	108	0.090	55–59	114	0.095
65–74	193	0.161	60–64	103	0.086
75–84	240	0.200	65–69	156	0.130
85–94	207	0.172	70–74	223	0.185
95–104	150	0.125	75–79	250	0.208
105–114	85	0.071	80–84	171	0.143
115–124	44	0.037	85–90	109	0.091
125–134	21	0.017		1200	1.000
135–145	3	0.003			
	1200	1.000			

Note: All times are given in minutes.

Following the procedures outlined in Section 8.2, we consecutively add together the probabilities of each individual time between arrivals interval,

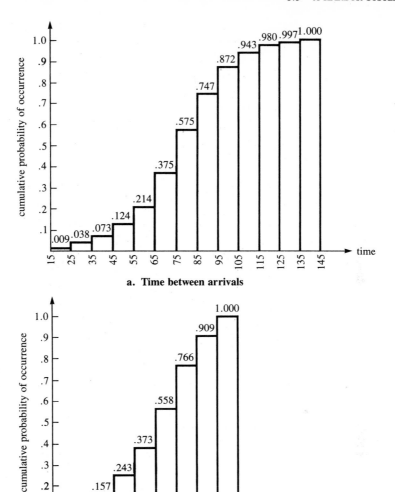

FIGURE 8-9 Cumulative histograms of the time between ship arrivals and the unloading times, from the data in Table 8-14.

as well as the probabilities of each individual unloading time interval. These computations result in the cumulative histograms depicted in Figure 8-9.

Next we use random numbers uniformly distributed over the interval $0 \leqslant x \leqslant 1$ to duplicate the various interarrival times and unloading times based on the cumulative histograms. We then use the midpoints of each interval and construct linear splines through adjacent data points. (We ask you to complete this construction in Problem 1.) Since it is very easy to calculate the inverse splines directly, we do so and summarize our results in Tables 8-15 and 8-16.

TABLE 8-15 The inverse linear splines provide for the time between arrivals of successive ships as a function of a random number in the interval $[0, 1]$

Random number interval	Corresponding arrival time	Inverse linear spline
$0 \leqslant x < .009$	$15 \leqslant b < 20$	$b = 555.6x + 15.0000$
$0.009 \leqslant x < .038$	$20 \leqslant b < 30$	$b = 344.8x + 16.8966$
$0.038 \leqslant x < .073$	$30 \leqslant b < 40$	$b = 285.7x + 19.1429$
$0.073 \leqslant x < .124$	$40 \leqslant b < 50$	$b = 196.1x + 25.6863$
$0.124 \leqslant x < .214$	$50 \leqslant b < 60$	$b = 111.1x + 36.2222$
$0.214 \leqslant x < .375$	$60 \leqslant b < 70$	$b = 62.1x + 46.7080$
$0.375 \leqslant x < .575$	$70 \leqslant b < 80$	$b = 50.0x + 51.2500$
$0.575 \leqslant x < .747$	$80 \leqslant b < 90$	$b = 58.1x + 46.5698$
$0.747 \leqslant x < .872$	$90 \leqslant b < 100$	$b = 80.0x + 30.2400$
$0.872 \leqslant x < .943$	$100 \leqslant b < 110$	$b = 140.8x - 22.8169$
$0.943 \leqslant x < .980$	$110 \leqslant b < 120$	$b = 270.3x - 144.8649$
$0.980 \leqslant x < .997$	$120 \leqslant b < 130$	$b = 588.2x - 456.4706$
$0.997 \leqslant x \leqslant 1.000$	$130 \leqslant b \leqslant 145$	$b = 5000.0x - 4855$

TABLE 8-16 The inverse linear splines provide for the unloading time of a ship as a function of a random number in the interval $[0, 1]$

Random number interval	Corresponding unloading time	Inverse linear spline
$0 \leqslant x < 0.017$	$45 \leqslant u < 47.5$	$u = 147x + 45.000$
$0.017 \leqslant x < 0.062$	$47.5 \leqslant u < 52.5$	$u = 111x + 45.611$
$0.062 \leqslant x < 0.157$	$52.5 \leqslant u < 57.5$	$u = 53x + 49.237$
$0.157 \leqslant x < 0.243$	$57.5 \leqslant u < 62.5$	$u = 58x + 48.372$
$0.243 \leqslant x < 0.373$	$62.5 \leqslant u < 67.5$	$u = 38.46x + 53.154$
$0.373 \leqslant x < 0.558$	$67.5 \leqslant u < 72.5$	$u = 27x + 57.419$
$0.558 \leqslant x < 0.766$	$72.5 \leqslant u < 77.5$	$u = 24x + 59.087$
$0.766 \leqslant x < 0.909$	$77.5 \leqslant u < 82.5$	$u = 35x + 50.717$
$0.909 \leqslant x \leqslant 1.000$	$82.5 \leqslant u \leqslant 90$	$u = 82.41x + 7.582$

Finally, we incorporate our linear spline submodels into the simulation model for the harbor system by generating $between_i$ and $unload_i$ for $i = 1, 2, \ldots, n$ in Steps 1 and 5 of our algorithm, according to the rules displayed in Tables 8-15 and 8-16. Employing these submodels, Table 8-17 gives the results of six independent simulation runs of 100 ships each.

TABLE 8-17 Harbor system simulation results for 100 ships

Average time of a ship in the harbor	108	95	125	78	123	101
Maximum time of a ship in the harbor	237	188	218	133	250	191
Average waiting time of a ship	38	25	54	9	53	31
Maximum waiting time of a ship	156	118	137	65	167	124
Percentage of time dock facilities are idle	0.09	0.09	0.08	0.12	0.06	0.10

Note: Based on the data exhibited in Table 8-14. All times are given in minutes.

8.3 PROBLEMS

1. Using the data from Table 8-14 and the cumulative histograms from Figure 8-9, construct cumulative plots of the time between arrivals and unloading time submodels (as in Figure 8-6). Calculate equations for the linear spline over each random number interval. Compare your results with the inverse splines given in Tables 8-15 and 8-16.

2. Modify the ship harbor system algorithm to keep track of the number of ships waiting in a queue.

3. Most small harbors have a maximum number of ships N_{max} that can be accommodated in the harbor area while they wait to be unloaded. If a ship cannot get into the harbor, assume it goes elsewhere to unload its cargo. Refine the ship harbor system algorithm to take these considerations into account.

4. Suppose the owners of the harbor docking facility decide to construct a second facility in order to accommodate the unloading of more ships. When a ship enters the harbor, it is directed to go to the nearest available facility, which is facility 1 if both facilities are available. Using the same assumptions for interarrival times between successive ships and unloading times as in the initial text example, modify the ship harbor algorithm for a system with two unloading facilities.

5. Construct a Monte Carlo simulation of a baseball game between your two favorite teams. Initially, use the batting averages of the various players and simulate their probability of getting a single, double, triple, or home run, or making an out. Later, consider double plays. Discuss how you would model the effect of various pitchers, stealing of bases, and so forth.

6. Establish odds for the various horses in a horse race. Construct a Monte Carlo simulation that advances the horses in phases (10 to 15) so that the distance advanced by each horse during each phase is based upon the odds you established for the horse. The first horse to reach the finish line is the winner unless more than one horse finishes at the end of a "turn." In case of ties, the horse traveling the greatest total distance is the winner.

8.3 PROJECTS

1. Write a computer program to implement the ship harbor simulation algorithm.

2. Write a computer program to implement your simulation of a baseball game in Problem 5.

3. Write a computer program to implement your simulation of a horse race in Problem 6. Run your simulation for 1000 trials and count the number of times each horse wins. How well do the results coincide with the odds you established?

MODEL

8.4 MORNING RUSH HOUR*

Consider an office building with 12 floors in a metropolitan area of some city. During the morning rush hour—from 7:50 A.M. to 9:10 A.M.—workers enter the lobby of the building and take an elevator to their floor of destination. There are four elevators servicing the building. The time between arrivals of the customers at the building varies in a probabilistic manner every 0–30 sec, and upon arrival each customer selects the first available elevator (numbered 1–4). When a person enters an elevator and selects the floor of destination, the elevator waits 15 sec before closing its doors. If another person arrives within the 15-sec interval, the waiting cycle is repeated. If no person arrives within the 15-sec interval, the elevator departs to deliver all of its passengers. We assume no other passengers are picked up along the way. After delivering its last passenger, the elevator returns to the main floor, picking up no passengers on the way down. The maximum occupancy of an elevator is 12 passengers. When a person arrives in the lobby and no elevator is available (because all four elevators are transporting their load of passengers), a queue begins to form in the lobby as the people wait for an elevator to return to the main floor.

The management of the building wants to provide good elevator service to its customers and is interested in exactly what service it is now giving. Some customers claim that they have to wait too long in the lobby before an elevator returns. Others complain that they spend too much time riding the elevator, and still others say that there is considerable congestion in the lobby during the morning rush hour. What really is the situation? Can the management resolve these complaints by a more effective means of scheduling or utilizing the elevators? You may wish to reread the discussion in Example 4 of Section 2.2 to review some of the different formulations of the elevator problem.

We wish to simulate the elevator system using an algorithm for computer implementation that will give answers to the following questions:

1. How many customers are actually being serviced in a typical morning rush hour?

* Optional section

2. If the *waiting time* of a person is the time the person stands in a queue—the time from arrival at the lobby until entry into an available elevator—what is the average and maximum time a person waits in a queue?
3. What is the length of the longest queue? The answer to this question will provide the management with information about congestion in the lobby.
4. If the *delivery time* is the time it takes a customer to reach his or her floor of destination after arrival in the lobby, including any waiting time for an elevator to become available, what is the average and maximum delivery time?
5. What is the average and maximum time a customer actually spends in the elevator?
6. How many stops are made by each elevator, and what percent of the total morning rush hour time is each elevator actually in use?

Before presenting an algorithm to provide answers to these questions, we will define the terms used in the algorithm and explain some of its underlying logic. Because the algorithm is rather complex, this approach should be more revealing than using some hypothetical numbers and taking you step by step through the algorithm. (This is a difficult program if you are not using GPSS or another simulation language.)

During the simulation there is a TIME clock that keeps track of the time (given in seconds). Initially, the value of TIME is 0 sec (at 7:50 A.M.), and the simulation ends when TIME reaches 4800 sec (at 9:10 A.M.). Each customer is assigned a number according to the order of his or her arrival: the first customer is labeled 1, the second customer 2, and so forth. Whenever another customer arrives at the lobby, the time between the customer's arrival and the time when the immediately preceding customer arrived is added to the TIME clock. This time between successive arrivals of customers i and $i - 1$ is labeled $between_i$ in the algorithm, and the arrival time of customer i is labeled $arrive_i$. Initially, for the customer arrival submodel, we assume that all values between 0 and 30 have an equal likelihood of occurring.

All four elevators have their own availability times, called $return_j$ for elevator j. If elevator j is currently available at the main floor, its time is the current time, so $return_j$ = TIME. If an elevator is in transit, its availability time is the time at which it will return to the main floor. Passengers enter an available elevator in the numerical order of the elevators: first elevator 1 (if it is available), next elevator 2 (if it is available), and so on. Maximum occupancy of an elevator is 12 passengers.

Whenever another customer arrives in the lobby of the building, two possible situations exist. Either an elevator is available for receiving passengers, or no elevator is available and a queue is forming as customers wait for one to become available. Once an elevator becomes available, it is "tagged" for loading and passengers can enter *only* that elevator until either it is fully occupied (with 12 passengers) or the 15-sec time delay is exceeded before the arrival of another customer. After loading, the elevator departs to deliver all of its passengers. Even if fully loaded with 12 passengers, it is assumed the elevator waits 15 sec to load the last passenger, allow floor selection, and get under way.

To keep track of which floors have been selected during the loading period of an elevator, and the number of times a particular floor has been selected, the algorithm sets up two one-dimensional arrays (having a component for each of the floors 1–12). (Although no one selects floor 1, the indexing is simplified with its inclusion.) These arrays are called $selvec_j$ and $flrvec_j$ for the "tagged" elevator j. If a customer selects floor 5, for instance, then a 1 is entered into the 5th component position of $selvec_j$ and also into the 5th component of $flrvec_j$. If another customer selects floor 5, then the 5th component of $flrvec_j$ is updated to a 2, and so forth. For example, suppose the passengers in elevator j have selected floor 3 twice, floor 5 twice, and floors 7, 8, and 12 once. Then for this elevator, $selvec_j = (0, 0, 1, 0, 1, 0, 1, 1, 0, 0, 0, 1)$ and $flrvec_j = (0, 0, 2, 0, 2, 0, 1, 1, 0, 0, 0, 1)$. These arrays are then used to calculate the transport time of elevator j so we can determine when it will return to the main floor. As stated previously, that return time is designated $return_j$. The arrays are also used to calculate the delivery times of each passenger in elevator j. Initially, we assume that a customer chooses a floor with equal likelihood. We assume that it takes 10 sec for an elevator to travel between floors, 10 sec to open *and* close its doors, and 3 sec for each passenger to disembark. We also assume that it takes 3 sec for each passenger in a queue to enter the next available elevator.

Summary of Elevator Simulation Algorithm Terms

$between_i$	time between successive arrivals of customers i and $i - 1$ (a random integer varying between 0 and 30 sec)
$arrive_i$	time of arrival from start of clock at $t = 0$ for customer i (calculated only if customer enters a queue waiting for an elevator)
$floor_i$	floor selected by customer i (a random integer varying between 2 and 12)
$elevator_i$	time customer i spends in an elevator
$wait_i$	time customer i waits before stepping into an elevator (calculated only if customer enters a queue waiting for an elevator)
$delivery_i$	time required to deliver customer i to destination floor from time of arrival, including any waiting time
$selvec_j$	binary 0, 1 one-dimensional array representing the floors selected for elevator j, not counting the number of times a particular floor has been selected
$flrvec_j$	integer one-dimensional array representing the number of times each floor has been selected for elevator j for the group of passengers currently being transported to their respective floors

$occup_j$ number of current occupants of elevator j

$return_j$ time from start of clock at $t = 0$ that elevator j returns to the main floor and is available for receiving passengers

$first_j$ an index, the customer number of the first passenger who enters elevator j after it returns to the main floor

$quecust$ customer number of the first person waiting in the queue

$queue$ total length of current queue of customers waiting for an elevator to become available

$startque$ clock time at which the (possibly updated) current queue commences to form

$stop_j$ total number of stops made by elevator j during the entire simulation

$eldel_j$ total time elevator j spends in delivering its current load of passengers

$operate_j$ total time elevator j operates during the entire simulation

$limit$ customer number of the last person to enter an available elevator before it commences transport

max largest index of a nonzero entry in the array $selvec_j$ (highest floor selected)

$remain$ number of customers left in the queue after loading next available elevator

$quetotal$ total number of customers who spent time waiting

$TIME$ current clock time in seconds, starting at $t = 0$

$DELTIME$ average delivery time of a customer to reach destination floor from time of arrival, including any waiting time

$ELEVTIME$ average time a person spends in an elevator

$MAXDEL$ maximum time required for a customer to reach his or her floor of destination from time of arrival

$MAXELEV$ maximum time a customer spends in an elevator

$QUELEN$ number of customers waiting in the longest queue

$QUETIME$ average time a customer who must wait spends in a queue

$MAXQUE$ longest time a customer spends in a queue

Elevator System Simulation Algorithm

Input   None required.

Output   Number of passengers serviced, DELTIME, ELEVTIME, MAXDEL, MAXELEV, QUELEN, QUETIME, MAXQUE, $stop_j$, and the percent time each elevator is in use.

Step 1 Initially set the following parameters to zero: DELTIME, ELEVTIME, MAXDEL, MAXELEV, QUELEN, QUETIME, MAXQUE, quetotal, remain.

Step 2 For the first customer, generate time between successive arrivals and floor destination, and initialize delivery time:

$$i = 1$$
$$\text{Generate between}_i \text{ and floor}_i$$
$$\text{delivery}_i = 15$$

Step 3 Initialize the clock time, elevator available clock times, elevator stops, and elevator operating times. Also, initialize all customer waiting times.

$$\text{TIME} = \text{between}_i$$
$$\text{For } k = 1 \text{ to 4: return}_k = \text{TIME} \quad \text{and} \quad \text{stop}_k = \text{operate}_k = 0$$
$$\text{For } k = 1 \text{ to 400: wait}_k = 0$$

(∗ The number 400 is an upper-bound guess for the total number of customers ∗)

Step 4 While TIME ≤ 4800, do Steps 5–32:

Step 5 Select the first available elevator:

If TIME ≥ return_1, then $j = 1$ else
If TIME ≥ return_2, then $j = 2$ else
If TIME ≥ return_3, then $j = 3$ else
If TIME ≥ return_4, then $j = 4$

ELSE (∗ no elevator is currently available ∗) GOTO Step 19.

Step 6 Set as an index the customer number of first person to occupy tagged elevator, and initialize the elevator occupancy floor selection vectors:

$$\text{first}_j = i, \qquad \text{occup}_j = 0$$
$$\text{For } k = 1 \text{ to 12: selvec}_j[k] = \text{flrvec}_j[k] = 0$$

Step 7 Load current customer on elevator j by setting the floor selection vectors and incrementing elevator occupancy:

$$\text{selvec}_j[\text{floor}_i] = 1$$
$$\text{flrvec}_j[\text{floor}_i] = \text{flrvec}_j[\text{floor}_i] + 1$$
$$\text{occup}_j = \text{occup}_j + 1$$

Step 8 Get next customer and update clock time:

$$i = i + 1$$
$$\text{Generate between}_i \text{ and floor}_i$$
$$\text{TIME} = \text{TIME} + \text{between}_i$$
$$\text{delivery}_i = 15$$

Step 9   Set all available elevators to current clock time:

For $k = 1$ to 4:

If TIME $\geq$ return$_k$, then return$_k$ = TIME.
Else leave return$_k$ as is.

Step 10   If between$_i \leq 15$ and occup$_j < 12$, then increase the delivery times for each customer on the tagged elevator j:

For $k = $ first$_j$ to $i - 1$:

delivery$_k$ = delivery$_k$ + between$_i$

and GOTO Step 7 to load current customer on the elevator and get the next customer.

Else (∗ send off the tagged elevator ∗):
Set limit $= i - 1$ and GOTO Step 11.

The sequence of Steps 11–18 implements delivery of all passengers on the currently tagged elevator j:

Step 11   For $k = $ first$_j$ to limit, do Steps 12–16.

Step 12   Calculate time customer k spends in elevator:

$$N = \text{floor}_k - 1 \text{ (an index)}$$

elevator$_k$ = travel time up to floor + time to drop off previous
customers + customer k drop off time
+ open/close door times on previous floors
+ open door on current floor

$$= 10N + 3 \sum_{m=1}^{N} \text{flrvec}_j[m] + 3 + 10 \sum_{m=1}^{N} \text{selvec}_j[m] + 5$$

Step 13   Calculate delivery time for customer k:

delivery$_k$ = delivery$_k$ + elevator$_k$

Step 14   Sum to total delivery time for averaging:

DELTIME = DELTIME + delivery$_k$

Step 15   If delivery$_k >$ MAXDEL, then MAXDEL = delivery$_k$.
Else leave MAXDEL as is.

Step 16   If elevator$_k >$ MAXELEV, then MAXELEV = elevator$_k$.
Else leave MAXELEV as is.

Step 17   Calculate total number of stops for elevator j, its time in transit, and the time at which it returns to the main floor:

$$\text{stop}_j = \text{stop}_j + \sum_{m=1}^{12} \text{selvec}_j[m]$$

$\text{Max} = $ index of largest nonzero entry in selvec_j (i.e., the highest floor visited)

$\text{eldel}_j = $ travel time + passenger drop-off time + doors time

$$= 20(\text{Max} - 1) + 3 \sum_{m=1}^{12} \text{flrvec}_j[m] + 10 \sum_{m=1}^{12} \text{selvec}_j[m]$$

$\text{return}_j = \text{TIME} + \text{eldel}_j$

$\text{operate}_j = \text{operate}_j + \text{eldel}_j$

Step 18   GOTO Step 5.

The sequence of Steps 19–32 is taken when no elevator is currently available and a queue of customers waiting for elevator service is set up.

Step 19   Initialize queue:

$\text{quecust} = i$ (number of first customer in queue)

$\text{startque} = \text{TIME}$ (starting time of queue)

$\text{queue} = 1$

$\text{arrive}_i = \text{TIME}$

Step 20   Get the next customer and update clock time:

$$i = i + 1$$

Generate between_i and floor_i

$\text{TIME} = \text{TIME} + \text{between}_i$

$\text{arrive}_i = \text{TIME}$

$\text{queue} = \text{queue} + 1$

Step 21   Check for elevator availability:

If $\text{TIME} \geqslant \text{return}_1$, then $j = 1$ and GOTO Step 22 else

If $\text{TIME} \geqslant \text{return}_2$, then $j = 2$ and GOTO Step 22 else

If $\text{TIME} \geqslant \text{return}_3$, then $j = 3$ and GOTO Step 22 else

If $\text{TIME} \geqslant \text{return}_4$, then $j = 4$ and GOTO Step 22.

ELSE (* no elevator is available yet *) GOTO Step 20.

Step 22   Elevator j is available. Initialize the floor selection vectors and assess the length of the queue:

For $k = 1$ to 12: $\text{selvec}_j[k] = \text{flrvec}_j[k] = 0$

$\text{remain} = \text{queue} - 12$

Step 23   If $\text{remain} \leqslant 0$, then $R = i$ and $\text{occup}_j = \text{queue}$.
Else $R = \text{quecust} + 11$ and $\text{occup}_j = 12$.

Step 24   Load customers onto elevator j:

For $k = \text{quecust}$ to R:

$\text{selvec}_j[\text{floor}_k] = 1$ and $\text{flrvec}_j[\text{floor}_k] = \text{flrvec}_j[\text{floor}_k] + 1$

Step 25 If queue ≥ QUELEN, then QUELEN = queue.
Else leave QUELEN as is.

Step 26 Update queuing totals:

$$quetotal = quetotal + occup_j$$

$$QUETIME = QUETIME + \sum_{m=quecust}^{R} [TIME - arrive_m]$$

Step 27 If (TIME − startque) ≥ MAXQUE, then MAXQUE = TIME − startque.
Else leave MAXQUE as is.

Step 28 Set index giving number of first customer to occupy tagged elevator:

$$first_j = quecust$$

Step 29 Calculate delivery and waiting times for each passenger on the tagged elevator:

For $k = first_j$ to R:
$$delivery_k = 15 + (TIME - arrive_k)$$
$$wait_k = TIME - arrive_k$$

Step 30 If remain ≤ 0, then set queue = 0 and GOTO Step 8 to get next customer. Else set limit = R and, for $k = first_j$ to limit, do Steps 12–17. When finished, GOTO Step 31.

Step 31 Update queue length and check for elevator availability:

$$queue = remain$$
$$quecust = R + 1$$
$$startque = arrive_{R+1}$$

Step 32 GOTO Step 20.

The sequence of Steps 33–36 calculates output values for the morning rush-hour elevator simulation.

Step 33 Output the following values:
$N = i - queue$, the total number of customers served
DELTIME = DELTIME/N, average delivery time
MAXDEL, maximum delivery time of a customer

Step 34 Output the average time spent in an elevator and the maximum time spent in an elevator:

$$ELEVTIME = \sum_{m=1}^{limit} \frac{elevator[m]}{limit} \quad \text{and MAXELEV}$$

Step 35 Output the number of customers waiting in the longest queue, the average time a customer who waits in line spends in a queue, and the longest time spent in a queue:

$$\text{QUELEN}$$

$$\text{QUETIME} = \frac{\text{QUETIME}}{\text{quetotal}}$$

$$\text{MAXQUE}$$

Step 36   Output the total number of stops for each elevator and the percent time each elevator is in transport:

$$\text{For } k = 1 \text{ to } 4 \text{: display stop}_k \text{ and } \frac{\text{operate}_k}{4800}$$

STOP

Note   For ease of presentation in the elevator simulation, TIME is updated only as the next customer arrives in the lobby. Therefore TIME is not an actual clock being updated every second. It is possible, when a queue has formed, that an elevator returns to the main floor during Step 20, before the next customer arrives. However, loading of the available elevator does not commence until that customer actually arrives. For this reason, the times spent waiting in a queue are slightly on the high side. In Problem 2 you are asked to modify Steps 20–32 in the algorithm so that loading commences immediately upon the return of the first available elevator.

TABLE 8-18 Results of elevator simulation for 15 consecutive days

Simulation number	Numbers of customers serviced	Average delivery time	Maximum delivery time	Average time in elevator	Maximum time in elevator	Number of customers in longest queue	Average time in a queue	Longest time in a queue	Total number of stops for each elevator				Percent of total time each elevator is in transport			
									1	2	3	4	1	2	3	4
1	328	147	412	89	208	12	40	166	67	76	56	52	84	87	80	75
2	322	146	409	88	211	18	43	176	74	62	52	61	88	80	78	77
3	309	139	385	87	201	12	37	161	62	61	61	62	85	83	80	80
4	331	149	371	89	205	13	42	149	72	69	68	44	85	82	87	73
5	320	146	404	87	208	15	48	178	72	52	58	58	86	78	80	74
6	313	153	405	91	211	13	45	146	69	62	72	47	85	82	82	74
7	328	138	341	88	195	10	35	120	59	66	70	61	82	81	84	82
8	312	147	377	86	198	12	46	163	69	60	61	43	86	82	81	73
9	329	139	352	87	208	11	37	155	58	63	70	57	83	86	82	78
10	314	143	325	88	205	9	35	128	65	68	57	65	86	84	78	83
11	317	137	344	85	202	10	38	129	64	75	64	56	87	85	81	77
12	341	153	396	90	211	18	45	177	83	63	63	53	91	82	78	77
13	318	136	345	80	208	11	36	140	58	64	63	55	85	82	83	73
14	319	140	356	88	208	13	33	135	64	67	58	65	84	83	82	81
15	323	147	386	91	218	15	39	166	76	64	70	54	86	81	87	76
Averages (rounded)	322	144	374	88	206	13	40	153	67	65	63	56	86	83	82	77

Note: All times are measured in seconds and rounded to the nearest second.

Table 8-18 gives the results of 15 independent simulations, representing three weeks of the morning rush hour, according to the preceding algorithm.

8.4 PROBLEMS

1. Consider an intersection of two one-way streets controlled by a traffic light. Assume that between 5 and 15 cars (varying probabilistically) arrive at the intersection every 10 seconds going in direction 1, and that between 6 and 24 cars arrive every 10 seconds in direction 2. Suppose that 36 cars per 10 sec can cross the intersection in direction 1 and that 20 cars per 10 sec can cross the intersection in direction 2, if the traffic light is green. No turning is allowed. Initially, assume the traffic light is green for 30 sec and red for 70 sec in direction 1. Write a simulation algorithm to answer the following questions for a 60-min time period:
   a. How many cars pass through the intersection in direction 1 during the hour?
   b. What is the average waiting time of a car stopped when the traffic signal is red in direction 1? The maximum waiting time?
   c. What is the average length of the queue of cars stopped for a red light in direction 1? The maximum length?
   d. What is the average number of cars passing through the intersection in direction 1 during the time when the traffic light is green? What is the maximum number?
   e. Answer questions a–d for direction 2.
   How would you use your simulation to determine the switching period for which the total waiting time in both directions is as small as possible? (You will have to modify it to account for the waiting times in direction 2.)
2. Modify Steps 20–32 in the elevator simulation algorithm so that loading of the first available elevator commences immediately upon its return. Thus, if $TIME > return_j$, so that elevator j is available for loading, then loading commences at time $return_j$ rather than TIME. Consider how you will now process customer i in Step 20 who has not yet quite arrived on the scene.

8.4 PROJECTS

1. Find a building in your local area that has from 4 to 12 floors that are serviced by 1 to 4 elevators. Collect data for the interarrival times (and, possibly, floor destinations) of the customers during a busy hour (like the morning rush hour), and build the interarrival and destination submodels based on your data (by constructing the cumulative histograms, as discussed in Section 8.2). Write a computer program incorporating your submodels into the elevator system algorithm to obtain results like those given in Table 8-18.

2. Write a computer program to implement your algorithm in Problem 1 of this section simulating the traffic intersection.

3. Pick a busy intersection of two one-way streets in your local area and collect data on how many cars arrive at the intersection in each direction every 10 sec. Determine also how many cars can cross the intersection in each direction, if the traffic light is green. Determine the switching times of the traffic light. Develop arrival and intersection-crossing submodels for each direction, based on your data. Write a computer program incorporating your submodels into the algorithm you developed in Problem 1 of this section. What are the best switching times of the traffic light in order to minimize the total waiting time for cars in both directions?

4. Write a summary report on the article, "A First Course in Continuous Simulation," by Richard Bronson, *The Two-Year College Mathematics Journal 13*, no. 5 (1982):300–310.

8.4 FURTHER READING

Forrester, Jay W. *World Dynamics.* Cambridge, Mass.: Wright-Allen, 1971.

Guetzkow, Harold, P. Kotler, and R. L. Schultz. *Simulation in Social and Administrative Science.* Englewood Cliffs, N.J.: Prentice-Hall, 1972.

Shannon, Robert E. *Systems Simulation: The Art and Science.* Englewood Cliffs, N.J.: Prentice-Hall, 1975.

PART FOUR

MODELING DYNAMIC BEHAVIOR

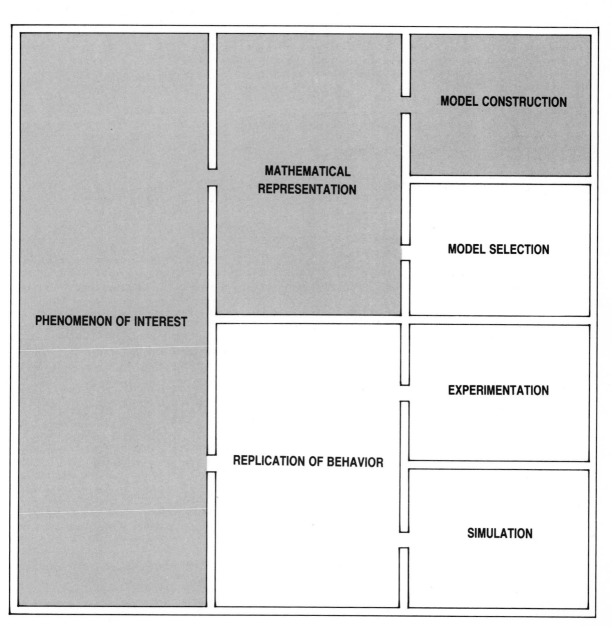

PHENOMENON OF INTEREST

MATHEMATICAL
REPRESENTATION

MODEL CONSTRUCTION

MODEL SELECTION

REPLICATION OF BEHAVIOR

EXPERIMENTATION

SIMULATION

MODELING USING THE DERIVATIVE

INTRODUCTION

Quite often we have information relating a rate of change of a dependent variable with respect to one or more independent variables and are interested in discovering the function relating the variables. For example, if P represents the number of people in a large population at some time t, then it is reasonable to assume that the rate of change of the population with respect to time depends on the current size of P as well as other factors that are discussed in Section 9.1. For ecological, economical, and other important reasons, it is desirable to determine a relationship between P and t in order to make predictions about P. If the present population size is denoted by $P(t)$ and the population size at time $t + \Delta t$ is $P(t + \Delta t)$, then the change in population ΔP during that time period Δt is given by

$$\Delta P = P(t + \Delta t) - P(t) \qquad (9.1)$$

The factors affecting the population growth are developed in detail in Section 9.1. For now, let's merely assume a simple proportionality: $\Delta P \propto P$. For example, if immigration, emigration, age, and gender are all neglected, we may assume that during a unit time period a certain percentage of the population reproduces while a percentage dies. Suppose that the constant of proportionality k is expressed as a percentage per unit time. Then our proportionality assumption gives

$$\Delta P = P(t + \Delta t) - P(t) = kP \, \Delta t \qquad (9.2)$$

difference equation Equation (9.2) is called a **difference equation.** Note that we are treating a discrete set of time periods rather than allowing t to vary *continuously* over

302

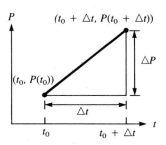

discrete mathematics
finite mathematics

FIGURE 9-1 The size of the population at future time intervals of length Δt gives a discrete set of points.

some interval. Difference equations belong to an important area of mathematics, called **discrete** or **finite mathematics.** In this situation the discrete set of times may give the population in future years at those distinct times (perhaps after the spring spawn in a fish population). Referring to Figure 9-1, observe that the horizontal distance between the points $(t_0, P(t_0))$ and $(t_0 + \Delta t, P(t_0 + \Delta t))$ is Δt, which may represent the time between spawning periods in a fish population growth problem or the length of a fiscal period in a budget growth problem. The time t_0 refers to a particular time. The vertical distance, ΔP in this case, represents the change in the dependent variable.

Assume now that t does vary continuously so that we can take advantage of the calculus. Division of Equation (9.2) by Δt gives

$$\frac{\Delta P}{\Delta t} = \frac{P(t + \Delta t) - P(t)}{\Delta t} = kP \qquad (9.3)$$

We may interpret $\Delta P/\Delta t$ physically as the *average rate of change* in P during the time period Δt. For example, $\Delta P/\Delta t$ may represent the average daily growth of the budget. However, in other scenarios it may have no physical interpretation: if fish spawn only in the spring, it is somewhat meaningless to talk about the average daily growth in the fish population. Again in Figure 9-1, $\Delta P/\Delta t$ can be interpreted geometrically as the slope of the line segment connecting the point $(t_0, P(t_0))$ and $(t_0 + \Delta t, P(t_0 + \Delta t))$. Next allow Δt to approach zero. The definition of the derivative gives the *differential equation*

$$\lim_{\Delta t \to 0} \frac{\Delta P}{\Delta t} = \frac{dP}{dt} = kP$$

where dP/dt represents the *instantaneous rate of change*. In many situations the instantaneous rate of change has an identifiable physical interpretation such as in the flow of heat from a space capsule after entering the ocean or the reading of a car speedometer as the car accelerates. However, in the case of a fish population that has a discrete spawning period or the budget process that has a discrete fiscal period, the instantaneous change may be somewhat meaningless. These latter scenarios are more appropriately modeled using difference equations, but it is occasionally advantageous to approximate a difference equation with a differential equation. Because of the scope of this book, we study the derivative for modeling dynamic scenarios.*

The derivative is used in two distinct roles:

1. To represent the instantaneous rate of change in "continuous" problems.
2. To approximate an average rate of change in "discrete" problems.

The advantage of approximating an average rate of change by a derivative is that the calculus often helps in uncovering a functional relationship between the variables under investigation. For instance, the solution to the model (9.3) is $P = P_0 e^{kt}$, where P_0 is the population at time $t = t_0$. However,

---

* An excellent introduction to difference equations is the UMAP module "Difference Equations With Applications," by Donald R. Sherbert, listed in the projects for Section 9.1.

many differential equations cannot be solved so easily using analytic techniques. In such cases the solutions are approximated using discrete methods. An introduction to numerical techniques is presented in Section 9.4. In case the solution being approximated is to a differential equation that is itself an approximation to a difference equation, the modeler should consider using a discrete method with the finite difference equation directly.

The interpretation of the derivative as an instantaneous rate of change is useful in many modeling applications. The geometrical interpretation of the derivative as the slope of the line tangent to the curve is useful for constructing numerical solutions. Let's review more carefully these important concepts from the calculus.

The Derivative as a Rate of Change

The origins of the derivative lie in humankind's curiosity about motion and our need to develop a deeper understanding of motion. The search for the laws governing planetary motion, the study of the pendulum and its application to clock building, and the laws governing the flight of a cannonball were the kinds of problems stimulating the minds of mathematicians and scientists in the 16th and 17th centuries. Such problems motivated the development of the calculus.

To remind ourselves of one interpretation of the derivative, consider a particle whose distance s from a fixed position depends upon time t. Let the graph in Figure 9-2 represent the distance s as a function of time t, and let (t_1, s_1) and (t_2, s_2) denote two points on the graph.

Define $\Delta t = t_2 - t_1$ and $\Delta s = s_2 - s_1$, and form the ratio $\Delta s/\Delta t$. Note that this ratio represents a rate: an increment of distance traveled Δs over some increment of time Δt. That is, the ratio $\Delta s/\Delta t$ represents the average velocity during the time period in question. Now remember how the derivative ds/dt evaluated at $t = t_1$ is defined:

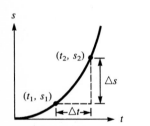

FIGURE 9-2 Graph of distance s as a function of time t.

$$\left.\frac{ds}{dt}\right|_{t=t_1} = \lim_{\Delta t \to 0} \frac{\Delta s}{\Delta t} \tag{9.4}$$

Physically, what occurs as $\Delta t \to 0$? Using the interpretation of average velocity, we can see that at each stage of using a smaller Δt we are computing the average velocity over smaller and smaller intervals with left endpoint at t_1 until, in the limit, we have the instantaneous velocity at $t = t_1$. If you think of the motion of a moving vehicle, this instantaneous velocity would correspond to the exact reading of its (perfect) speedometer at the instant t_1.

More generally, if $y = f(x)$ is a differentiable function, then the derivative dy/dx at any given point can be interpreted as the *instantaneous rate of change* of y with respect to x at that point. Interpreting the derivative as an instantaneous rate of change is useful in many modeling applications.

The Derivative as the Slope of the Tangent Line

Let's consider another interpretation of the derivative. As scholars sought knowledge about the laws of planetary motion, their chief need was to ob-

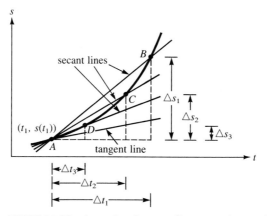

FIGURE 9-3 The slope of each secant line approximates the slope of the tangent line to the curve at the point A.

serve and measure the heavenly bodies. However, the construction of lenses for use in telescopes was a difficult task. To grind a lens to the correct curvature to achieve the desired light refraction requires knowing the tangent to the curve describing the lens surface.

Let's examine the geometrical implications of the limit in Equation (9.4). We consider $s(t)$ now simply as a curve. Let's examine a set of secant lines each emanating from the point $A = (t_1, s(t_1))$ on the curve. To each secant there corresponds a pair of increments $(\Delta t_i, \Delta s_i)$ as shown in Figure 9-3. The lines AB, AC, and AD are secant lines. As $\Delta t \to 0$, these secant lines approach the line tangent to the curve at the point A. Since the slope of each secant is $\Delta s/\Delta t$, we may interpret the derivative as *the slope of the line tangent to the curve $s(t)$ at the point A*. The interpretation of the derivative evaluated at a point as the slope of the line tangent to the curve at that point is useful in constructing numerical approximations to solutions of differential equations. Numerical approximations are discussed in Section 9.4.

MODEL

9.1 POPULATION GROWTH

Interest in how populations tend to grow was stimulated in the late 18th century when Thomas Malthus (1766–1834) published *An Essay on the Principle of Population as it Affects the Future Improvement of Society*. In his book Malthus put forth an exponential growth model for human population and concluded that eventually the population would exceed the capacity to grow an adequate food supply. Although the assumptions of the Malthusian model leave out factors important to population growth (so the model has proven to be inaccurate for technologically developed countries), nevertheless it is instructive to examine this model as a basis for later refinement.

Problem identification   Suppose you know the population at some given time, say it is P_0 at time $t = t_0$, and you are interested in predicting the population P at some future time $t = t_1$. In other words, you want to *find a population function $P(t)$ for $t_0 \leqslant t \leqslant t_1$ satisfying $P(t_0) = P_0$.*

Assumptions   Consider some factors that pertain to population growth. Two obvious factors are the *birthrate* and the *deathrate*. The birthrate and deathrate are determined by different factors. The birthrate is influenced by infant mortality rate, attitudes toward and availability of contraceptives, attitudes toward abortion, health care during pregnancy, and so forth. The deathrate is affected by sanitation and public health, wars, pollution, medicines, diet, psychological stress and anxiety, and so forth. Other factors that influence population growth in a given region are immigration and emigration, living space restrictions, availability of food and water, epidemics, and so forth. For our model, let's neglect all these latter factors. (If we are dissatisfied with our results, we can include these factors later in a more refined model, possibly in a simulation model.) Now we'll consider only the birthrate and deathrate. Since knowledge and technology have helped humankind diminish the deathrate below the birthrate, human populations have tended to grow.

Let's begin by assuming that during a small unit time period a percentage b (given as a decimal equivalent) of the population is newly born. Similarly, a percentage c of the population dies. In other words, the new population $P(t + \Delta t)$ is the old population $P(t)$ plus the number of births minus the number of deaths during the time period Δt. Symbolically,

$$P(t + \Delta t) = P(t) + bP(t)\Delta t - cP(t)\Delta t$$

or

$$\frac{\Delta P}{\Delta t} = bP - cP = kP$$

Thus, we are really assuming that the average rate of change of the population over an interval of time is proportional to the size of the population. Using the instantaneous rate of change to approximate the average rate of change, we have the following model:

$$\frac{dP}{dt} = kP, \qquad P(t_0) = P_0, \qquad t_0 \leqslant t \leqslant t_1 \tag{9.5}$$

where (for growth) k is a positive constant.

Solving the model   We can separate the variables and rewrite Equation (9.5) by moving all terms involving P and dP to one side of the equation and all terms in t and dt to the other. This gives

$$\frac{dP}{P} = k\,dt$$

Integration of both sides of this last equation yields

$$\ln P = kt + C \qquad\qquad\qquad\qquad \textbf{(9.6)}$$

for some constant C. Applying the condition $P(t_0) = P_0$ to Equation (9.6) to find C results in

$$\ln P_0 = kt_0 + C$$

or

$$C = \ln P_0 - kt_0$$

Then substitution for C into (9.6) gives

$$\ln P = kt + \ln P_0 - kt_0$$

or, simplifying algebraically,

$$\ln \frac{P}{P_0} = k(t - t_0)$$

Finally, by taking the exponential of both sides of the preceding equation and multiplying the result by P_0, we obtain the solution:

$$P(t) = P_0 e^{k(t - t_0)} \qquad\qquad\qquad\qquad \textbf{(9.7)}$$

Malthusian model of population growth   Equation (9.7), which is known as the **Malthusian model of population growth,** predicts that population grows exponentially with time.

Verifying the model   Since $\ln (P/P_0) = k(t - t_0)$, our model predicts that if we plot $\ln P/P_0$ versus $t - t_0$, a straight line passing through the origin with slope k should result. However, if we plot the population data for the United States for several years, the model does not fit very well, especially in the later years. In fact, the 1970 census for the population of the United States was 203,211,926, and in 1950 it was 150,697,000. Substituting these values into Equation (9.7) and dividing the first result by the second gives

$$\frac{203{,}211{,}926}{150{,}697{,}000} = e^{k(1970 - 1950)}$$

Thus,

$$k = \left(\frac{1}{20}\right) \ln \frac{203{,}211{,}926}{150{,}697{,}000} \approx 0.015$$

That is, during the 20-year period from 1950 to 1970, population in the United States was increasing at the average rate of 1.5% per year. We can use this information together with Equation (9.7) to predict the population for 1980. In this case, $t_0 = 1970$, $P_0 = 203,211,926$, and $k = 0.015$ yields

$$P(1980) = 203,211,926e^{0.015(1980 - 1970)} = 236,098,574$$

The 1980 census for the population of the United States was 226,505,000 (rounded to the nearest thousand). Thus, our prediction is off the mark by roughly 4%. We can probably live with that magnitude error, but let's look into the distant future. Our model predicts that the population of the United States will be 28,688 billion in the year 2300, a population that exceeds current estimates of the maximum sustainable population of the entire planet! We are forced to conclude that our model is unreasonable over the long run.

Some populations do grow exponentially provided that the population is not too large. However, in most populations individual members eventually compete with one another for food, living space, and other natural resources. Let's refine our Malthusian model of population growth to reflect this competition.

Refining the model to reflect limited growth   Recall our graphical model for the growth in a fish population in the absence of fishing presented in Section 5.5. In this limited growth model we assume that there is some maximum sustainable population M (we called $M = N_u$ in Section 5.5). Moreover, the proportionality factor k, measuring the rate of population growth in Equation (9.5), is now no longer constant, but a function of the population. As the population increases and gets closer to the maximum population M, the rate k decreases. One simple submodel for k is the linear one

$$k = r(M - P), \qquad r > 0$$

where r is a constant. Substitution into Equation (9.5) leads to

$$\frac{dP}{dt} = r(M - P)P \tag{9.8}$$

or

$$\frac{dP}{P(M - P)} = r\,dt \tag{9.9}$$

Again we assume the *initial condition* $P(t_0) = P_0$. (The model (9.8) was first introduced by the Dutch mathematical biologist Pierre-François Verhulst, 1804–1849), and is referred to as *logistic growth*. It follows from elementary algebra that

$$\frac{1}{P(M - P)} = \frac{1}{M}\left(\frac{1}{P} + \frac{1}{M - P}\right)$$

Thus, Equation (9.9) can be rewritten as

$$\frac{dP}{P} + \frac{dP}{M - P} = rM \, dt$$

which integrates to

$$\ln P - \ln |M - P| = rMt + C \qquad \textbf{(9.10)}$$

for some arbitrary constant C. Using the initial condition, we evaluate C in the case $P < M$:

$$C = \ln \frac{P_0}{M - P_0} - rMt_0$$

Substituting into (9.10) and simplifying gives

$$\ln \frac{P}{M - P} - \ln \frac{P_0}{M - P_0} = rM(t - t_0)$$

or

$$\ln \frac{P(M - P_0)}{P_0(M - P)} = rM(t - t_0)$$

Exponentiating both sides of this equation gives

$$\frac{P(M - P_0)}{P_0(M - P)} = e^{rM(t - t_0)}$$

or

$$P_0(M - P)e^{rM(t - t_0)} = P(M - P_0)$$

Then,

$$P_0 M e^{rM(t - t_0)} = P(M - P_0) + P_0 P e^{rM(t - t_0)}$$

so that solving for the population P gives

$$P(t) = \frac{P_0 M e^{rM(t - t_0)}}{M - P_0 + P_0 e^{rM(t - t_0)}}$$

In order to estimate P as $t \to \infty$, we rewrite this last equation as

$$P(t) = \frac{M P_0}{[P_0 + (M - P_0)e^{-rM(t - t_0)}]} \qquad \textbf{(9.11)}$$

Notice from Equation (9.11) that $P(t)$ approaches M as t tends to infinity. Moreover, from Equation (9.8) we calculate the second derivative

$$P'' = rMP' - 2rPP' = rP'(M - 2P)$$

so $P'' = 0$ when $P = M/2$. In words, when the population P reaches half the limiting population M, the growth dP/dt is most rapid and then starts to diminish toward zero. One advantage of recognizing that the maximum rate of growth occurs at $P = M/2$ is that the information can be used to estimate

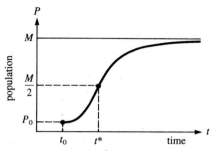

FIGURE 9-4 Graph of the limited growth model.

logistic curve M. In a situation where the modeler is satisfied that the growth involved is essentially logistic, if the point of maximum rate of growth has been reached, then $M/2$ can be estimated. The graph of the limited growth equation (9.11) is depicted in Figure 9-4 for the case $P < M$ (see Problem 2 for the case $P > M$). Such a curve is called a **logistic curve.**

Verifying the limited growth model   Let's test our model (9.11) against some real-world data. Equation (9.10) suggests a straight-line relationship of $\ln [P/(M - P)]$ versus t. Let's test this model using the data given in Table 9-1 for the growth of yeast in a culture. In order to plot $\ln [P/(M - P)]$ versus t, we need an estimate for the limiting population M. From the data in Table 9-1 we see that the population never exceeds 661.8. We estimate $M \approx$

TABLE 9-1 Growth of yeast in a culture

Time in hours	Observed yeast biomass	Biomass calculated from logistic Equation (9.13)	Percent error
0	9.6	8.9	−7.3
1	18.3	15.3	−16.4
2	29.0	26.0	−10.3
3	47.2	43.8	−7.2
4	71.1	72.5	2.0
5	119.1	116.3	−2.4
6	174.6	178.7	2.3
7	257.3	258.7	0.5
8	350.7	348.9	−0.5
9	441.0	436.7	−1.0
10	513.3	510.9	−4.7
11	559.7	566.4	1.2
12	594.8	604.3	1.6
13	629.4	628.6	−0.1
14	640.8	643.5	0.4
15	651.1	652.4	0.2
16	655.9	657.7	0.3
17	659.6	660.8	0.2
18	661.8	662.5	0.1

Data from R. Pearl, "The Growth of Population," *Quart. Rev. Biol. 2* (1927): 532–548.

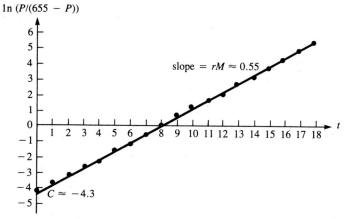

ln $(P/(655 - P))$

slope $= rM \approx 0.55$

$C \approx -4.3$

FIGURE 9-5 Plot of ln $[P/(665 - P)]$ versus t for the data in Table 9-1.

665 and plot ln $[P/(665 - P)]$ versus t. The graph is shown in Figure 9-5 and does approximate a straight line. Thus we accept the assumptions of logistic growth for bacteria. Now Equation (9.10) gives

$$\ln \frac{P}{M - P} = rMt + C$$

and from the graph in Figure 9-5 we can estimate the slope $rM \approx 0.55$, so that $r \approx 0.0008271$ from our estimate for $M \approx 665$.

It is often convenient to express the logistic equation (9.11) in another form. To this end, let t^* denote the time when the population P reaches half the limiting value; that is, $P(t^*) = M/2$. It follows from Equation (9.11) that

$$t^* = t_0 - \frac{1}{rM} \ln \frac{P_0}{M - P_0}$$

(see Problem 1a). Solving this last equation for t_0, substituting the result into (9.11), and simplifying algebraically gives

$$P(t) = \frac{M}{1 + e^{-rM(t - t^*)}} \qquad \textbf{(9.12)}$$

(see Problem 1b).

We can estimate t^* for the yeast culture data presented in Table 9-1 using Equation (9.10) and our graph in Figure 9-5:

$$t^* = -\frac{C}{rM} \approx \frac{4.3}{0.55} \approx 7.82$$

This calculation gives the logistic equation

$$P(t) = \frac{665}{1 + 73.8e^{-0.55t}} \qquad \textbf{(9.13)}$$

by substituting $M = 665$, $r = 0.0008271$, and $t^* = 7.82$ in (9.12).

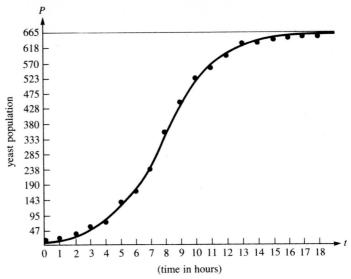

FIGURE 9-6 Logistic curve showing the growth of yeast in a culture based on the data from Table 9-1 and the model (9.13). The small circles indicate the observed values.

The logistic model is known to agree quite well for populations of organisms that have very simple life histories, as for instance, yeasts growing in a culture where space is limited. Table 9-1 shows the calculations for the logistic equation (9.13), and you can see from the calculated error that there is very good agreement with the original data. A plot of the curve is shown in Figure 9-6.

Next let's consider some data for human populations. A logistic equation for the population growth of the United States was formulated by Pearl and Reed in 1920. One form of their logistic curve is given by

$$P(t) = \frac{197,273,522}{1 + e^{-0.03134(t - 1914.32)}} \qquad (9.14)$$

where $M = 197,273,522$, $r = 1.5887 \times 10^{-10}$, and $t^* = 1914.32$ were determined using the census figures for the years 1790, 1850, and 1910 (we ask you to estimate M, r, and t^* in Problem 4).

Table 9-2 compares the values predicted in 1920 by the logistic equation (9.14) with the observed values of the population of the United States. The predicted values agree quite well with the observations up to the year 1950, but as you can see, the predicted values are much too small for the years 1970 and 1980. This should not be too surprising since our model fails to take into account such factors as immigration into the United States, wars, advances in medical technology, and so forth. In populations of higher plants and animals, which have complicated life histories and long periods of individual development, there are likely to be numerous responses that greatly modify the population growth.

TABLE 9-2 Population of the United States from 1790 to 1980, with predictions from Equation (9.14)

Year	Observed population	Predicted population	Percent error
1790	3,929,000	3,929,000	0.0
1800	5,308,000	5,336,000	0.5
1810	7,240,000	7,227,000	−0.2
1820	9,638,000	9,756,000	1.2
1830	12,866,000	13,108,000	1.9
1840	17,069,000	17,505,000	2.6
1850	23,192,000	23,191,000	−0.0
1860	31,443,000	30,410,000	−3.3
1870	38,558,000	39,370,000	2.1
1880	50,156,000	50,175,000	0.0
1890	62,948,000	62,767,000	−0.3
1900	75,995,000	76,867,000	1.1
1910	91,972,000	91,970,000	−0.0
1920	105,711,000	107,393,000	1.6
1930	122,755,000	122,396,000	−0.3
1940	131,669,000	136,317,000	3.5
1950	150,697,000	148,677,000	−1.3
1960	179,323,000	159,230,000	−11.2
1970	203,212,000	167,943,000	−17.4
1980	226,505,000	174,941,000	−22.8

9.1 PROBLEMS

1. **a.** Show that the population P in the logistic equation reaches half the maximum population M at time t^* given by

$$t^* = t_0 - (1/rM) \ln [P_0/(M - P_0)]$$

   b. Derive the form given by Equation (9.12) for population growth according to the logistic law.
   c. Derive the equation $\ln [P(M - P)] = rMt - rMt^*$ from Equation (9.12).
2. Consider the solution of Equation (9.8). Evaluate the constant C in (9.10) in the case that $P > M$ for all t. Sketch the solutions in this case. Sketch also a solution curve for the case that $M/2 < P < M$.
3. The following data were obtained for the growth of a sheep population introduced into a new environment on the island of Tasmania. (Adapted from J. Davidson, "On the Growth of the Sheep Population in Tasmania," *Trans. Roy. Soc. S. Australia 62*(1938): 342–346.)

t (year)	1814	1824	1834	1844	1854	1864
$P(t)$	125	275	830	1200	1750	1650

a. Make an estimate of M by graphing $P(t)$.

b. Plot $\ln [P/(M - P)]$ against t. If a logistic curve seems reasonable, estimate rM and t^*.

4. Using the data for the U.S. population in Table 9-2, estimate M, r, and t^* using the same technique as in the text. Assume you are making the prediction in 1951 using previous census. Use the data from 1960 to 1980 to check your model.

5. The modern philosopher Jean Jacques Rousseau formulated a simple model of population growth for 18th-century England based on the following assumptions:

> The birthrate in London is less than that in rural England.
> The deathrate in London is greater than that in rural England.
> As England industrializes, more and more people migrate from the countryside to London.

Rousseau then reasoned that since London's birthrate was lower and its deathrate higher and rural people tend to migrate there, the population of England would eventually decline to zero. Criticize Rousseau's conclusion.

6. Consider the spreading of a highly communicable disease on an isolated island with population size N. A portion of the population travels abroad and returns to the island infected with the disease. You would like to predict the number of people X who will have been infected by some time t. Consider the following model:

$$\frac{dX}{dt} = kX(N - X)$$

a. List two major assumptions implicit in the preceding model. How reasonable are the assumptions you list?

b. Graph dX/dt versus X.

c. Graph X versus t if the initial number of infections is $X_1 < N/2$. Graph X versus t if the initial number of infections is $X_2 > N/2$.

d. Solve the model given earlier for X as a function of t.

e. From Part d, find the limit of X as t approaches infinity.

f. Consider an island with a population of 5000. At various times during the epidemic the number of people infected was recorded as follows:

t (days)	2	6	10
X (people infected)	1887	4087	4853
$\ln (X/(N - X))$	−.5	1.5	3.5

Do the data collected support the given model?

g. Use the results in Part f to estimate the constants in the model, and predict the number of people who will be infected by $t = 12$ days.

7. Assume we are considering the survival of whales and that if the number of whales falls below a minimum survival level m the species will become

extinct. Assume also that the population is limited by the carrying capacity M of the environment. That is, if the whale population is above M, then it will experience a decline because the environment cannot sustain that large a population level.

a. Discuss the following model for the whale population:

$$\frac{dP}{dt} = k(M - P)(P - m)$$

where $P(t)$ denotes the whale population at time t and k is a positive constant.

b. Graph dP/dt versus P and P versus t. Consider the cases where the initial population $P(0) = P_0$ satisfies $P_0 < m$, $m < P_0 < M$, and $M < P_0$.

c. Solve the model in Part a, assuming that $m < P < M$ for all time. Show that the limit of P as t approaches infinity is M.

d. Discuss how you would test the model in Part a? How would you determine M and m?

e. Assuming that the model reasonably estimates the whale population, what implications are suggested for fishing? What controls would you suggest?

8. Sociologists recognize a phenomenon called "social diffusion," which is the spreading of a piece of information, a technological innovation, or a cultural fad among a population. The members of the population can be divided into two classes: those who have the information and those who do not. In a fixed population whose size is known, it is reasonable to assume that the rate of diffusion is proportional to the number who have the information times the number yet to receive it. If X denotes the number of individuals who have the information in a population of N people, then a mathematical model for social diffusion is given by $dX/dt = kX(N - X)$, where t represents time and k is a positive constant.

a. Solve the model and show that it leads to a logistic curve.

b. At what time is the information spreading fastest?

c. How many people will eventually receive the information?

9.1 PROJECTS

1. Complete the requirements of the UMAP module, "The Cobb-Douglas Production Function," by Robert Geitz, UMAP 509. A mathematical model relating the output of an economic system to labor and capital is constructed from the assumptions that (a) marginal productivity of labor is proportional to the amount of production per unit of labor, (b) marginal productivity of capital is proportional to the amount of production per unit of capital, and (c) if either labor or capital tends to zero, then so does production.

2. Complete the UMAP module "The Diffusion of Innovation in Family Planning," by Kathryn N. Harmon, UMAP 303. This module gives an interesting application of finite difference equations to study the process through which public policies are diffused in order to understand how national governments might adopt family planning policies.

3. Complete the UMAP module "Difference Equations With Applications," by Donald R. Sherbert, UMAP 322. This module presents a good introduction to solving first and second order linear difference equations, including the method of undetermined coefficients for nonhomogeneous equations. Applications to problems in population and economic modeling are included.

9.1 FURTHER READING

Frauenthal, James C. *Introduction to Population Modeling.* Lexington, Mass.: COMAP, 1979.

Hutchinson, G. Evelyn. *An Introduction to Population Ecology.* New Haven, Conn.: Yale University Press, 1978.

Levins, R. "The Strategy of Model Building in Population Biology." *American Scientist 54*(1966): 421–431.

Lotka, A. J. *Elements of Mathematical Biology.* New York: Dover, 1956.

Odum, E. P. *Fundamentals of Ecology.* Philadelphia: Saunders, 1971.

Pearl, R., L. J. Reed. "On the Rate of Growth of the Population of the United States since 1790." *Proceedings of the National Academy of Science 6*(1920): 275–288.

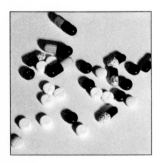

MODEL

9.2 PRESCRIBING DRUG DOSAGE*

The problem of how much of a dosage to prescribe for a drug and how often the dosage should be administered is an important one in pharmacology. For most drugs there is a concentration below which the drug is ineffective and a concentration above which the drug is dangerous.

---

* This section is adapted from UMAP Unit 72, based on the work of Brindell Horelick and Sinan Koont. The adaptation is presented with the permission of COMAP, Inc./UMAP, 271 Lincoln St., Lexington, Mass. 02173.

Problem identification *How can the doses and the time between doses be adjusted to maintain a safe but effective concentration of the drug in the blood?*

The concentration in the blood resulting from a single dose of a drug normally decreases with time as the drug is eliminated from the body (see Figure 9-7). However, we are interested in what happens to the concentration of the drug in the blood as doses are given at regular intervals. If H denotes the highest safe level of the drug and L its lowest effective level, it would be desirable to prescribe a dosage C_0 with time T between doses so that the concentration of the drug in the bloodstream remains between L and H over each dose period.

Let's consider several ways in which the drugs might be administered. In Figure 9-8a the time between doses is such that effectively there is no buildup of the drug in the system. In other words, the "residual concentration" from previous doses is approximately zero. On the other hand, in Figure 9-8b the interval between doses relative to the amount administered and the decay rate of the concentration is such that a residual concentration exists at each time the drug is taken (after the first dose). Furthermore, as depicted in the graph, this residual level seems to be approaching a limit. We will be concerned with determining if this situation is indeed the case and, if so, what that limit must be. Our ultimate goal in prescribing drugs is to determine dose *amounts* and *intervals* between doses such that the lowest effective level L is reached quickly and thereafter the concentration is maintained between the lowest effective level L and the highest safe level H, as depicted in Figure 9-9. We begin by determining the limiting residual level, which depends

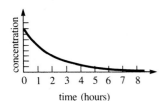

FIGURE 9-7 The concentration of a drug in the bloodstream decreases with time.

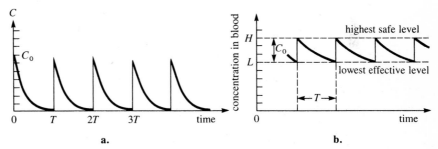

FIGURE 9-8 Residual buildup depends upon the time interval between administration of drug doses.

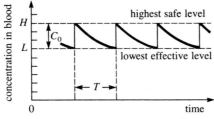

FIGURE 9-9 Safe but effective levels of the drug in the blood. C_0 is the change in concentration produced by one dose and T is the time interval between doses.

upon our assumptions for the rate of assimilation of the drug in the blood-stream and the rate of decay after assimilation.

Assumptions   In order to solve the problem we have identified, let's consider the factors that determine the concentration $C(t)$ of the drug in the blood-stream at any time t. We begin with

$$C(t) = f(\text{decay rate, assimilation rate, dosage amount, dosage interval}, \ldots)$$

and various other factors including body weight and blood volume. To sim-plify our assumptions, let's assume that body weight and blood volume are constants (say an average over some specific age group), and that concen-tration level is the critical factor in determining the effect of a drug.

Next we determine submodels for decay rate and assimilation rate.

Submodel for decay rate   Consider the elimination of the drug from the bloodstream. Probably this is a discrete phenomenon, but let's approximate it by a continuous function. Clinical experiments have revealed that the de-crease in the concentration of a drug in the bloodstream will be proportional to the concentration itself. Mathematically, this assumption means that if we assume the concentration of drug in the blood at time t is a differentiable function $C(t)$, then

$$C'(t) = -kC(t) \tag{9.15}$$

elimination constant   In this formula k is a positive constant, called the **elimination constant** of the drug. Notice $C'(t)$ is negative, as it should be if it is to describe a decreasing concentration. Usually the quantities in Equation (9.15) are measured as follows: the time t is given in hours, $C(t)$ is milligrams per milliliter of blood (mg/ml), $C'(t)$ is mg ml^{-1} hr^{-1}, and k is hr^{-1}.

Assume now that the concentrations H and L can be determined experi-mentally for a given population, such as an age group. (We will be saying more about this assumption in the following discussion.) Then set the drug concentration for a single dose at the level:

$$C_0 = H - L \tag{9.16}$$

If we assume that C_0 is the concentration at $t = 0$, then we have the model:

$$\frac{dC}{dt} = -kC, \qquad C(0) = C_0 \tag{9.17}$$

The variables can be separated in Equation (9.17) and the model solved in the same way as the Malthusian model of population growth presented in the preceding section. Solution of the model gives

$$C(t) = C_0 e^{-kt} \tag{9.18}$$

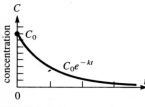

FIGURE 9-10 Exponential model for decay of drug con-centration with time.

To obtain the concentration at time $t > 0$, multiply the *initial concentration* C_0 by e^{-kt}. The graph of $C(t)$ looks like the one in Figure 9-10.

Submodel for assimilation rate   Having made an assumption about how drug concentrations decrease with time, let's consider how they increase again when drugs are administered. Our initial assumption is that when a drug is taken, it is diffused so rapidly throughout the blood that the graph of the concentration for the absorption period is, for all practical purposes, vertical. That is, we assume an instantaneous rise in concentration whenever a drug is administered. This assumption may not be as reasonable for a drug taken by mouth as it is for a drug that is injected directly into the bloodstream. Now let's see how the drug accumulates in the bloodstream with repeated doses.

Drug accumulation with repeated doses   Consider what happens to the concentration $C(t)$ when a dose that is capable of raising the concentration by C_0 mg/ml each time it is given is administered regularly at fixed time intervals of length T.

Suppose at time $t = 0$ the first dose is administered. According to our model (9.18), after T hours have elapsed, the residual $R_1 = C_0 e^{-kT}$ remains in the blood, and then the second dose is administered. Because of our assumption concerning the increase in drug concentration as previously discussed, the level of concentration instantaneously jumps to $C_1 = C_0 + C_0 e^{-kT}$. Then after T hours elapse again, the residual $R_2 = C_1 e^{-kT} = C_0 e^{-kT} + C_0 e^{-2kT}$ remains in the blood. This possibility of accumulation of the drug in the blood is depicted in the graph in Figure 9-11.

Next, we determine a formula for the nth residual R_n. If we let C_{i-1} be the concentration at the beginning of the ith interval and R_i the *residual concentration* at the end of it, we can easily obtain the following Table 9-3.

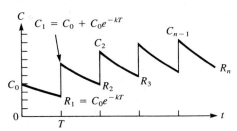

FIGURE 9-11 One possible effect of repeating equal doses.

TABLE 9-3 Calculation of residual concentration of drug

i	C_{i-1}	R_i
1	C_0 — multiply by e^{-kT} → $C_0 e^{-kT}$	
	← add C_0 —	
2	$C_0 + C_0 e^{-kT}$	$C_0 e^{-kT} + C_0 e^{-2kT}$
3	$C_0 + C_0 e^{-kT} + C_0 e^{-2kT}$	$C_0 e^{-kT} + C_0 e^{-2kT} + C_0 e^{-3kT}$
⋮	⋮	⋮
n		$C_0 e^{-kT} + \ldots + C_0 e^{-nkT}$

From the table:

$$R_n = C_0e^{-kT} + C_0e^{-2kT} + \cdots + C_0e^{-nkT} \qquad (9.19)$$
$$= C_0e^{-kT}(1 + r + r^2 + \cdots + r^{n-1})$$

where $r = e^{-kT}$. Algebraically it is easy to verify that

$$1 + r + r^2 + \cdots + r^{n-1} = \frac{1 - r^n}{1 - r}$$

so substitution for r gives the result:

$$R_n = \frac{C_0e^{-kT}(1 - e^{-nkT})}{1 - e^{-kT}} \qquad (9.20)$$

Notice that the number e^{-nkT} is close to 0 when n is large. In fact, the larger n becomes, the closer e^{-nkT} gets to 0. As a result, the sequence of R_n's has a limiting value, which we call R:

$$R = \lim_{n \to \infty} R_n = \frac{C_0e^{-kT}}{1 - e^{-kT}}$$

or,

$$R = \frac{C_0}{e^{kT} - 1} \qquad (9.21)$$

In summary, if a dose that is capable of raising the concentration by C_0 mg/ml is repeated at intervals of T hours, then the limiting value R of the residual concentrations is given by the formula (9.21). The number k in the formula is the elimination constant of the drug.

Determining the dose schedule   From Table 9-3 the concentration C_{n-1} at the beginning of the nth interval is given by

$$C_{n-1} = C_0 + R_{n-1} \qquad (9.22)$$

If the desired dosage level is required to approach the highest safe level H as depicted in Figure 9-9, then we want C_{n-1} to approach H as n becomes large. That is,

$$H = \lim_{n \to \infty} C_{n-1} = \lim_{n \to \infty} (C_0 + R_{n-1}) = C_0 + R$$

Combining this last result with $C_0 = H - L$ yields

$$R = L \qquad (9.23)$$

A meaningful way to examine what happens to the residual concentration R for different intervals T between doses is to look at R in comparison with C_0, the change in concentration due to each dose. To make this comparison, we form the dimensionless ratio:

$$\frac{R}{C_0} = \frac{1}{e^{kT} - 1} \qquad (9.24)$$

Equation (9.24) says that R/C_0 will be close to 0 whenever the time T between doses is long enough to make $e^{kT} - 1$ sufficiently large. As for the intermediate values of R_n, we can see from Table 9-3 that each R_n is obtained from the previous R_{n-1} by adding a positive quantity $C_0 e^{-nkT}$. This means that all the R_n's are positive, because R_1 is positive. It also means that R is larger than each of the R_n's. In symbols,

$$0 < R_n < R$$

for all n.

The implication of this for drug dosage is that whenever R is small, the R_n's are even smaller. In particular, whenever T is long enough to make $e^{kT} - 1$ significantly large, the residual concentration from each dose is almost nil. The various administrations of the drug are then essentially independent, and the graph of $C(t)$ looks like the one depicted in Figure 9-12.

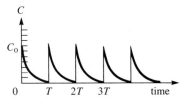

FIGURE 9-12 Drug concentration for long intervals between doses.

On the other hand, suppose the length of time T between doses is so short that e^{kT} is not very much larger than 1, so that R/C_0 is significantly greater than 1. As R_n becomes larger, the concentration C_n after each dose becomes larger. The loss during the time period after each dose increases with larger C_n from Equation (9.17). Finally, the drop in concentration after each dose becomes imperceptibly close to the rise in concentration C_0 due to each dose. When this condition prevails (the loss in concentration equalling the gain), the concentration will oscillate between R at the end of each period and $R + C_0$ at the start of each period. This situation is depicted in Figure 9-13.

Suppose a drug is ineffective below the concentration L and harmful above some higher concentration H, as discussed previously. Assume now that L

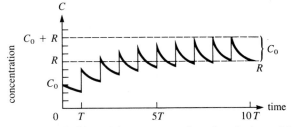

FIGURE 9-13 Buildup of drug concentration when the interval between doses is short.

and H are "safe" guidelines so that a person would not suffer a severe overdose if the drug concentration rises somewhat above H, and that it is not necessary to begin the build-up process all over again if the concentration falls slightly below L. Then for patient convenience we might opt for the strategy of maximizing the time between drug doses by setting $R = L$ and $C_0 = H - L$, as we have indicated previously. Then substitution of $R = L$ and $C_0 = H - L$ in Equation (9.21) yields

$$L = \frac{H - L}{e^{kT} - 1}$$

We then solve the preceding equation for e^{kT} to obtain

$$e^{kT} = H/L$$

Taking the logarithm of both sides of this last equation and dividing the result by k gives the desired dose schedule:

$$T = \frac{1}{k} \ln \frac{H}{L} \qquad \textbf{(9.25)}$$

To reach an effective level rapidly, administer a dose, often called a *loading dose*, that will immediately produce a blood concentration of H mg/ml. (For example, this loading dose might equal $2C_0$.) This medication can be followed every $T = (1/k) \ln (H/L)$ hours by a dose that raises the concentration by $C_0 = H - L$ mg/ml.

Verifying the model   Our model for prescribing a safe and effective dosage of drug concentration appears to be a good one. It is in accord with the common medical practice of prescribing an initial dose several times larger than the succeeding periodic doses. Also the model is based on the assumption that the decrease in the concentration of the drug in the bloodstream is proportional to the concentration itself, which has been verified clinically. Moreover, the elimination constant k, which is the positive constant of proportionality in that relationship, is an easily measured parameter (see Problem 1). The model also provides quantitatively for the prediction of concentration levels under varying conditions for dose rates, using Equation (9.21). Thus, the drug may be tested to determine experimentally the lowest effective level L and the highest safe level H, with appropriate safety factors to allow for inaccuracies in the modeling process. Then formulas (9.16) and (9.25) can be used to prescribe a safe and effective dosage of the drug (assuming the loading dose is several times larger than C_0). So our model is useful.

One deficiency in the model is the assumption of an instantaneous rise in concentration whenever a drug is administered. A drug, such as aspirin, taken orally requires a finite time to diffuse into the bloodstream; the assumption, therefore, is not realistic for such a drug. For such cases the graph of concentration versus time for a single dose might resemble the graph shown in Figure 9-14.

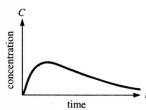

FIGURE 9-14 The concentration of a drug in the bloodstream for a single dose taken orally.

9.2 PROBLEMS

1. Discuss how the elimination constant k in Equation (9.15) could be obtained experimentally for a given drug.

2. **a.** If $k = 0.05$ hr^{-1} and the highest safe concentration is e times the lowest effective concentration, find the length of time between repeated doses that will ensure safe but effective concentrations.

   b. Does (a) give enough information to determine the size of each dose?

3. Suppose that $k = 0.01$ hr^{-1} and $T = 10$ hr. Find the smallest n such that $R_n > 0.5 R$.

4. Given $H = 2$ mg/ml, $L = 0.5$ mg/ml, and $k = 0.02$ hr^{-1}, suppose that concentrations below L are not only ineffective but also harmful. Determine a scheme for administering this drug (in terms of concentration and times of dosage).

5. Suppose that $k = 0.2$ hr^{-1} and that the smallest effective concentration is 0.03 mg/ml. A single dose that produces a concentration of 0.1 mg/ml is administered. Approximately how many hours will the drug remain effective?

6. Suggest other phenomena for which the model described in the text might be used.

7. Sketch how a series of doses might accumulate based on the concentration curve given in Figure 9-14.

8. A patient is given a dosage Q of a drug at regular intervals of time T. The concentration of the drug in the blood has been shown experimentally to obey the law:

$$\frac{dC}{dt} = -ke^C$$

   a. If the first dose is administered at $t = 0$ hr, show that after T hr have elapsed, the residual

$$R_1 = -\ln (kT + e^{-Q})$$

   remains in the blood.

   b. Assuming an instantaneous rise in concentration whenever the drug is administered, show that after the second dose and T hr have elapsed again, the residual

$$R_2 = -\ln \left[kT(1 + e^{-Q}) + e^{-2Q} \right]$$

   remains in the blood.

   c. Show that the limiting value R of the residual concentrations for doses of Q mg/ml repeated at intervals of T hr is given by the formula:

$$R = -\ln \frac{kT}{1 - e^{-Q}}$$

d. Assuming the drug is ineffective below a concentration L and harmful above some higher concentration H, show that the dose schedule T for a safe and effective concentration of the drug in the blood satisfies the formula

$$T = \frac{1}{k}(e^{-L} - e^{-H})$$

where k is a positive constant.

9.2 PROJECTS

1. Write a summary report on the article "Case Studies in Cancer and Its Treatment by Radiotherapy," by J. R. Usher and D. A. Abercrombie, *International Journal of Mathematics Education in Science and Technology 12*, no. 6 (1981), pp. 661–682. Present your report to the class.

In Projects 2–5 complete the requirements of the designated UMAP module.

2. "Selection in Genetics," by Brindell Horelick and Sinan Koont, UMAP 70. This module introduces genetic terminology and basic results about genotype distribution in successive generations. A recurrence relation is obtained from which the nth generation frequency of a recessive gene can be determined. Calculus is used to derive a technique for approximating the number of generations required for this frequency to fall below any given positive value.

3. "Epidemics," by Brindell Horelick and Sinan Koont, UMAP 73. This unit poses two problems: (1) At what rate must infected persons be removed from a population to keep an epidemic under control? (2) What portion of a community will become infected during an epidemic? Threshold removal rate is discussed, and the extent of an epidemic is discussed when the removal rate is slightly below threshold.

4. "Tracer Methods in Permeability," by Brindell Horelick and Sinan Koont, UMAP 74. This module describes a technique for measuring the permeability of red corpuscle surfaces to K^{42} ions, using radioactive tracers. Students learn how radioactive tracers can be used to monitor substances in the body, and learn some of the limitations and strengths of the model described in this unit.

5. "Modeling the Nervous System. Reaction Time and the Central Nervous System," by Brindell Horelick and Sinan Koont, UMAP 67. The process by which the central nervous system reacts to a stimulus is modeled, and the predictions of the model are compared with experimental data. Students learn what conclusions can be drawn from the model about reaction time, and are given an opportunity to discuss the merits of various assumptions about the relation between intensity of excitation and stimulus intensity.

9.3 SCENARIOS REVISITED

Example 1 Braking Distance

In our model for vehicular total stopping distance (see Section 3.3) one of the submodels is braking distance:

$$\text{braking distance} = h(\text{weight, speed})$$

Using an argument based on the result that the work done by the braking system must equal the change in kinetic energy, we found that the braking distance d_b is proportional to the square of the velocity. We now use an argument based on the derivative to establish that same result.

Let's assume that the braking system is designed in such a way that the maximum braking force increases in proportion to the mass of the car. Basically, this means that if the force per unit area applied by the braking hydraulic system remains constant, then the surface area in contact with the brakes would have to increase in proportion to the mass of the car. From an engineering standpoint this assumption seems reasonable.

The implication of the assumption is that the deceleration felt by the passengers is constant, which is probably a plausible design criterion. If it is further assumed that under a panic stop the maximum braking force F is applied continuously, then we obtain

$$F = -km$$

for some positive proportionality constant k. Since F is the only force acting on the car under our assumptions, this gives

$$ma = m\frac{dv}{dt} = -km$$

(the negative sign signals deceleration). Thus,

$$\frac{dv}{dt} = -k$$

which integrates to

$$v = -kt + C_1$$

If v_0 denotes the velocity at $t = 0$ when the brakes are initially applied, substitution gives $C_1 = v_0$ so that

$$v = -kt + v_0 \tag{9.26}$$

If t_s denotes the time it takes for the car to stop after the brakes have been applied, then $v = 0$ when $t = t_s$ substituted into Equation (9.26) gives

$$t_s = \frac{v_0}{k} \tag{9.27}$$

If x represents the distance traveled by the car after the brakes are applied, then x is the integral of $v = dx/dt$. Thus, from (9.26)

$$x = -0.5 \, kt^2 + v_0 t + C_2$$

When $t = 0$, $x = 0$, which implies $C_2 = 0$ so that

$$x = -0.5 \, kt^2 + v_0 t \qquad \textbf{(9.28)}$$

Next, let d_b denote the braking distance; that is, $x = d_b$ when $t = t_s$. Substitution of these results into (9.28) yields

$$d_b = -0.5 \, kt_s^2 + v_0 t_s$$

Using (9.27) in this last equation, we have

$$d_b = \frac{-v_0^2}{2k} + \frac{v_0^2}{k} = \frac{v_0^2}{2k} \qquad \textbf{(9.29)}$$

Therefore, d_b is proportional to the square of the velocity, in accordance with the submodel obtained in Section 3.3.

In Chapter 4 we tested the submodel $d_b \propto v^2$ against some data and found reasonable agreement. The constant of proportionality was estimated to be 0.054 ft·hr²/mi², which corresponds to a value of k in (9.29) of approximately 19.9 ft/sec² (see Problem 1). If we interpret k as the deceleration felt by a passenger in the vehicle (since $F = -km$ by assumption), we will find it useful to interpret this constant as $0.6 \, g$ (where g is the acceleration of gravity).

Example 2 Elapsed Time of a Tape Recorder*

Consider again the problem of relating the counter on a tape recorder to the amount of playing time that has elapsed, as presented in Section 3.5. We assumed that during the play or recording of a tape the linear speed of the tape across the read/write heads is constant. If l denotes the length of tape played, this assumption translates to

$$\frac{dl}{dt} = s \qquad \textbf{(9.30)}$$

where s is a constant.

Our second assumption is that the counter reading c is proportional to the number of revolutions $\theta/2\pi$ of the take-up drive. Thus,

$$c = m \frac{\theta}{2\pi} \qquad \textbf{(9.31)}$$

---

* The differential equations model constructed in this section was suggested by Dan Kalman, "A Model for Playing Time," *Mathematics Magazine 54*, no. 5 (1981): 247–250.

where θ is the turning angle of the take-up reel measured in radians. We tested the submodel (9.31) in Chapter 3 and found that $m \approx 11/10$ from Figure 3-22.

At the beginning of the play, when $t = 0$, we assume further that $l = 0$, $\theta = 0$, and that $r = r_0$ is the radius of the hub of the empty take-up reel (see Figure 9-15).

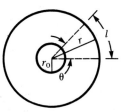

FIGURE 9-15 When the take-up reel turns through an angle of θ radians, a length l of tape is played, where $l = r\theta$.

Formulating the model   The rate of change of l is related to r and θ as follows. A small change $\Delta\theta$ in the angle of displacement increases l by the arc length of the outermost wind of tape between θ and $\theta + \Delta\theta$. Since $\Delta\theta$ is small, there is no significant effect on the radius r of the take-up reel. Since the arc length along a circle is its radius times the central angle, we have

$$\Delta l = r\,\Delta\theta$$

It follows that

$$\frac{\Delta l}{\Delta t} = r\,\frac{\Delta\theta}{\Delta t}$$

and taking limits as Δt goes to 0 gives

$$\frac{dl}{dt} = r\,\frac{d\theta}{dt}$$

From our assumption $dl/dt = s$, this last equation becomes

$$\frac{d\theta}{dt} = \frac{s}{r} \qquad\qquad \textbf{(9.32)}$$

where s is the constant linear speed of the tape across the playback head.

Next we consider how the radius r of the take-up reel varies with time. Note that an increase of 2π in θ adds one layer of tape onto the take-up reel. If h denotes the thickness of the tape, then one revolution of the tape increases r by h. Assuming this increase is distributed evenly throughout the wind gives

$$\Delta r = h\,\frac{\Delta\theta}{2\pi}$$

That is, the change in the radius of the take-up reel is the thickness of the tape times the number of revolutions. For $\Delta\theta$ less than 2π, division of the preceding equation by Δt and passage to the limit as Δt approaches zero leads to

$$\frac{dr}{dt} = \left(\frac{h}{2\pi}\right)\left(\frac{d\theta}{dt}\right) \qquad\qquad \textbf{(9.33)}$$

Together, Equations (9.32) and (9.33) form a system of ordinary differential equations subject to the initial conditions $\theta(0) = 0$ and $r(0) = r_0$. This system is solvable by elementary means.

Solving the model Substitution of (9.32) into (9.33) yields the differential equation:

$$\frac{dr}{dt} = \left(\frac{h}{2\pi}\right)\left(\frac{s}{r}\right)$$

which separates into

$$r\,dr = \left(\frac{hs}{2\pi}\right)dt$$

Integrating both sides of this last equation gives

$$\frac{1}{2}r^2 = \left(\frac{hs}{2\pi}\right)t + C_1$$

where C_1 is a constant of integration. The initial condition $r = r_0$ when $t = 0$ implies that $C_1 = (1/2)\,r_0{}^2$ so that

$$\frac{1}{2}r^2 = \left(\frac{hs}{2\pi}\right)t + \frac{1}{2}r_0{}^2$$

Solving this last equation algebraically for t, we have

$$t = \frac{\pi}{hs}[r^2 - r_0{}^2] \tag{9.34}$$

The previous equation gives an expression for the playback time, but our problem is to relate the playing time to the counter number. Thus, we need an expression for the radius r of the tape currently on the take-up reel in terms of the angular displacement θ undergone in advancing the tape across the playback head, and then we can relate θ to c by Equation (9.31). We observed that $\Delta r = h(\Delta\theta/2\pi)$ so that $\Delta r/\Delta\theta = h/2\pi$ and passage to the limit as $\Delta\theta \to 0$ gives

$$\frac{dr}{d\theta} = \frac{h}{2\pi}$$

Integrating this last expression with respect to θ gives

$$r = \frac{h}{2\pi}\theta + C_2$$

where C_2 is a constant of integration. Since $\theta = 0$ and $r = r_0$ when $t = 0$, we find that $C_2 = r_0$, which leads to

$$r = \frac{h\theta}{2\pi} + r_0 \tag{9.35}$$

Next, substitute the expression for r from (9.35) into (9.34) and simplify algebraically to obtain (see Problem 3)

$$t = \frac{r_0}{s}\theta + \frac{h}{4\pi s}\theta^2 \tag{9.36}$$

Finally, we use our assumption (9.31) that the counter number c varies directly with the angular displacement according to the rule $c = m(\theta/2\pi)$. Substitution for θ in (9.36) gives our desired result:

$$t = \frac{\pi}{ms}\left(2r_0 c + \frac{h}{m}c^2\right) \tag{9.37}$$

Note that the model (9.37) agrees with the model (3.19) constructed in Chapter 3.

For the tape recorder we used to collect the data in Section 4.4, we found that $2r_0 \approx 2.355$ in., $m \approx 11/10$ (from Figure 3-22), $s = 3.75$ in. per sec, and $h = h_{\text{eff}} \approx 0.001503$ in. Thus,

$$\frac{2\pi r_0}{ms} \approx 1.794 \quad \text{and} \quad \frac{\pi h}{m^2 s} \approx 0.00104$$

yielding the model:

$$t = 1.794c + 0.00104c^2 \tag{9.38}$$

Model (9.38) is in close agreement with our result when we fit a curve of the form $t = ac + bc^2$ to the data in Section 4.4.

9.3 PROBLEMS

1. **a.** Using the estimate that $d_b = 0.054v^2$, where 0.054 has dimension ft·hr^2/mi^2, show that the constant k in (9.29) has the value 19.9 ft/sec^2.
    b. Using the data in Table 3-1, plot d_b in ft versus $v^2/2$ in ft^2/sec^2 to estimate $1/k$ directly.
2.* Consider launching a satellite into orbit using a single-stage rocket. The rocket is continuously losing mass, which is being propelled away from it at significant speeds. We are interested in predicting the maximum speed the rocket can attain.
    a. Assume the rocket of mass m is moving with speed v. In a small increment of time Δt it loses a small mass Δm_p, which leaves the rocket with speed u in a direction opposite to v. Here Δm_p is the small propellant mass. The resulting speed of the rocket is $v + \Delta v$. Neglect all external forces (such as gravity, atmospheric drag, and so forth) and assume Newton's second law of motion

$$\text{force} = \frac{d}{dt}(\text{momentum of system})$$

where momentum is mass times velocity. Derive the model:

$$\frac{dv}{dt} = \left(\frac{-c}{m}\right)\frac{dm}{dt}$$

---

* This problem was suggested by D. N. Burghes and M. S. Borrie, *Modelling with Differential Equations* (West Sussex, England: Ellis Horwood, 1981).

where $c = u + v$ is the relative exhaust speed (the speed of the burnt gases relative to the rocket).

b. Assume initially at time $t = 0$ the velocity $v = 0$ and the mass of the rocket is $m = M + P$, where P is the mass of the payload satellite and $M = \epsilon M + (1 - \epsilon)M$ $(0 < \epsilon < 1)$ is the initial fuel mass ϵM plus the mass $(1 - \epsilon)M$ of the rocket casings and instruments. Solve the model in Part a to obtain the speed:

$$v = -c \ln \frac{m}{M + P}$$

c. Show that when all the fuel is burned, the speed of the rocket is given by

$$v_f = -c \ln \left[1 - \frac{\epsilon}{1 + \beta} \right]$$

where $\beta = P/M$ is the ratio of the payload mass to the rocket mass.

d. Find v_f if $c = 3$ km/sec, $\epsilon = 0.8$, and $\beta = 1/100$.
   (These are fairly typical values in satellite launchings.)

e. Suppose scientists plan to launch a satellite in a circular orbit h km above the earth's surface. Assume that the gravitational pull toward the center of the earth is given by Newton's inverse square law of attraction:

$$\frac{\gamma m M_e}{(h + R_e)^2}$$

where γ is the universal gravitational constant, m the mass of the satellite, M_e the earth's mass, and R_e is the radius of the earth. Assume that this force must be balanced by the centrifugal force $mv^2/(h + R_e)$, where v is the speed of the satellite. What speed must be attained by a rocket in order to launch a satellite into an orbit 100 km above the earth's surface? From your computation in Part d, can a single-stage rocket launch a satellite into an orbit of that height?

3. Derive Equation (9.36) by substituting (9.35) into (9.34) and simplifying algebraically.

4. The Gross National Product (GNP) represents the sum of consumption purchases of goods and services, government purchases of goods and services, and gross private investment (which is the increase in inventories plus buildings constructed and equipment acquired). Assume that the GNP is increasing at the rate of 3% per year, and that the national debt is increasing at a rate proportional to the GNP.

a. Construct a system of two ordinary differential equations modeling the GNP and national debt.

b. Solve the system in Part a, assuming the GNP is M_0 and the national debt is N_0 at year 0.

c. Does the national debt eventually outstrip the GNP? Consider the ratio of the national debt to the GNP.

9.3 PROJECTS

Complete the requirements of the indicated UMAP module.

1. "Kinetics of Single Reactant Reactions," by Brindell Horelick and Sinan Koont, UMAP 232. The unit discusses reaction orders of irreversible single reactant reactions. The equation $a'(t) = -k(a(t))^n$ is solved for selected values of n; reaction orders of various reactions are found from experimental data, and the notion of half-life is discussed. Some background knowledge of chemistry is required.
2. "Radioactive Chains: Parents and Daughters," by Brindell Horelick and Sinan Koont, UMAP 234. When a radioactive substance A decays into a substance B, A and B are called parent and daughter. It may happen that B itself is radioactive, and is the parent of a new daughter C, and so on. There are three radioactive chains that together account for all naturally occurring radioactive substances beyond Thallium on the periodic table. This unit develops models for calculating the amounts of the substances in radioactive chains and discusses transient and secular states of equilibrium between parent and daughter.
3. "The Relationship Between Directional Heading of an Automobile and Steering Wheel Deflection," by John E. Prussing, UMAP 506. This unit develops a model relating the compass heading and the steering wheel deflection using basic geometric and kinematic principles.

9.4 NUMERICAL APPROXIMATION METHODS*

In the models developed in the preceding sections of this chapter, we found an equation relating a derivative to some function of the independent and dependent variables; that is,

$$\frac{dy}{dx} = g(x, y)$$

where g is some function in which either x or y may not appear explicitly. Moreover, we were given some starting value; that is, $y(x_0) = y_0$. Finally, we were interested in the values of y for a specific set of x values; that is, $x_0 \leqslant x \leqslant b$. In summary we determined models of the general form:

$$\frac{dy}{dx} = g(x, y), \qquad y(x_0) = y_0, \qquad x_0 \leqslant x \leqslant b$$

* This section is adapted from UMAP Unit 625, which was written by the authors and David H. Cameron. The adaptation is presented with the permission of COMAP, Inc./UMAP, 271 Lincoln St., Lexington, Mass. 02173.

We call first-order ordinary differential equations with the preceding conditions *first-order initial value problems*. As seen from our previous models, they constitute an important class of problems. We now discuss the three parts of the model.

First-Order Initial Value Problems

The differential equation $dy/dx = g(x, y)$ As discussed in our models, we are interested in finding a function $y = f(x)$ whose derivative satisfies an equation $dy/dx = g(x, y)$. Thus, although we do not know f, we can compute its derivative given particular values of x and y. As a result, we can find the slope of the tangent line to the solution curve $y = f(x)$ at specified points (x, y).

The initial value $y(x_0) = y_0$ The initial value equation states that at the initial point x_0, we know the y value is $f(x_0) = y_0$. Geometrically, this means that the point (x_0, y_0) lies on the solution curve (Figure 9-16). Thus, we know where our solution curve begins. Moreover, from the differential equation $dy/dx = g(x, y)$ we know that the slope of the solution curve at (x_0, y_0) is the number $g(x_0, y_0)$. This is also depicted in Figure 9-16.

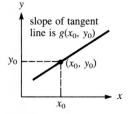

FIGURE 9-16 The solution curve passes through the point (x_0, y_0) and has slope $g(x_0, y_0)$.

The interval $x_0 \leqslant x \leqslant b$ The condition $x_0 \leqslant x \leqslant b$ gives the particular interval of the x axis with which we are concerned. Thus, we would like to relate y with x over the interval $x_0 \leqslant x \leqslant b$ by finding the solution function $y = f(x)$ passing through the point (x_0, y_0) with slope $g(x_0, y_0)$ there (Figure 9-17). Note that the function $y = f(x)$ is continuous over $x_0 \leqslant x \leqslant b$ because its derivative exists there.

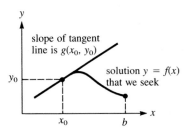

FIGURE 9-17 The solution $y = f(x)$ to the initial value problem is a continuous function over the interval from x_0 to b.

Approximating Solutions to Initial Value Problems

We shall now study a method that utilizes the three parts of the initial value problem together with the geometrical interpretation of the derivative to construct a sequence of discrete points in the plane that numerically approximate the points on the actual solution curve $y = f(x)$. We begin with an example illustrating the method.

Example: Interest compounded continuously Suppose at the end of year 1 we have $1000 invested at 7% annual interest compounded continuously. We would like to know how much money we will have at the end of year 20. Letting $Q(t)$ represent the amount of money at any time t, we have

$$\frac{dQ}{dt} = 0.07Q, \qquad Q(1) = 1000, \qquad 1 \leqslant t \leqslant 20$$

We would like to find the function $Q(t)$ solving this initial value problem. Using a dashed curve to represent the unknown function Q, we sketch the preceding information in Figure 9-18.

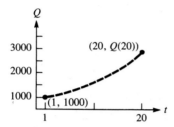

FIGURE 9-18 The curve satisfying $\dfrac{dQ}{dt} = 0.07Q$, $Q(1) = 1000$, $1 \leqslant t \leqslant 20$.

We know the derivative of $Q(t)$ for known values of Q and t. In particular, we know that at the point $(1, 1000)$ the derivative is $dQ/dt = 0.07(1000) = 70$. Since the derivative can be interpreted as the slope of the line tangent to the curve, we sketch a tangent line to our unknown function Q at $t = 1$ with slope 70. This situation is depicted in Figure 9-19. Since we do not know $Q(t)$, we cannot determine exactly the value $Q(20)$. However, we can approximate it by the value *on the tangent line* when $t = 20$. Now the equation of the tangent line T in point-slope form is given by

$$T - 1000 = 70(t - 1)$$

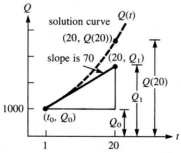

FIGURE 9-19 The point $(20, Q_1)$ on the tangent line approximates the actual solution point $(20, Q(20))$.

In other words,

$$T = Q_0 + \left.\frac{dQ}{dt}\right|_{t=t_0} \Delta t$$

where $Q_0 = 1000$, $\left.dQ/dt\right|_{t=t_0} = 70$, and $\Delta t = t - 1$. When $t = 20$, we see that

$$T(20) = 1000 + 70(20 - 1) = 2330$$

Then we make the approximation

$$Q(20) \approx 2330 = Q_1$$

to the value of the unknown function at $t = 20$. Thus, starting with $1000 at the end of 1 year, we estimate that we will have $2330 at the end of year 20 if interest is compounded continuously at an annual rate of 7%.

Estimating with two steps   We emphasize that we have used the known starting value $Q(1) = 1000$ to calculate the estimate $Q(20) \approx 2330$. How can we improve the approximation and get a more accurate picture of the solution curve? We have assumed that the derivative Q' is the constant 70 over the interval $1 \leqslant t \leqslant 20$, but we know it actually changes as Q and t change. Perhaps our estimate of $Q(20)$ will be more accurate if we make another estimate at an intermediate point. Setting $\Delta t = (20 - 1)/2 = 9.5$ in that same problem, we obtain

$$Q(10.5) \approx Q_1 = Q_0 + \left.\frac{dQ}{dt}\right|_{t=1} \Delta t = 1000 + 70(9.5) = 1665$$

This is depicted in Figure 9-20.

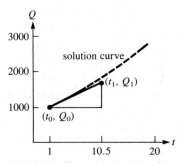

FIGURE 9-20 The point (t_1, Q_1) estimates the solution curve at the halfway value $t = 10.5$.

Next we use the estimate $Q(10.5) \approx 1665$ to approximate the derivative at $t = 10.5$ from the formula:

$$\left.\frac{dQ}{dt}\right|_{t=10.5} = 0.07\, Q(10.5)$$

Note an important difference from our first calculation for $Q'(1)$. We *know* the value of the derivative at $t = 1$ exactly because we know $Q(1) = 1000$, but we must *estimate* the derivative at $t = 10.5$ because $Q(10.5) \approx 1665$ is only an estimate. Now we calculate our estimate $Q(20)$:

$$Q(20) \approx Q_2 = Q_1 + \frac{dQ}{dt}\bigg|_{t=10.5} \Delta t$$

$$= Q_1 + 0.07 \, Q(10.5) \, \Delta t$$

$$\approx 1665 + (0.07)(1665)(9.5) = 2772.23$$

This two-step process is shown in Figure 9-21. You will see shortly that the approximation $Q(20) \approx 2772.23$ is closer to the actual value of the solution at $t = 20$.

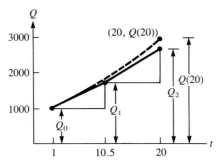

FIGURE 9-21 The value Q_2 approximates the solution $Q(20)$ in a two-step process.

Euler's Algorithm

Let us generalize our procedure. Given an initial value problem:

$$\frac{dy}{dx} = g(x, y), \qquad y(x_0) = y_0, \qquad x_0 \leqslant x \leqslant b$$

one can approximate the solution using the following procedure:

Step 1 First divide the interval $x_0 \leqslant x \leqslant b$ into n subintervals using the equally spaced points:

$$x_1 = x_0 + \Delta x, \qquad x_2 = x_1 + \Delta x, \ldots,$$
$$x_n = x_{n-1} + \Delta x = b$$

Step 2 Next, obtain the sequence of approximations:

$$y_1 = y_0 + g(x_0, y_0) \, \Delta x$$
$$y_2 = y_1 + g(x_1, y_1) \, \Delta x$$
$$\vdots$$
$$y_n = y_{n-1} + g(x_{n-1}, y_{n-1}) \, \Delta x$$

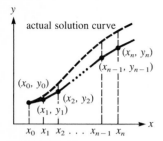

FIGURE 9-22 The points (x_i, y_i) approximate the solution curve.

In this way a table of approximate values to the solution is built up in a step-by-step fashion. The situation is depicted in Figure 9-22. Notice that an error is produced at each step. As these errors accumulate with more and more steps, the approximations $y_1, y_2, \ldots, y_n$ get further and further away from the actual solution curve.

Improving the estimate of the graph Euler's algorithm can easily be coded for computer implementation to facilitate reductions in step size. Since there was a significant change in the estimate to $Q(20)$ (from 2330 to 2772.23) when we reduced the step size Δt in the compound interest example, we would not be too confident in our results at this point. Using a calculator program, we applied Euler's algorithm with step sizes $\Delta t = 19$, $\Delta t = 9.5$, and $\Delta t = 1.0$ and obtained approximations for Q every integer value of t between 1 and 20. To get an idea of what the unknown solution function looks like, we plotted the points obtained from the various step sizes on a single graph. The graph is displayed in Figure 9-23.

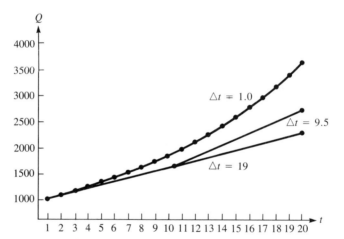

FIGURE 9-23 Plot of the approximate solution for the various step sizes.

It might be tempting to reduce the step size even further in order to obtain greater accuracy. However, each additional calculation not only requires additional computer time, but more importantly introduces round-off error. Since these errors accumulate, an ideal method would improve the accuracy of the approximations yet minimize the number of calculations. For this reason Euler's method may prove unsatisfactory. More refined numerical methods for solving the initial value problem are investigated in courses in the numerical methods of differential equations. We will not pursue those methods here.

Finding the solution analytically Just as we did for the population problem in Section 9.1, we can separate the variables and integrate the initial value problem

$$\frac{dQ}{dt} = .07Q, \qquad Q(1) = 1000, \qquad 1 \leqslant t \leqslant 20$$

to obtain

$$Q = C_1 e^{0.07t}$$

Finally we apply the initial condition that $Q(1) = 1000$ to evaluate the constant C_1: substituting $t = 1$ into the last equation gives

$$1000 = Q(1) = C_1 e^{0.07}$$

so $C_1 = 1000/e^{0.07} \approx 932.39382$. We now have our desired solution function:

$$Q = \left(\frac{1000}{e^{0.07}}\right) e^{0.07t}$$

If we evaluate this solution at $t = 20$, we obtain $Q(20) = 3781.04$. Thus our approximations for $Q(20)$ (2330 with $\Delta t = 19$, 2772.23 with $\Delta t = 9.5$, 3616.53 with $\Delta t = 1$) became more accurate as Δt decreased. Note that even with $\Delta t = 1$, an error of 164.51 results. We can further reduce this error by decreasing Δt. With the aid of a calculator, $\Delta t = 0.01$ in Euler's method gives the approximation $Q(20) \approx 3763.56$.

Separation of variables The method of direct integration employed in this chapter only works in those cases in which the dependent and independent variables can be algebraically separated. Any such differential equation can be written in the form:

$$p(y) \, dy = q(x) \, dx$$

For example, given the differential equation:

$$u(x)v(y) \, dx + q(x)p(y) \, dy = 0$$

we can arrange the equation in the following form:

$$\frac{p(y)}{v(y)} \, dy = \frac{-u(x)}{q(x)} \, dx$$

Since the right-hand side is a function of x and the left-hand side is a function of y only, the solution is obtained by simply integrating both sides directly. You should remember from calculus that even if we are successful in separating the variables, it may not be possible to find the integrals in closed form. The method just described is called *separation of variables* and is one of the most elementary methods available for solving a differential equation.

9.4 PROBLEMS

1. When interest is compounded, the interest earned is added to the principal amount so that it may also earn interest. For a one-year period, the principal amount P is given by

$$P = \left(1 + \frac{i}{n}\right)^n P(0)$$

where i is the annual interest rate (given as a decimal) and n is the number of times during the year that the interest is compounded.

   To lure depositors, banks offer to compound interest at different intervals: semiannually, quarterly, or daily. A certain bank advertises that it compounds interest *continuously*. If $100 is deposited initially, formulate a mathematical model describing the growth of the initial deposit during the first year. Assume an annual interest rate of 10%.

2. Use the differential equation model formulated in the preceding problem to answer the following.
   a. From the derivative evaluated at $t = 0$, determine an equation of the tangent line T passing through the point (0, 100).
   b. Estimate $Q(1)$ by finding $T(1)$, where $Q(t)$ denotes the amount of money in the bank at time t (assuming no withdrawals).
   c. Estimate $Q(1)$ using a step size of $\Delta t = 0.5$.
   d. Estimate $Q(1)$ using a step size of $\Delta t = 0.25$.
   e. Plot the estimates you obtained for $\Delta t = 1.0$, 0.5, and 0.25 to approximate the graph of $Q(t)$.

3. a. For the differential equation model obtained in Problem 1, find $Q(t)$ by separating the variables and integrating.
   b. Evaluate $Q(1)$.
   c. Compare your previous estimates of $Q(1)$ with its actual value.
   d. Find the effective annual interest rate when an annual rate of 10 percent is compounded continuously.
   e. Compare the effective annual interest rate computed in Part d with interest compounded:
      i. Semiannually: $(1 + .10/2)^2$
      ii. Quarterly: $(1 + .10/4)^4$
      iii. Daily: $(1 + .10/365)^{365}$
   f. Estimate the limit of $(1 + .10/n)^n$ as $n \to \infty$ by evaluating the expression for $n = 1,000$; 10,000; 100,000.
   g. What is $\lim_{n \to \infty} (1 + 0.1/n)^n$?

9.4 PROJECTS

Complete the requirements of the indicated UMAP module.

1. "Feldman's Model," by Brindell Horelick and Sinan Koont, UMAP 75. This unit develops a version of G. A. Feldman's model of growth in a planned economy in which all of the means of production are owned by the state. Originally the model was developed by Feldman in connection with planning the economy of the Soviet Union. Students compute numerical values for rates of output, national income, their rates of change, and the propensity to save, and discuss the effects of changes in the parameters of the model and in the units of measurement.

2. "The Digestive Process of Sheep," by Brindell Horelick and Sinan Koont, UMAP 69. This unit introduces a differential equation model for the digestive processes of sheep. The model is tested and fit using collected data and the least-squares criterion.

INTERACTIVE DYNAMIC SYSTEMS

INTRODUCTION

Interactive situations occur in the study of economics, ecology, electrical circuits, mechanical systems, celestial mechanics, control systems, and so forth. For example, the study of the dynamics of population growth of various plants and animals is an important ecological application of mathematics. Different species interact in a variety of ways. One animal may serve as the primary food source for another, commonly referred to as a "predator-prey" relationship. Two species may depend upon one another for mutual support, such as a bee's using a plant's nectar as food while simultaneously pollinating that plant; such a relationship is referred to as "mutualism." Another possibility occurs when two or more species compete against one another for a common food source or even compete for survival (like a military confrontation between two armies). In this chapter we develop some elementary models to explain these interactive situations, and we analyze the models using graphical techniques.

In modeling interactive situations involving the dynamics of population growth, we are interested in the answers to certain questions concerning the species under investigation. For instance, will one species eventually dominate the other and drive it to extinction? Can the species coexist? If so, will their populations reach equilibrium levels, or will they vary in some predictable fashion? Moreover, how sensitive are the answers to the preceding questions relative to the initial population levels or to external perturbations (such as natural disasters, development of chemical or biological agents used to control the populations, and so forth)?

Because we are modeling the rates of change with respect to time, the models invariably involve differential equations (or in a discrete analysis, difference equations). Even with very simple assumptions, these equations are often nonlinear and generally cannot be solved analytically, although numerical techniques exist. Nevertheless, qualitative information about the behavior of the variables often can be obtained by simple graphical analysis. We will demonstrate how graphical analysis can be used to answer questions like those posed in the preceding paragraph. We will also point out limitations for such an analysis and conditions requiring a more sophisticated mathematical analysis.

10.1 GRAPHING SOLUTIONS TO FIRST-ORDER DIFFERENTIAL EQUATIONS

Consider again the first-order differential equation

$$\frac{dy}{dx} = g(x, y)$$

which we investigated in the last chapter. Each time an initial value $y(x_0) = y_0$ is specified, the solution curve must pass through the point (x_0, y_0) and have the slope value $g(x_0, y_0)$ there. Graphically, we can draw a short line segment with the proper slope through each point (x, y) in the plane. The resultant configuration is known as the **direction field** of the first-order differential equation and is illustrated in Figure 10-1. Thus, the direction field gives a visual indication of the family of possible solutions to the differential equation. A solution curve is tangent to the direction line at each point through which the curve passes. One solution curve is depicted for the family of curves indicated by the direction field in Figure 10-1.

direction field

Often the solutions to a first-order differential equation can be expressed in the form $y = f(x)$, where each solution is distinguished by a different constant resulting from integration. For example, if we separate the variables in the differential equation

$$\frac{dy}{dx} = \frac{2y}{x}$$

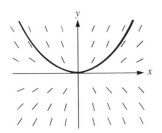

FIGURE 10-1 The direction field of a first-order differential equation assigns to each point in the plane the slope $y' = g(x, y)$.

we obtain

$$\frac{dy}{y} = 2\frac{dx}{x}$$

and integration of both sides then gives

$$\ln|y| = 2\ln|x| + \ln C$$

(where the constant of integration is named $\ln C$ for computational convenience). Applying the exponential to both sides of this last equation yields

$$|y| = Cx^2, \qquad C > 0 \tag{10.1}$$

Equation (10.1) represents a family of parabolas, each parabola being distinguished by a different value of the positive constant C. If $y \geqslant 0$, the curves

$y = Cx^2$ open upward; if $y < 0$, the curves $y = -Cx^2$ open downward. This family is depicted in Figure 10-2. Excluding the origin, exactly one of these parabolas passes through each point in the plane. Thus, by specifying an initial condition $y(x_0) = y_0$, we select a *unique* solution curve passing through the point (x_0, y_0).

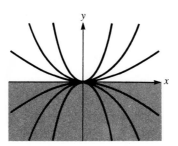

FIGURE 10-2 The family of parabolas $|y| = Cx^2$, $C > 0$.

The question of whether more than one solution curve can pass through a specific point (x_0, y_0) is an important one, but a full discussion is beyond the scope of this book. For uniqueness to occur, certain conditions must be met by the function $g(x, y)$ defining the differential equation. Whenever uniqueness does apply, you know that two solutions cannot cross at the point in question. In Figure 10-2 there is a unique solution parabola through each point in the plane with the exception of the origin; through the origin there are infinitely many solutions. Notice that the constant function $y = 0$ is a solution to the differential equation. This fact is readily seen when the equation is written in the form

$$xy' = 2y$$

which is clearly satisfied by $y = 0$. However, the solution $y = 0$ is not one of the curves represented by the family (10.1). The difficulty lies with the lack of continuity of the function $g(x, y) = 2y/x$ at the origin.*

Rest Points

Qualitative information concerning the solutions can often be obtained from the differential equation. For example, consider the differential equation describing the Malthusian model of population growth discussed in Chapter 9:

$$\frac{dP}{dt} = kP, \qquad k > 0$$

---

* A good discussion of the uniqueness problem appears in W. E. Boyce and R. C. DiPrima, *Elementary Differential Equations and Boundary Value Problems*, 3rd ed. (New York: Wiley, 1977), Section 2.11.

Thus, the Malthusian model assumes a simple proportionality between growth rate and population (see Figure 10-3a). Since $P > 0$ and $k > 0$, you can see that $dP/dt > 0$, which means that the population curve $P(t)$ is everywhere increasing. Moreover, the smaller the value of k, the less rapid is the growth in the population over time. For a fixed value of k, as P increases so does its rate of change dP/dt. Thus, the solution curves must appear qualitatively as depicted in Figure 10-3b. Assuming k is constant, at each population level P the derivative dP/dt is constant. Therefore, all the solution curves are horizontal translates of one another. The initial population $P(0) = P_0$ distinguishes these various curves.

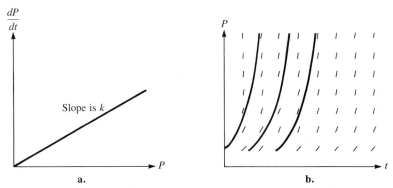

a.   b.

FIGURE 10-3 Graphs of dP/dt versus P and the solution curves for $dP/dt = kP$, where $k > 0$ is fixed.

Next, recall the limited growth model studied in Chapter 9:

$$\frac{dP}{dt} = r(M - P)P \tag{10.2}$$

The graph of dP/dt versus P is the parabola shown in Figure 10-4. Since the second derivative

$$P'' = rP'(M - 2P) \tag{10.3}$$

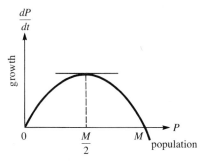

FIGURE 10-4 $dP/dt = r(M - P)P$, $r > 0$, $M > 0$.

is zero when $P = M/2$, positive for $P < M/2$, and negative for $P > M/2$, we conclude that dP/dt is at a maximum at that population level. Note that for $0 < P < M$, the derivative dP/dt is positive and the population $P(t)$ is increasing; for $M < P$, dP/dt is negative and $P(t)$ is decreasing. Moreover, if $P < M/2$, the second derivative P'' is positive and the population curve is concave upward; for $P > M/2$, P'' is negative and the population curve is concave downward.

The parabola in Figure 10-4 has zeros at the population levels $P = 0$ and $P = M$. At those levels, $dP/dt = 0$, so no change in the population P can occur. That is, if the population level is at 0, it will remain there for all time; if it is at the level M, it will remain there for all time. Points for which the derivative dP/dt is zero are called **rest points, equilibrium points, stationary points,** or **critical points** of the differential equation. The behavior of solutions near rest points is of significant interest to us. Let's examine what happens to the population when P is near the rest points $P = M$ and $P = 0$.

Suppose the population P is slightly less than M. Then dP/dt is positive, so the population increases and gets closer to M. On the other hand, if $P > M$, $dP/dt < 0$, so the population will decrease toward M. Thus no matter what positive value is assigned to the starting population, it will tend to the limiting value M as time tends to infinity. We say that M is an **asymptotically stable** rest point, because whenever the population level is perturbed away from that level it tends to return there again. However, if the starting population is not at the level M, the population $P(t)$ cannot reach M in a finite amount of time. This fact follows from the property that $P = M$ is a solution to the limited growth model (10.2) and two solutions cannot cross.

Next, consider the rest point $P = 0$ in Figure 10-4. If P is perturbed slightly away from 0 so that $M > P > 0$, then dP/dt is positive and the population increases toward M. In this situation we say that $P = 0$ is an **unstable** rest point, because any population not starting at that level tends to move away from it.

In general, equilibrium solutions to differential equations are classified as stable or unstable according to whether, graphically, nearby solutions stay close to or converge to the equilibrium, or diverge away from the equilibrium, respectively, as the independent variable t tends to infinity.

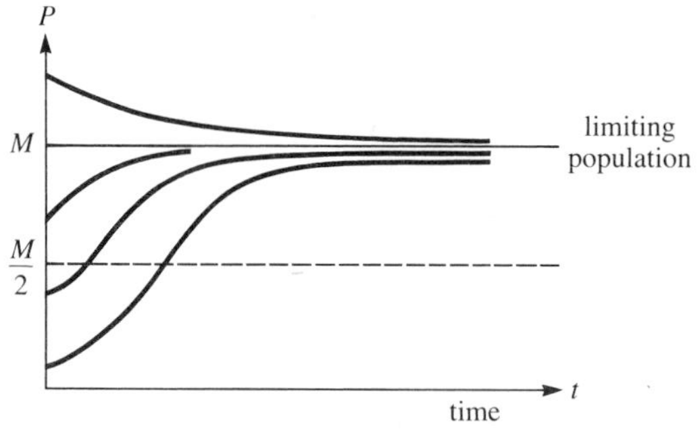

(margin terms:) **rest points** **equilibrium points** **stationary points** **critical points** **asymptotically stable** **unstable**

FIGURE 10-5 Population curves for the logistic model $dP/dt = r(M - P)P$, $r > 0$, $M > 0$.

At $P = M/2$, P'' is zero and the derivative dP/dt is at a maximum (see Figure 10-4). Thus the population P is increasing most rapidly when $P = M/2$ and a point of inflection occurs in the graph there. These features give each solution curve its characteristic "S" or *sigmoid* shape. From the information we have obtained, the family of solutions to the limited growth model must appear (approximately) as shown in Figure 10-5. Notice again that at each population level P, the derivative dP/dt is constant, so that the solution curves are horizontal translates of one another. The curves are distinguished by the initial population level $P(0) = P_0$. In particular, note the solution curve when $M/2 < P_0 < M$ in Figure 10-5.

Systems of First-Order Differential Equations

In modeling the playing time for a tape recorder in the last chapter, we obtained a system of two first-order differential equations. That system was so simple we were able to convert it to a single separable equation by substitution. The separable equation was then solved by integration. Usually, however, it is not so easy to solve a system of differential equations; in fact, it is rare that we can find an analytic solution when the equations are nonlinear, although numerical solution methods exist. It is worthwhile, therefore, to consider a qualitative graphical analysis for solutions to a system of differential equations analogous to our preceding development for single equations. We restrict our discussion to special systems involving only two first-order differential equations.

The system

$$\frac{dx}{dt} = f(x, y)$$

$$\frac{dy}{dt} = g(x, y)$$

(10.4)

autonomous   is called an **autonomous** system of differential equations. In such a system the independent variable t is absent (that is, t does not appear explicitly on the right side of Equations 10.4). To emphasize the physical significance of autonomous systems, think of the independent variable t as denoting time, and the dependent variables as giving position (x, y) in the Cartesian plane. Thus, autonomous systems are not time dependent. In order that the system be suitably well behaved, we assume throughout our discussion that the functions f and g, together with their first partial derivatives $\partial f/\partial x$, $\partial f/\partial y$, $\partial g/\partial x$, and $\partial g/\partial y$ are all continuous over a suitable region of the xy-plane.

solution   It is useful to think of a solution to the autonomous system (10.4) as a curve in the xy-plane. That is, a **solution** to (10.4) is a pair of parametric equations, $x = x(t)$ and $y = y(t)$, whose derivatives satisfy the system. The

trajectory   solution curve whose coordinates are $(x(t), y(t))$, as t varies over time, is called
path   a **trajectory, path,** or **orbit** of the system. The xy-plane is referred to as the
orbit
phase plane   **phase plane.** It is convenient to think of a trajectory as the path of a moving particle and we will appeal to this idea throughout this chapter. Note that

as the particle moves through the phase plane with increasing t, the direction it moves from a point (x, y) depends only on the coordinates (x, y) and not on the time of arrival.

If (x, y) is a point in the phase plane for which $f(x, y) = 0$ and $g(x, y) = 0$ simultaneously, then both the derivatives dx/dt and dy/dt are zero. Hence there is no motion in either the x or the y direction, and the particle is stationary. Such a point is called a **rest point,** or **equilibrium point,** of the system. Notice that whenever (x_0, y_0) is a rest point of the system (10.4), the equations $x = x_0$ and $y = y_0$ give a solution to the system. In fact, this constant solution is the only one passing through the point (x_0, y_0) in the phase plane. The trajectory associated with this solution is simply the rest point (x_0, y_0) itself. Hence, the particle is "at rest" there. A trajectory $x = x(t)$, $y = y(t)$ is said to approach the rest point (x_0, y_0) if $x(t) \to x_0$ and $y(t) \to y_0$ as $t \to \infty$. In applications it is of interest to see what happens to a trajectory when it comes near a rest point.

rest point

equilibrium point

The idea of stability is central to any discussion of the behavior of trajectories near a rest point. Roughly, the rest point (x_0, y_0) is **stable** if any trajectory that starts "close" to the point stays close to it for all future time. It is **asymptotically stable** if it is stable and if any trajectory that starts close to (x_0, y_0) approaches that point as t tends to infinity. If it is not stable, the rest point is said to be **unstable.** These notions will be clarified when we examine specific modeling applications further on in the chapter. Our goal here is to have a language with which to discuss qualitatively our differential equations models. It is not our intent to study the theoretical aspects of stability, which would require greater mathematical precision than we have presented here.

stable

asymptotically stable

unstable

The following results are useful in investigating solutions to the autonomous system (10.4). We offer these results without proof.

1. There is at most one trajectory through any point in the phase plane.
2. A trajectory that starts at a point other than a rest point cannot reach a rest point in a finite amount of time.
3. No trajectory can cross itself unless it is a closed curve. If it is a closed curve, it is a periodic solution.

The implications of these three properties are that from a starting point that is not a rest point, the resulting motion:

a. will move along the same trajectory regardless of the starting time;
b. cannot return to the starting point unless the motion is periodic;
c. can never cross another trajectory; and
d. can only approach (never reach) a rest point.

Therefore, the resulting motion of a particle along a trajectory behaves in one of three possible ways: (i) the particle approaches a rest point, or (ii) the particle moves along or approaches asymptotically a closed path, or (iii) at least one of the trajectory components, $x(t)$ or $y(t)$, becomes arbitrarily large as t tends to infinity. We will apply these ideas to the models we develop in the next several sections.

10.1 PROBLEMS

1. Construct a direction field for the following differential equations, and sketch in a solution curve.

   a. $dy/dx = y$ **b.** $dy/dx = x$
   c. $dy/dx = x + y$ **d.** $dy/dx = x - y$
   e. $dy/dx = xy$ **f.** $dy/dx = 1/y$

   Sketch a number of solutions to the following equations, showing the correct slope, concavity, and any points of inflection.

2. $dy/dx = (y + 2)(y - 3)$
3. $dy/dx = y^2 - 4$
4. $dy/dx = y^3 - y$
5. $dy/dx = x - 2y$
6. Analyze graphically the equation $dy/dt = ry$, when $r < 0$. What happens to any solution curve as t becomes large?
7. Develop graphically the following models. First graph dP/dt versus P, and then obtain various graphs of P versus t by selecting different initial values $P(0)$ (as in our population example in the text). Identify and discuss the nature of the equilibrium points in each model.

   a. $dP/dt = a - bP$, $a, b > 0$
   b. $dP/dt = P(a - bP)$, $a, b > 0$
   c. $dP/dt = k(M - P)(P - m)$, $k, M, m > 0$
   d. $dP/dt = kP(M - P)(P - m)$, $k, M, m > 0$

8. Sketch a number of trajectories corresponding to the following autonomous systems, and indicate the direction of motion for increasing t. Identify and classify any rest points according to being stable, asymptotically stable, or unstable.

   a. $dx/dt = x$, $dy/dt = y$
   b. $dx/dt = -x$, $dy/dt = 2y$
   c. $dx/dt = y$, $dy/dt = -2x$
   d. $dx/dt = -x + 1$, $dy/dt = -2y$

9. The Department of Fish and Game in a certain state is planning to issue deer hunting permits. It is known that if the deer population falls below a certain level m, then the deer will become extinct. It is also known that if the deer population goes above the maximum carrying capacity M, the population will decrease to M.

   a. Discuss the reasonableness of the following model for the growth rate of the deer population as a function of time

   $$\frac{dP}{dt} = kP(M - P)(P - m)$$

   where P is the population of the deer and k is a constant of proportionality. Include a graph of dP/dt versus P as part of your discussion.

   b. Explain how this growth rate model differs from the logistic model $dP/dt = kP(M - P)$. Is it better or worse than the logistic model?

   c. Show that if $P > M$ for all t, then the limit of $P(t)$ as $t \to \infty$ is M.

   d. Discuss what happens if $P < m$ for all t.

   e. Assuming that $m < P < M$ for all t, explain briefly the steps you would use to solve the differential equation. Do not attempt to solve the differential equation.

   f. Graphically discuss the solutions to the differential equation. What are the equilibrium points of the model? Explain the dependence of the equilibrium level of P on the initial conditions. How many deer hunting permits should be issued?

10.1 PROJECTS

1. Complete the requirements of the UMAP module, "Whales and Krill: A Mathematical Model," by Raymond N. Greenwell, UMAP 610. A predator–prey system involving whales and krill is modeled by a system of differential equations. Although the equations are not solvable, information is extracted using dimensional analysis and the study of equilibrium points. The concept of maximum sustainable yield is introduced and used to draw conclusions about fishing strategies. You will learn to construct a differential equations model, remove dimensions from a set of equations, find equilibrium points of a system of differential equations and learn their significance, and practice manipulative skills in algebra and calculus. (This is a good project to assign before the various topics just mentioned are formally presented in the text.)

MODEL

10.2 A COMPETITIVE HUNTER MODEL*

Up to now we have seen how single species growth can be modeled as the Malthusian model or the limited growth model. Let's turn our attention to how two different species might compete for common resources.

* This section is adapted from UMAP Unit 628, based on the work of Stanley C. Leja and one of the authors. The adaptation is presented with the permission of COMAP, Inc./UMAP, 271 Lincoln St. Lexington, Mass. 02173.

Problem identification   Imagine a small pond that is mature enough to support wildlife. We desire to stock the pond with game fish, say trout and bass. Let $x(t)$ denote the population of the trout at any time t, and let $y(t)$ denote the bass population. *Is coexistence of the two species in the pond possible? If so, how sensitive is the final solution of population levels to the initial stockage levels and external perturbations?*

Assumptions   The level of the trout population $x(t)$ depends on many variables: the initial level x_0, the capability of the environment to support trout, the amount of competition for limited resources, the existence of predators, and so forth. Initially, we assume that the environment can support an unlimited number of trout, so that in isolation:

$$\frac{dx}{dt} = ax \quad \text{for} \quad a > 0$$

(Later we may find it desirable to refine the model and use a limited growth assumption.) Next, we modify the preceding differential equation to take into account the competition of the trout with the bass population for living space and a common food supply. The effect of the bass population is to decrease the growth rate of the trout population. This decrease is roughly proportional to the number of possible interactions between the two species, so one submodel is to assume that the decrease is proportional to the product of x and y. These considerations are modeled by the equation:

$$\frac{dx}{dt} = ax - bxy = (a - by)x \tag{10.5}$$

The intrinsic growth rate $k = a - by$ decreases as the level of the bass population increases. The constants a and b indicate the degrees of "self-regulation" of the trout population and its "competition" with the bass population, respectively. These coefficients must be determined experimentally or by analyzing historical data.

   The situation for the bass population is analyzed in the same manner. Thus we obtain the following autonomous system of two first-order differential equations for our model:

$$\frac{dx}{dt} = (a - by)x$$
$$\frac{dy}{dt} = (m - nx)y \tag{10.6}$$

where $x(0) = x_0$, $y(0) = y_0$, and a, b, m, and n are all positive constants. This model is useful in studying the growth patterns of species exhibiting competitive behavior like the trout and bass.

Graphical analysis of the model   One of our concerns is whether the trout and bass populations reach equilibrium levels. If so, then we will know whether coexistence of the two species in the pond is possible. The only way

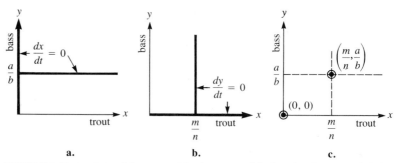

FIGURE 10-6 Rest points of the competitive hunter model given by the system (10.6).

such a state can be achieved is that both populations stop growing; that is, $dx/dt = 0$ and $dy/dt = 0$. Thus, we seek the rest points or equilibrium points of the system (10.6).

Setting the right sides of (10.6) equal to zero and solving for x and y simultaneously, we find the rest points $(x, y) = (0, 0)$ and $(x, y) = (m/n, a/b)$ in the phase plane. Along the vertical line $x = m/n$ and the x-axis in the phase plane, the growth dy/dt in the bass population is zero; along the horizontal line $y = a/b$ and the y-axis, the growth dx/dt in the trout population is zero. If the initial stockage is at these rest point levels, there would be no growth in either population. These features are depicted in Figure 10-6.

Considering the approximations necessary in any model, it is inconceivable that we would estimate precisely the values for the constants a, b, m, and n in our system (10.6). So the pertinent behavior we need to investigate is what happens to the solution trajectories in the vicinity of the rest points $(0, 0)$ and $(m/n, a/b)$. Specifically, are these points stable or unstable?

To investigate this question graphically, let's analyze the directions of dx/dt and dy/dt in the phase plane. (Although $x(t)$ and $y(t)$ represent the trout and bass populations, respectively, it is helpful to think of the trajectories as paths of a moving particle, in accord with our discussion in the preceding section.) Whenever dx/dt is positive, the horizontal component $x(t)$ of the trajectory is increasing and the particle is moving toward the right; whenever dx/dt is negative, the particle is moving to the left. Likewise, if dy/dt is positive, the component $y(t)$ is increasing and the particle is moving upward; if dy/dt is negative, the particle is moving downward. In our system (10.6), the vertical line $x = m/n$ divides the phase plane into two half-planes. In the left half-plane, dy/dt is positive and in the right half-plane it is negative. The directions of the associated trajectories are indicated in Figure 10-7. Likewise, the horizontal line $y = a/b$ determines the half-planes where dx/dt is positive or negative. The directions of the associated trajectories are indicated in Figure 10-8. Along the line $y = a/b$ itself, $dx/dt = 0$. Therefore, any trajectory crossing this line will do so vertically. Similarly, along the line $x = m/n$, $dy/dt = 0$, so the line will be crossed horizontally. Finally, along the y-axis motion must be vertical, and along the x-axis motion must be

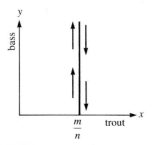

FIGURE 10-7 To the left of $x = m/n$ the trajectories move upward, and to the right they move downward.

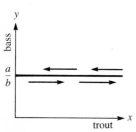

FIGURE 10-8 Above the line $y = a/b$ the trajectories move to the left, and below the line they move to the right.

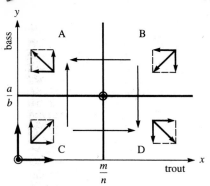

FIGURE 10-9 Composite graphical analysis of the trajectory directions in the four regions determined by $x = m/n$ and $y = a/b$.

horizontal. Combining all this information together into a single graph gives the four distinct regions A, B, C, D with their respective trajectory directions as depicted in Figure 10-9.

Analyzing the motion in the vicinity of $(0, 0)$, we can see that all motion is away from that rest point: the motion is upward and toward the right. In the vicinity of the rest point $(m/n, a/b)$ the behavior depends on the region in which the trajectory begins. If the trajectory starts in region B, for instance, then it will move downward and leftward toward the rest point. However, as it gets nearer to the rest point, the derivatives dx/dt and dy/dt approach zero. Depending on where the trajectory begins and the relative sizes of the constants a, b, m, and n, either the trajectory will continue moving downward and into region D as it swings past the rest point, or the trajectory will move leftward into region A. Once it enters either one of these latter two regions, the trajectory moves away from the rest point. Thus both rest points are unstable. Note that there is exactly one trajectory from region B that gets arbitrarily close to the rest point, slowing down as it gets nearer and nearer to $(m/n, a/b)$. Trajectories that start above this path will cross the line $x = m/n$; those below will cross $y = a/b$. These features are suggested in Figure 10-10.

Model interpretation The graphical analysis conducted so far leads us to the preliminary conclusion that, under the assumptions of our model, reaching equilibrium levels of both species is highly unlikely. Furthermore, the initial stockage levels turn out to be important in determining which of the two species might survive. Perturbations of the system may also affect the outcome of the competition. Thus, mutual coexistence of the species is highly improbable. This phenomenon is known as the *Principle of Competitive Exclusion*, or *Gause's principle*. Moreover, the initial conditions completely determine the outcome as depicted in Figure 10-11. You can see from that graph that any perturbation causing a switch from one region (say below the line joining the two rest points) to the other region (above the line) would change the outcome. Actually the curve separating the starting points in which the bass win from those in which the trout win may not be a straight

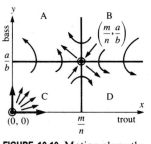

FIGURE 10-10 Motion along the trajectories near the rest points $(0, 0)$ and $(m/n, a/b)$.

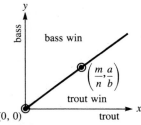

FIGURE 10-11 Qualitative results of analyzing the competitive hunter model.

FIGURE 10-12 Trajectory direction near a rest point.

line as depicted in the figure. One of the limitations of our graphical analysis is that we have not determined that separating curve precisely. If we are satisfied with our model, we may very well want to determine that separating boundary.

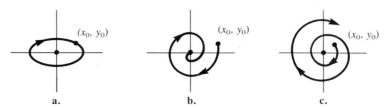

FIGURE 10-13 Three possible trajectory motions: (a) periodic motion, (b) motion toward an asymptotically stable rest point, and (c) motion near an unstable rest point.

Limitations of a graphical analysis It is not always possible to determine the nature of the motion near a rest point using only graphical analysis. To understand this limitation, consider the rest point and the direction of motion of the trajectories shown in Figure 10-12. The information given in the figure is insufficient to distinguish between the three possible motions shown in Figure 10-13. Moreover, even if you have determined by some other means that Figure 10-13c correctly portrays the motion near the rest point, you might be tempted to deduce that the motion will grow without bound in both the x and y directions. However, consider the system given by

$$\frac{dx}{dt} = y + x - x(x^2 + y^2)$$

$$\frac{dy}{dt} = -x + y - y(x^2 + y^2)$$

(10.7)

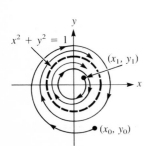

FIGURE 10-14 The solution $x^2 + y^2 = 1$ is a limit cycle.

limit cycle

It can be shown that $(0, 0)$ is the only rest point for (10.7). Yet any trajectory starting on the unit circle $x^2 + y^2 = 1$ will traverse the unit circle in a periodic solution because in that case $dy/dx = -x/y$ (see Problem 2). Moreover, if a trajectory starts "inside" the circle (provided it does not start at the origin), it will spiral outward asymptotically, getting closer and closer to the circular path as t tends to infinity. Likewise, if the trajectory starts "outside" the circular region, it will spiral inward and again approach the circular path asymptotically. The solution $x^2 + y^2 = 1$ is called a **limit cycle.** The trajectory behavior is sketched in Figure 10-14. Thus, if the system (10.7) models population behavior for two competing species, we would have to conclude that the population levels will eventually be periodic. This example illustrates that the results of a graphical analysis are useful for determining the motion in the immediate vicinity of an equilibrium point only. (Here we have assumed that negative values for x and y have a physical meaning in Figure 10-14, or that the point $(0, 0)$ represents the translation of a rest point from the first quadrant to the origin.)

10.2 PROBLEMS

1. List three important considerations that are ignored in the development of the competitive hunter model presented in the text.
2. For the system (10.7), show that any trajectory starting on the unit circle $x^2 + y^2 = 1$ will traverse the unit circle in a periodic solution. First introduce polar coordinates and rewrite the system as $dr/dt = r(1 - r^2)$ and $d\theta/dt = 1$.
3. Develop a model for the growth of trout and bass, assuming that in isolation trout demonstrate exponential decay (so that $a < 0$ in Equation (10.6)) and that the bass population actually grows logistically with a population limit M. Analyze graphically the motion in the vicinity of the rest points in your model. Is coexistence possible?
4. How might the competitive hunter model (10.6) be validated? Include a discussion of how the various constants a, b, m, and n might be estimated. How could state conservation authorities use the model to ensure the survival of both species?
5. Consider the competitive hunter model defined by

$$\frac{dx}{dt} = a(1 - x/k_1)x - bxy$$

$$\frac{dy}{dt} = m(1 - y/k_2)y - nxy$$

where x represents the trout population and y the bass population.
   a. What assumptions are implicitly being made about the growth of trout and bass in the absence of competition?
   b. Interpret the constants a, b, m, n, k_1, k_2 in terms of the physical problem.
   c. Perform a graphical analysis and answer the following questions:
      i.   What are the possible equilibrium levels?
      ii. Is coexistence possible?
      iii. Pick several typical starting points and sketch typical trajectories in the phase plane.
      iv. Interpret the outcomes predicted by your graphical analysis in terms of the constants a, b, m, n, k_1, and k_2.
   Note: When you get to Step i, you should realize that at least five cases exist. You will need to analyze all five cases. One case is when the lines are coincident.
6. Consider the following economic model. Let P be the price of a single item on the market. Let Q be the quantity of the item available on the

market. Both P and Q are functions of time. If one considers price and quantity as two interacting species, the following model might be proposed:

$$\frac{dP}{dt} = aP(b/Q - P)$$

$$\frac{dQ}{dt} = cQ(fP - Q)$$

where a, b, c, and f are positive constants. Justify and discuss the adequacy of the model.

a. If $a = 1$, $b = 20{,}000$, $c = 1$, and $f = 30$, find the equilibrium points of this system. Classify each equilibrium point with respect to its stability, if possible. If a point cannot be readily classified, give some explanation.

b. Perform a graphical stability analysis to determine what will happen to the levels of P and Q as time increases.

c. Give an economic interpretation of the curves that determine the equilibrium points.

10.2 PROJECTS

Complete the requirements of the referenced UMAP module.

1. "The Budgetary Process: Incrementalism" (UMAP 332), "Competition" (UMAP 333), by Thomas W. Likens. The politics of budgeting revolve around the allocation of limited resources to agencies and groups competing for them. UMAP 332 develops a model to explain how levels of appropriation change from one period to the next if new budgets are determined by Congress and federal agencies by making marginal adjustments in the status quo. The model assumes that the share received by one agency will not affect or depend on the share received by another agency. In UMAP 333 the model is refined to address the conflictive nature of politics and the necessary interdependence of budgetary decisions.

2. "The Growth of Partisan Support I: Model and Estimation" (UMAP 304), "The Growth of Partisan Support II: Model Analytics" (UMAP 305), by Carol Weitzel Kohfeld. UMAP 304 presents a simple model of political mobilization, refined to include the interaction between supporters of a particular party and recruitable non-supporters. UMAP 305 investigates the mathematical properties of the first-order quadratic-difference equation model. The model is tested using data from three U.S. counties. An understanding of linear first-order difference equations with constant coefficients is required.

3. "Random Walks: An Introduction to Stochastic Processes," by Ron Barnes, UMAP 520. Random walks are introduced by an example of a gambling game. The associated finite difference equation is developed and solved. The concept of expected gain and the duration of a game are introduced and their usefulness is demonstrated. Generalizations to Markov chains and continuous processes are discussed. Applications in the life sciences and genetics are noted.

10.2 FURTHER READING

Tuchinsky, Philip M. *Man in Competition with the Spruce Budworm*, UMAP Expository Monograph. The population of tiny caterpillars periodically explodes in the evergreen forests of Eastern Canada and Maine. They devour the trees' needles and cause great damage to forests that are central to the economy of the region. The province of New Brunswick is using mathematical models of the budworm/forest interaction in an effort to plan for and control the damage. The monograph surveys the ecological situation and examines the computer simulation and differential equation models that are currently in use.

MODEL

10.3 A PREDATOR–PREY MODEL*

In this section we study a model of population growth of two species in which one species is the primary food source for the other. One example of such a situation occurs in the Southern Ocean, where the baleen whales eat the Antarctic krill, *Euphausia superboa*, as their principal food source. Another example is wolves and rabbits in a closed forest; the wolves eat the rabbits for their principal food source and the rabbits eat vegetation in the forest. Still other examples include sea otters as predators and abalone as prey; and the ladybird beetle, *Novius cardinalis*, as predator and the cottony cushion insect, *Icerya purchasi*, as prey.

Problem identification   Let's take a closer look at the situation of the baleen whales and Antarctic krill. The whales eat the krill and the krill live on the plankton in the sea. If the whales eat too many krill, so that the krill cease to be abundant, the food supply of the whales is greatly reduced. Then the

* Optional section

whales will starve or leave the area in search of a new supply of krill. As the population of baleen whales dwindles, the krill population makes a comeback since not so many of them are being eaten. As the krill population increases, the food supply for the whales grows and, consequently, so does the baleen whale population. And more baleen whales are eating more and more krill again. *In the pristine environment, does this cycle continue indefinitely or does one of the species eventually die out?* The baleen whales in the Southern Ocean have been overexploited to the extent that their current population is around one-sixth its estimated pristine level. Thus there appears to be a surplus of Antarctic krill. (Already something like 100,000 tons of krill are being harvested annually.) What effect does exploitation of the whales have on the balance between the whale and krill populations? What are the implications that a krill fishery may hold for the depleted stocks of baleen whales and for other species like seals, seabirds, penguins, and fish that depend on krill for their main source of food? The ability to answer such questions is important to management of multispecies fisheries. Let's see what answers can be obtained from a graphical modeling approach.

Assumptions Let $x(t)$ denote the Antarctic krill population at any time t, and let $y(t)$ denote the population of baleen whales in the Southern Ocean. The level of the krill population depends on a number of factors including the ability of the ocean to support them, the existence of competitors for the plankton they ingest, the presence and levels of predators, and so forth. As a rough, first model, let's start by assuming that the ocean can support an unlimited number of krill so that

$$\frac{dx}{dt} = ax \qquad \text{for } a > 0$$

(Later we may want to refine the model with a limited growth assumption. This refinement is presented in UMAP 610 listed in the 10.1 Projects section.) Second, assume that the krill are eaten primarily by the baleen whales (so neglect any other predators). Then the growth rate of the krill is diminished in a way that is proportional to the number of interactions between them and the baleen whales. One interaction assumption leads to the differential equation:

$$\frac{dx}{dt} = ax - bxy = (a - by)x \qquad \textbf{(10.8)}$$

Notice that the intrinsic growth rate $k = a - by$ decreases as the level of the baleen whale population increases. The constants a and b indicate the degrees of "self-regulation" of the krill population and the "predatoriness" of the baleen whales, respectively. These coefficients must be determined experimentally or from historical data. So far our first equation (10.8) governing the growth of the krill population looks just like either of the equations in the competitive hunter model presented in the preceding section.

Next, consider the baleen whale population $y(t)$. In the absence of krill the whales have no food, so we will assume that their population declines at a rate proportional to their numbers. This assumption produces the exponential decay equation:

$$\frac{dy}{dt} = -my \qquad \text{for } m > 0$$

However, in the presence of krill the baleen whale population increases at a rate proportional to the interactions between the whales and their krill food supply. Thus, the preceding equation is modified to give

$$\frac{dy}{dt} = -my + nxy = (-m + nx)y \qquad \textbf{(10.9)}$$

Notice from (10.9) that the intrinsic growth rate $r = -m + nx$ of the whales increases as the level of the krill population increases. The positive coefficients m and n would be determined experimentally or from historical data. Putting the results (10.8) and (10.9) together gives the following autonomous system of differential equations for our predator–prey model:

$$\frac{dx}{dt} = (a - by)x$$

$$\textbf{(10.10)}$$

$$\frac{dy}{dt} = (-m + nx)y$$

where $x(0) = x_0$, $y(0) = y_0$, a, b, m, and n are all positive constants. The system (10.10) governs the interaction of the baleen whales and Antarctic krill populations under our unlimited growth assumptions and in the absence of other competitors and predators.

Graphical analysis of the model Let's determine whether the krill and whale populations reach equilibrium levels. The rest points or equilibrium levels occur when $dx/dt = dy/dt = 0$. Setting the right sides of (10.10) equal to zero and solving for x and y simultaneously gives the rest points $(x, y) = (0, 0)$ and $(x, y) = (m/n, a/b)$. Along the vertical line $x = m/n$ and the x-axis in the phase plane, the growth dy/dt in the baleen whale population is zero; along the horizontal line $y = a/b$ and the y-axis, the growth dx/dt in the krill population is zero. These features are depicted in Figure 10-15.

Because the values for the constants a, b, m, and n in the system (10.10) will only be estimates, we need to investigate the behavior of the solution trajectories near the two rest points $(0, 0)$ and $(m/n, a/b)$. Thus, we analyze the directions of dx/dt and dy/dt in the phase plane. In the system (10.10), the vertical line $x = m/n$ divides the phase plane into two half-planes. In the left half-plane, dy/dt is negative, and in the right half-plane it is positive. In a similar way, the horizontal line $y = a/b$ determines two half-planes. In the upper half-plane, dx/dt is negative, and in the lower half-plane it is positive.

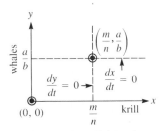

FIGURE 10-15 Rest points of the predator–prey model given by the system (10.10).

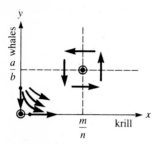

FIGURE 10-16 Trajectory directions in the predator–prey model.

The directions of the associated trajectories are indicated in Figure 10-16. Along the y-axis, motion must be vertical and toward the rest point $(0, 0)$, and along the x-axis, motion must be horizontal and away from the rest point $(0, 0)$.

From Figure 10-16 you can see that the rest point $(0, 0)$ is unstable. Entrance to the rest point is along the line $x = 0$, where there is no krill population. Thus the whale population declines to zero in the absence of its primary food supply. All other trajectories recede from the rest point. The rest point $(m/n, a/b)$ is more complicated to analyze. The information given by the figure is insufficient to distinguish between the three possible motions shown in Figure 10-13 of the preceding section. We cannot tell whether the motion is periodic, asymptotically stable, or unstable. Thus we must perform a further analysis.

An analytic solution of the model Since the number of baleen whales will depend upon the number of Antarctic krill available for food, we assume that y is a function of x. Then from the chain rule for derivatives, we have

$$\frac{dy}{dx} = \frac{dy/dt}{dx/dt}$$

or,

$$\frac{dy}{dx} = \frac{(-m + nx)y}{(a - by)x} \tag{10.11}$$

Equation (10.11) is a separable first-order differential equation and may be rewritten as

$$\left(\frac{a}{y} - b\right) dy = \left(n - \frac{m}{x}\right) dx \tag{10.12}$$

Integration of each side of (10.12) yields

$$a \ln y - by = nx - m \ln x + k_1$$

or

$$a \ln y + m \ln x - by - nx = k_1$$

where k_1 is a constant.

Using properties of the natural logarithm and exponential functions, this last equation can be rewritten as

$$\frac{y^a x^m}{e^{by + nx}} = K \tag{10.13}$$

where K is a constant. Equation (10.13) defines the solution trajectories in the phase plane. We now show that these trajectories are closed and represent periodic motion.

The predator–prey trajectories are periodic Equation (10.13) can be rewritten as

$$\left(\frac{y^a}{e^{by}}\right) = K\left(\frac{e^{nx}}{x^m}\right) \tag{10.14}$$

Let's determine the behavior of the function $f(y) = y^a/e^{by}$. Using the first derivative test (see Problem 1), we can easily show that $f(y)$ has a relative maximum at $y = a/b$ and no other critical points. For simplicity of notation, call this maximum value M_y. Moreover, $f(0) = 0$ and from l'Hôpital's rule $f(y)$ approaches 0 as y tends to infinity. Similar arguments apply to the function $g(x) = x^m/e^{nx}$, which achieves its maximum value M_x at $x = m/n$. The graphs of the functions f and g are depicted in Figure 10-17.

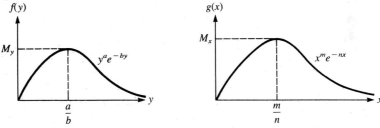

FIGURE 10-17 The graphs of the functions $f(y) = y^a/e^{by}$ and $g(x) = x^m/e^{nx}$.

From Figure 10-17, the largest value for $y^a e^{-by} x^m e^{-nx}$ is $M_y M_x$. That is, Equation (10.14) has no solutions if $K > M_y M_x$ and exactly one solution, $x = m/n$ and $y = a/b$, when $K = M_y M_x$. Let's consider what happens when $K < M_y M_x$.

Suppose $K = sM_y$, where $s < M_x$ is a positive constant. Then the equation:

$$x^m e^{-nx} = s$$

has exactly two solutions $x_m < m/n$, and $x_M > m/n$ (see Figure 10-18). Now if $x < x_m$, then $x^m e^{-nx} < s$ so that $se^{nx}x^{-m} > 1$ and

$$f(y) = y^a e^{-by} = Ke^{nx}x^{-m} = sM_y e^{nx}x^{-m} > M_y$$

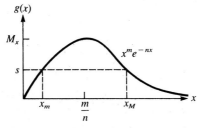

FIGURE 10-18 The equation $x^m e^{-nx} = s$ has exactly two solutions for $s < M_x$.

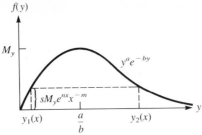

FIGURE 10-19 When $x_m < x < x_M$, there are exactly two solutions for y in Equation (10.14).

Therefore, there is no solution for y in Equation (10.14) when $x < x_m$. Likewise there is no solution when $x > x_M$. If $x = x_m$ or $x = x_M$, then Equation (10.14) has exactly the one solution $y = a/b$.

Finally, if x lies between x_m and x_M, Equation (10.14) has exactly two solutions. The smaller solution $y_1(x)$ is less than a/b, and the larger solution $y_2(x)$ is greater than a/b. This situation is depicted in Figure 10-19. Moreover, as x approaches either x_m or x_M, $f(y)$ approaches M_y so that both $y_1(x)$ and $y_2(x)$ approach a/b. It follows that the trajectories defined by Equation (10.14) are periodic and have the form depicted in Figure 10-20.

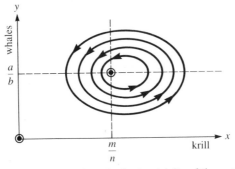

FIGURE 10-20 Trajectories in the vicinity of the rest point $(m/n, a/b)$ are periodic.

Model interpretation   What conclusions can be drawn from the trajectories in Figure 10-20? First, because the trajectories are *closed* curves, they predict that, under the assumptions of our model (10.10), neither the baleen whales nor the Antarctic krill will become extinct. (Remember, the model is based on the pristine situation.) The second observation is that along a single trajectory the two populations fluctuate between their maximum and minimum values. That is, starting with populations in the region where $x > m/n$ and $y > a/b$, the krill population will decline and the whale population increase until the krill population reaches the level $x = m/n$, at which point the whale population also begins to decline. Both populations continue to decline until the whale population reaches the level $y = a/b$ and the krill population begins to increase. And so on, around the trajectory. Recall from our discus-

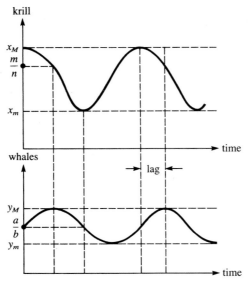

FIGURE 10-21 The whale population lags behind the krill population as both populations fluctuate cyclically between their maximum and minimum values.

sion in Section 10.1 that the trajectories never cross. A sketch of the population curves is shown in Figure 10-21. In that figure you can see the krill population fluctuates between its maximum and minimum values over one complete cycle. Notice that when the krill are plentiful, the whale population has its maximum rate of increase, but that the whale population reaches its maximum value after the krill population is on the decline. The predator *lags* behind the prey in a cyclic fashion.

The effects of harvesting For given initial population levels $x(0) = x_0$ and $y(0) = y_0$, the whale and krill populations will fluctuate with time around one of the closed trajectories depicted in Figure 10-20. Let T denote the time it takes to complete one full cycle and return to the starting point. Now the *average* levels of the krill and baleen whale populations over the time cycle are defined by the integrals

$$\bar{x} = \frac{1}{T} \int_0^T x(t)\,dt \quad \text{and} \quad \bar{y} = \frac{1}{T} \int_0^T y(t)\,dt$$

respectively. Now, from Equation (10.8),

$$\left(\frac{1}{x}\right)\left(\frac{dx}{dt}\right) = a - by$$

so that integration of both sides from $t = 0$ to $t = T$ leads to

$$\int_0^T \left(\frac{1}{x}\right)\left(\frac{dx}{dt}\right) dt = \int_0^T (a - by)\,dt$$

or,

$$\ln x(T) - \ln x(0) = aT - b \int_0^T y(t)\,dt$$

Because of the periodicity of the trajectory, $x(T) = x(0)$, and this last equation gives the average value:

$$\bar{y} = \frac{a}{b}$$

In an analogous manner, it can be shown that (see Problem 2)

$$\bar{x} = \frac{m}{n}$$

Therefore, the *average* levels of the predator and prey populations are in fact their *equilibrium* levels. Let's see what this means in terms of harvesting krill.

Let's assume that the effect of fishing for krill is to decrease its population level at a rate $rx(t)$. The constant r indicates the "intensity" of fishing and includes such factors as the number of fishing vessels at sea, the number of fishermen casting nets for krill, and so forth. Since less food is now available for the baleen whales, assume the whale population also decreases at a rate $ry(t)$. Incorporating these fishing assumptions into our model, we obtain the refined model:

$$\frac{dx}{dt} = (a - by)x - rx = [(a - r) - by]x$$

(10.15)

$$\frac{dy}{dt} = (-m + nx)y - ry = [-(m + r) + nx]y$$

The autonomous system (10.15) is of the same form as (10.10) (provided that $a - r > 0$) with a replaced by $a - r$ and m replaced by $m + r$. Thus, the new average population levels will be

$$\bar{x} = \frac{m + r}{n} \quad \text{and} \quad \bar{y} = \frac{a - r}{b}$$

Consequently, a moderate amount of harvesting krill (so that $r < a$) actually *increases* the average level of krill and *decreases* the average baleen whale population (under our assumptions for the model). The increase in krill population is beneficial to other species in the Southern Ocean (seals, seabirds, penguins, and fish) that depend on the krill for their main food source. The fact that some fishing increases the number of krill is known as *Volterra's principle*. The autonomous system (10.10) was first proposed by Lotka (1925) and Volterra (1931) as a simple model of predator–prey interaction.

The Lotka-Volterra model can be modified to reflect the situation in which both the predator and the prey are diminished by some kind of depleting force, such as when applying insecticide treatments that destroy both the insect predator and its insect prey. An example is given in Problem 3.

A number of biologists and ecologists argue that the Lotka-Volterra model (10.10) is unrealistic because the system is not asymptotically stable, whereas most observable natural predator–prey systems tend to equilibrium levels as time evolves. Nevertheless, regular population cycles, as suggested by the trajectories in Figure 10-20, do occur in nature. Some scientists have proposed models other than the Lotka-Volterra model that do exhibit oscillations that are asymptotically stable (so that the trajectories approach equilibrium solutions). One such model is given by

$$\frac{dx}{dt} = ax + bxy - rx^2$$

$$\frac{dy}{dt} = -my + nxy - sy^2$$

In this last autonomous system, the term rx^2 indicates the degree of internal competition of the prey for their limited resource (such as food and space), and the term sy^2 indicates the degree of competition among the predators for the finite amount of available prey. An analysis of this model is more difficult than that presented for the Lotka-Volterra model, but it can be shown that the trajectories of the model are not periodic and tend to equilibrium levels. The constants r and s are positive and would be determined by experimentation or historical data.

10.3 PROBLEMS

1. Apply the first and second derivative tests to the function $f(y) = y^a/e^{by}$ to show that $y = a/b$ is a unique critical point that yields the relative maximum $f(a/b)$. Show also that $f(y)$ approaches 0 as y tends to infinity.
2. Derive the result that the average value $\bar{x}$ of the prey population modeled by the Lotka-Volterra system (10.10) is given by the population level m/n.
3. In 1868 the accidental introduction into the United States of the cottony cushion insect (*Icerya purchasi*) from Australia threatened to destroy the American citrus industry. To counteract this situation, a natural Australian predator, a ladybird beetle (*Novius cardinalis*) was imported. The beetles kept the scale insects down to a relatively low level. When DDT was discovered to kill scale insects, farmers applied it in the hopes of reducing even further the scale insect population. However, DDT turned out to be fatal to the beetle as well and the overall effect of using the insecticide was to increase the numbers of the scale insect.

    Modify the Lotka-Volterra model to reflect a predator–prey system of two insect species where farmers apply (on a continuing basis) an insecticide that destroys both the insect predator and the insect prey at a common rate proportional to the numbers present. What conclusions do you reach concerning the effects of the application of the insecticide? Use graphical analysis to determine the effect of using the insecticide once on an irregular basis.
4. In a 1969 study, E. R. Leigh concluded that the fluctuations in the numbers of Canadian lynx and its primary food source, the hare, trapped

by the Hudson's Bay Company between 1847 and 1903 were periodic. The actual population levels of both species differed greatly from the predicted population levels obtained from the Lotka-Volterra predator–prey model.

Use the entire model-building process to modify the Lotka-Volterra model to arrive at a more realistic model for the growth rates of both species. Answer the following questions in the appropriate sections of the model-building process:

a. How have you modified the basic assumptions of the predator–prey model?

b. Why are your modifications an improvement to the basic model?

c. What are the equilibrium points for your model?

d. If it is possible, classify each equilibrium point as either stable or unstable.

e. Based on your equilibrium analysis, what values will the population levels of lynx and hare approach as t tends to infinity?

f. Explain how you would use your revised model to suggest hunting policies for Canadian lynx and hare. *Hint*: You are introducing a second predator—man—into the system.

5. Consider two species whose survival depends upon their mutual coopera- tion. An example would be a species of bee that feeds primarily on the nectar of one plant species and simultaneously pollinates that plant. One simple model of this mutualism is given by the autonomous system:

$$\frac{dx}{dt} = -ax + bxy$$

$$\frac{dy}{dt} = -my + nxy$$

a. What assumptions are implicitly being made about the growth of each species in the absence of cooperation?

b. Interpret the constants a, b, m, and n in terms of the physical problem.

c. What are the equilibrium levels?

d. Perform a graphical analysis and indicate the trajectory directions in the phase plane.

e. Find an analytic solution and sketch typical trajectories in the phase plane.

f. Interpret the outcomes predicted by your graphical analysis. Do you believe the model is realistic?

10.3 PROJECTS

1. Complete the requirements of the UMAP module, "Graphical Analysis of Some Difference Equations in Biology," by Martin Eisen, UMAP 553. The growth of many biological populations can be modeled by difference

equations. This module shows how the behavior of the solutions to certain equations can be predicted by graphical techniques.

2. Prepare a summary of the paper by May et al. listed in the references for this section. It extends the results of this section to include the situation of one prey and two predators.

10.3 FURTHER READING

Clark, Colin W. *Mathematical Bioeconomics: The Optimal Management of Renewable Resources.* New York: Wiley, 1976.

May, R. M. *Stability and Complexity in Model Ecosystems,* Monographs in Population Biology VI. Princeton, N.J.: Princeton University Press, 1973.

May, R. M., J. R. Beddington, C. W. Clark, S. J. Holt, R. M. Lewis, "Management of Multispecies Fisheries." *Science 205* (July 1979): 267–277.

May, R. M., ed. *Theoretical Ecology: Principles and Applications.* Philadelphia: Saunders, 1976.

10.4 TWO MILITARY EXAMPLES*

Example 1 The Nuclear Arms Race Revisited

In Chapter 1 we investigated the nuclear arms race in order to ascertain the effects that a particular strategy might have on the arms race, such as what is likely to occur if one of the two countries introduces mobile launching pads. Let's examine the economic aspects of the arms race.

Problem identification Consider again two countries engaged in a nuclear arms race. Let's attempt to assess qualitatively the effect of the arms race on the level of defense spending. *Specifically, we are interested in knowing if the arms race will lead to uncontrolled spending eventually dominated by the country with the greatest economic assets. Or will an equilibrium level of spending eventually be reached where each country spends a steady-state amount on defense?*

Assumptions Define the variable x as the annual defense expenditure for Country 1 and the variable y as the annual defense expenditure for Country 2. We assume that each nation is ready to defend itself and considers a defense budget to be necessary. Let's examine the situation from the point of view of Country 1. Its rate of spending depends on several factors. In the absence of any spending on the part of Country 2, or any grievance with that country, it is reasonable to assume that the defense spending would decrease at a rate proportional to the amount being spent. The proportionality constant is indicative of the requirement to maintain the current arsenal

---

* Optional section

(a percentage of the current spending) and acts as an economic restraint on defense spending (in the sense that government monies need to be spent in other areas like health and education). Thus,

$$\frac{dx}{dt} = -ax \qquad \text{for } a > 0$$

Qualitatively, this last equation says there will be no *growth* in defense spending. Now what happens when Country 1 perceives Country 2 engaging in defense spending? Country 1 will feel compelled to increase its defense budget in order to offset the defense buildup of its adversary and shore up its own security. Let's assume that the rate of increase for Country 1 is proportional to the amount Country 2 spends, where the proportionality coefficient is a measure of the perceived effectiveness of Country 2's weapons. This assumption seems reasonable, at least up to a point. As Country 2 adds weapons to its arsenal, Country 1 will perceive a need to add weapons to its own arsenal, where the numbers added are based on the assessment of the effectiveness of Country 2's weapons. Thus the previous equation is modified to become

$$\frac{dx}{dt} = -ax + by$$

We are assuming b is a constant, although this assumption is somewhat unrealistic. You would expect some kind of diminishing return of perceived effectiveness as Country 2 continues to add weapons. That is, if Country 2 has 100 weapons, Country 1 might perceive the need to add 40 weapons. But if Country 2 adds 200 weapons to its arsenal, Country 1 might perceive the need to add only 75 weapons. Thus, realistically, the constant b is more likely to be a decreasing function of y (and, in some instances, might be an increasing function). Later we may wish to refine our model to account for this probable diminishing return.

Finally, let's add in a constant term to reflect any underlying grievance felt by Country 1 toward Country 2. That is, even if the level of spending by both countries were zero, Country 1 would still feel compelled to be armed against Country 2, perhaps based on a fear of future aggressive action on the part of its adversary. If $y = 0$, then $c - ax$ represents a growth in defense spending for Country 1, and, until $c = ax$, growth will continue in order to achieve "deterrence." These assumptions then lead to the differential equation

$$\frac{dx}{dt} = -ax + by + c$$

where a, b, and c are nonnegative constants. The constant a indicates an economic restraint on defense spending, b indicates the intensity of rivalry with Country 2, and c indicates the deterrent or grievance factor. Although we are assuming c to be constant, it is more likely to be a function of both the variables x and y.

An entirely similar argument for Country 2 yields the differential equation

$$\frac{dy}{dt} = mx - ny + p$$

where m, n, and p are nonnegative constants, which are interpreted like b, a, and c, respectively. The preceding equations constitute our model for arms expenditures.

Graphical analysis of the model   We are interested in whether the defense expenditures reach equilibrium levels. If so, then we will know that the arms race will not lead to uncontrolled spending. This situation means that both defense budgets must stop growing so that $dx/dt = 0$ and $dy/dt = 0$. Thus we seek the rest points or equilibrium points of our model.

First, consider the case where neither country has a grievance against the other nor perceives any need for deterrence. Then $c = 0$ and $p = 0$, so our model becomes

$$\frac{dx}{dt} = -ax + by$$

$$\frac{dy}{dt} = mx - ny$$

(10.16)

For the autonomous system (10.16), $(x, y) = (0, 0)$ is a rest point. In this state there are no defense expenditures on either side and the two countries live in permanent *peace* with all conflicts resolved through nonmilitary means. (Such a peaceful state has existed between the United States and Canada since 1817). However, if grievances that are not resolved to the mutual satisfaction of both sides do arise, then the two countries will feel compelled to arm, leading to the equations:

$$\frac{dx}{dt} = c \quad \text{and} \quad \frac{dy}{dt} = p$$

Thus (x, y) will not remain at the rest point $(0, 0)$ if c and p are positive, so the rest point is unstable.

Now, consider our general model:

$$\frac{dx}{dt} = -ax + by + c$$

$$\frac{dy}{dt} = mx - ny + p$$

(10.17)

Setting the right sides equal to zero yields the linear system:

$$ax - by = c$$

$$mx - ny = -p$$

(10.18)

Each of the equations in (10.18) represents a straight line in the phase plane. If the determinant of the coefficients, $bm - an$, is not equal to zero, then

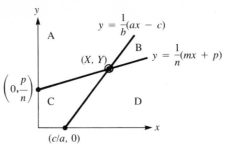

FIGURE 10-22 If $an - bm > 0$, the model (10.17) has a unique rest point (X, Y) in the first quadrant. Along the line $by = ax - c$, $dx/dt = 0$; along the line $ny = mx + p$, $dy/dt = 0$.

these two straight lines intersect at a unique rest point, denoted by (X, Y). It is easy to solve (10.18) to obtain this rest point:

$$X = \frac{bp + cn}{an - bm}$$

$$Y = \frac{ap + cm}{an - bm}$$

Assume that $an - bm > 0$, so that the rest point (X, Y) lies in the first quadrant of the phase plane. The situation is depicted in Figure 10-22. Shown in the figure are four regions labeled A, B, C, and D that are determined by the two intersecting lines. Let's examine the trajectory directions in each of these regions.

Any point (x, y) in region A lies above both of the lines represented by (10.18) so that $ny - mx - p > 0$ and $by - ax + c > 0$. It follows from (10.17) that for region A, $dy/dt < 0$ and $dx/dt > 0$. For a point (x, y) in region B, $ny - mx - p > 0$ and $by - ax + c < 0$, so that $dy/dt < 0$ and $dx/dt < 0$. Similarly, in region C, $dy/dt > 0$ and $dx/dt > 0$, and in region D, $dy/dt > 0$ and $dx/dt < 0$. These features are suggested in Figure 10-23.

An analysis of the trajectory motion in the vicinity of the rest point (X, Y) in Figure 10-23 reveals that every trajectory in the phase plane approaches the rest point. Thus, under our assumptions that $a/b > m/n$, so the intersection point (X, Y) lies in the first quadrant, the rest point (X, Y) is stable. Therefore, we conclude that defense spending for both countries will approach equilibrium (steady-state) levels of $x = X$ and $y = Y$. (In the problems for this section you are asked to investigate the case where $a/b < m/n$, and you will see that uncontrolled spending occurs.)

Let's examine the meaning of the inequality $a/b > m/n$ (which is equivalent to $an - bm > 0$). From the point of view of Country 1, the economic restraint a compared to the perceived intensity of rivalry b must be high against some specific ratio m/n for Country 2. The constant b has to do with perception and is, in part, a psychological factor. If b is lowered, the chances for the rest point to lie in the first quadrant are increased, with the beneficial result of steady-state levels of defense spending eventually likely for both

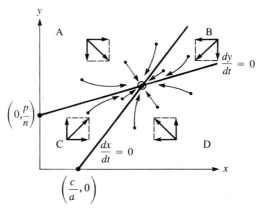

FIGURE 10-23 Composite graphical analysis of the trajectory directions in the four regions determined by the intersecting lines (10.18).

countries. The value of our model is not in its capacity to make predictions, but in its clarification of what can happen under different conditions regulating the parameters a, b, c, m, n, and p in the model. Certainly we can see that it is important for nations to reduce tension levels and perceived threats through mutual cooperation, respect, and disarmament policies if a runaway arms race is to be avoided. Moreover, a permanent peace can be achieved only if grievances are resolved to the mutual satisfaction of each country. However, note the effect of Taylor's "deterrence" strategy in Chapter 1. In that strategy $(0, 0)$ is no longer an equilibrium point so the possibility of total disarmament is ruled out. We have a stable equilibrium in this model because of the economic constraint $(c - ax)$. Even in the absence of defense spending on the part of Country 2 $(y = 0)$, Country 1 will increase its defense budget until the cost of maintaining its armaments becomes exorbitant $(c = ax)$.

MODEL

Example 2 Lanchester Combat Models

Consider the situation of combat between two homogeneous forces: a homogeneous X force (for example, tanks) opposed by another homogeneous Y force (for example, antitank weapons). We want to know if one force will eventually win out over the other, or will the combat end in a draw? Other questions of interest include the following: How do the force levels decrease over time in battle? How many survivors will the winner have? How long will the battle last? How do changes in the initial force levels and weapon-system parameters affect the battle's outcome? In this section we will consider one basic combat model and several of its refinements.

Assumptions   Let $x(t)$ and $y(t)$ denote the strengths of the forces X and Y at time t, respectively. Usually t is measured in hours or days from the beginning of the combat. Let's examine what is meant by the "strengths" of the two forces X and Y. The strength $x(t)$, for example, includes a number of factors. If X is a homogeneous tank force, its "strength" depends on the number of tanks in operation, the level of technology used in the tank design, the quality of workmanship in the manufacturing process, the level of training and the skills of the individuals operating the tank, and so forth. For our purposes, let's assume that the strength $x(t)$ is simply the number of tanks in operation at time t. Likewise, the strength $y(t)$ is the number of antitank weapons operational at time t.

In the actual state of affairs the numbers $x(t)$ and $y(t)$ are nonnegative integers. However, it is convenient to idealize the situation and assume that $x(t)$ and $y(t)$ are continuous functions of time. For instance, if there are 500 tanks at 1400 hours (two o'clock in the afternoon in military time) and 487 tanks at 1500 hours, then it is reasonable to assume that there are 497.4 tanks at 1412 hours (if we perform a linear interpolation between the data points). That is, 2.6 of the 13 tanks lost in the 1 hr of combat were lost in the first 12 min. We also assume that $x(t)$ and $y(t)$ are differentiable functions of t, so that they are "smooth" functions without any "corners" or "cusps" on their graphs. These idealizations enable us to model the strength functions by differential equations.

Now, what can we assume about how the force levels change as a result of combat? For the basic combat model, let's assume that the combat *casualty rate* for the X force is proportional to the strength of the Y force. Other factors affect the change in the X force over time such as reinforcements of the force by bringing in additional tanks (or troops if it is a troop force), or tank losses due to mechanical or electronic failures, or losses due to operator errors or desertions, and so forth. (Can you name some additional factors?) We will ignore all of these other factors in our initial model, so that the rate of change in $x(t)$ is given by

$$\frac{dx}{dt} = -ay, \qquad a > 0 \tag{10.19}$$

The positive constant a in Equation (10.19) is called the *antitank weapon kill rate or attrition-rate coefficient* and reflects the degree to which a single antitank weapon can destroy tanks. Thus we begin our analysis with the simplest assumption: namely, that the loss rate is proportional to the number of firers. Later we assume an interaction is necessary between firer and target (the firer must locate the target before firing), and we refine our model accordingly. In the refined model the attrition-rate coefficient is proportional to the number of targets.

Under similar assumptions, the rate of change in the Y force is given by

$$\frac{dy}{dt} = -bx, \qquad b > 0 \tag{10.20}$$

Here the constant b indicates the degree to which a single tank can destroy antitank weapons. The autonomous system given by (10.19) and (10.20), together with the initial strength levels $x(0) = x_0$ and $y(0) = y_0$, is called a **Lanchester-type combat model** named after F. W. Lanchester, who investigated air combat situations during World War I. The equations (10.19) and (10.20) constitute our basic model subject to the assumptions we have made. We assume throughout that $x \geqslant 0$ and $y \geqslant 0$, since negative force levels have no physical meaning.

Lanchester-type combat model

Analysis of the model Setting the right sides of (10.19) and (10.20) equal to zero, we see that $(0, 0)$ is a rest point for the basic combat model. The trajectory directions in the phase plane are determined from the observations that $dx/dt < 0$ and $dy/dt < 0$, when $x > 0$ and $y > 0$. Moreover, if $x = 0$, then $dy/dt = 0$ (and we assume also that $dx/dt = 0$ since $x < 0$ has no physical meaning). These considerations lead to the trajectory directions depicted in Figure 10-24. Note that our assumptions imply that a trajectory terminates when it reaches either coordinate axis. (Otherwise the rest point $(0, 0)$ would be unstable.)

It is easy to find an analytic solution to the basic model. From the chain rule,

$$\frac{dy}{dx} = \frac{dy/dt}{dx/dt}$$

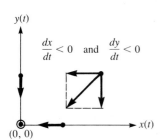

FIGURE 10-24 The rest point $(0, 0)$ for the basic Lanchester combat model.

and substitution of (10.19) and (10.20) gives

$$\frac{dy}{dx} = \frac{-bx}{-ay}$$

Separating the variables in this last equation yields

$$-ay \, dy = -bx \, dx \tag{10.21}$$

Integration of each side of (10.21) and employment of the initial force levels $x(0) = x_0$ and $y(0) = y_0$ produces the **Lanchester square law model:**

Lanchester square law model

$$a(y^2 - y_0{}^2) = b(x^2 - x_0{}^2) \tag{10.22}$$

Setting $C = ay_0{}^2 - bx_0{}^2$, we obtain the equation:

$$ay^2 - bx^2 = C \tag{10.23}$$

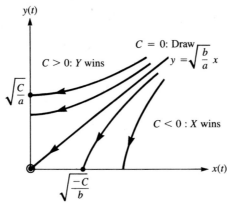

FIGURE 10-25 Trajectories of the basic Lanchester combat model. The trajectories are hyperbolas satisfying the Lanchester square law (10.22).

Typical trajectories in the phase plane represented by Equation (10.23) are depicted in Figure 10-25. The trajectories for $C \neq 0$ are hyperbolas, and when $C = 0$, the trajectory is the straight line $y = \sqrt{b/a}\, x$. When $C < 0$, the trajectory intersects the x-axis at $x = \sqrt{-C/b}$; then the X (tank) force wins because the Y force has been totally eliminated. On the other hand, if $C > 0$, the Y force wins with a final strength level of $y = \sqrt{C/a}$. These considerations are indicated in Figure 10-25.

Let's investigate the situation where the Y (antitank) force wins. Then the constant C must be positive, so that

$$\left(\frac{y_0}{x_0}\right)^2 > \frac{b}{a} \tag{10.24}$$

The inequality (10.24) gives a necessary and sufficient condition that the Y force wins under the assumptions of our model (so that reinforcements are not permitted, for example). From the inequality you can see that a doubling of the initial Y force level results in a fourfold advantage for that force, assuming the X force remains at the same initial level x_0 for constant a and b. This means that Country X must increase b (its technology) by a factor of 4 in order to keep pace with the increase in the size of Country Y's force if the strength x_0 is kept at the same level. Figure 10-26 depicts a typical graph showing the force level curves $x(t)$ and $y(t)$ when the inequality (10.24) is satisfied. Observe from the figure that it is not necessary for the initial Y force level y_0 to exceed the level x_0 of the X force in order to ensure victory for Y. The crucial relationship is given by the inequality (10.24).

The model given by Equations (10.19) and (10.20) can be converted to a single second-order differential equation in the dependent variable y as follows: Differentiate (10.20) to obtain

$$\frac{d^2 y}{dt^2} = -b \frac{dx}{dt}$$

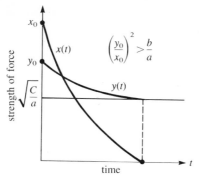

FIGURE 10-26 Force level curves $x(t)$ and $y(t)$ for the basic Lanchester combat model when $C > 0$ and the Y force wins.

Then substitute (10.19) into this last equation to get

$$\frac{d^2y}{dt^2} = aby$$

or

$$\frac{d^2y}{dt^2} - aby = 0 \qquad \textbf{(10.25)}$$

In the problem section you are asked to verify that the function

$$y(t) = y_0 \cosh\sqrt{ab}\,t - x_0\sqrt{a/b}\,\sinh\sqrt{ab}\,t \qquad \textbf{(10.26)}$$

satisfies the differential equation (10.25) subject to the condition that $y(0) = y_0$. Similarly, the solution for the X force level is

$$x(t) = x_0 \cosh\sqrt{ab}\,t - y_0\sqrt{a/b}\,\sinh\sqrt{ab}\,t \qquad \textbf{(10.27)}$$

subject to the initial level $x(0) = x_0$.

Equation (10.26) can be written in the more revealing form:

$$\frac{y(t)}{y_0} = \cosh\sqrt{ab}\,t - \left(\frac{x_0}{y_0}\right)\sqrt{\frac{a}{b}}\,\sinh\sqrt{ab}\,t \qquad \textbf{(10.28)}$$

This latter expression (10.28) says that Y's current force level divided by the initial one, that is, the *normalized force level*, depends on just two parameters: (1) a dimensionless engagement parameter $E = (x_0/y_0)\sqrt{a/b}$, and (2) a time parameter $T = \sqrt{ab}\,t$. The constant $\sqrt{ab}$ represents the intensity of battle and controls how quickly the battle is driven to its conclusion. The ratio a/b represents the relative effectiveness of individual combatants on the two opposing sides.

Refinements of the basic Lanchester combat model   In the basic Lanchester combat model

$$\frac{dx}{dt} = -ay$$

$$\frac{dy}{dt} = -bx$$

(10.29)

it has been assumed that the single-weapon attrition rates a and b are constant over time. However, in many circumstances the X force and the Y force are changing positions and the weapons' effectiveness depends on the distance between firer and target. Thus, $a = a(t)$ and $b = b(t)$ are time dependent. In that case the model (10.29) is no longer an autonomous system and it is much more difficult to extract analytically information from the model.

In some situations the single-weapon attrition rate a depends not only on time but also on the number of targets x. This situation occurs, for example, when target detection depends on the number of targets. In this case $a = a(t, x)$ is a function of time and also the number of targets. The model now becomes analytically intractable, but numerical methods can be used to generate force-level results.

One can go even further in enriching the basic model (10.29). For example, if $a = a(t, x/y)$, then the single-weapon attrition rate depends on time and also the force ratio x/y. In still another operational circumstance it is possible that $a = a(t, x, y)$, so that the attrition rate coefficient depends on time, the number of targets, and the number of firers.

When a weapon system employs "area" fire and enemy targets defend a constant area, the corresponding Lanchester attrition-rate coefficients depend on the number of targets. Let's assume, for instance, that the attrition rate is directly proportional to the number of targets. Then the basic model becomes

$$\frac{dx}{dt} = -gxy$$

$$\frac{dy}{dt} = -hyx$$

(10.30)

where g and h are positive constants, and $x(0) = x_0$ and $y(0) = y_0$ are the initial force levels. Assuming there are no operational losses and no reinforcements on either side, the model (10.30) reflects combat between two guerilla forces.

The system (10.30) is easily solvable. From the chain rule we obtain

$$\frac{dy}{dx} = \frac{h}{g}$$

and separation of variables leads to the equation:

$$g \, dy = h \, dx$$

Integration then yields the *linear combat law*:

$$g(y - y_0) = h(x - x_0) \tag{10.31}$$

Setting $K = g y_0 - h x_0$, we obtain the equation:

$$gy - hx = K \tag{10.32}$$

If $K > 0$, the Y force wins; if $K < 0$, the X force wins. The trajectories for the model (10.30) are depicted in Figure 10-27. Notice from (10.32) that the Y force wins provided that

$$\frac{y_0}{x_0} > \frac{h}{g} \tag{10.33}$$

In this case a doubling of the initial Y force simply doubles the advantage of that force, assuming the X force retains its same initial level x_0.

It can be shown (see Problem 3) that the X force level satisfying the model (10.30) is given by

$$x(t) = \begin{cases} x_0 \left[\dfrac{h x_0 - g y_0}{h x_0 - g y_0 e^{-(h x_0 - g y_0)t}} \right] & \text{for } h x_0 \neq g y_0 \\[2ex] \dfrac{x_0}{1 + h x_0 t} & \text{for } h x_0 = g y_0 \end{cases} \tag{10.34}$$

A similar result holds for the Y force level.

Let's consider a few more simple, yet natural, enrichments to the basic homogeneous-force combat model (10.29). For example, if the X force were to continuously commit more combatants to the battle at the rate $q(t) \geqslant 0$, then the rate of change of the X force level would become

$$\frac{dx}{dt} = -ay + q(t) \tag{10.35}$$

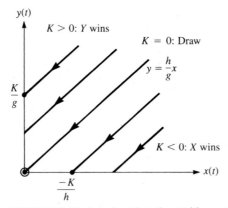

FIGURE 10-27 Trajectories when the attrition-rate coefficients are directly proportional to the number of targets.

Here $q(t) < 0$ would mean the continuous withdrawal of forces. Now, what might the function $q(t)$ look like? First, we would observe that replacements are drawn from a limited pool of manpower and weapon stocks. Then we could assume that these resources are committed to the battle at a constant rate m for as long as they last. If R denotes the total number of reserves that X can commit to battle, these considerations translate to the submodel:

$$q(t) = \begin{cases} m & \text{if } 0 \leqslant t \leqslant \dfrac{R}{m} \\[2ex] 0 & \text{if } t > \dfrac{R}{m} \end{cases} \tag{10.36}$$

Another consideration is that operational losses may occur. By *operational losses* we mean those due to noncombat mishaps such as disease, desertions, breakdown of machinery (like a tank), and so forth. Many of the factors involved in the operational loss rate, such as psychological factors inherent in desertions, would be difficult to identify precisely. However, we might assume that the operational loss rate is proportional to the strength of the force. In that case, assuming also there are reinforcements, the rate of change of the X force level would be

$$\frac{dx}{dt} = -cx - ay + q(t) \tag{10.37}$$

where a and c are positive constants, and $q(t) \geqslant 0$. Similar considerations apply to the Y force.

10.4 PROBLEMS

1. In our model (10.17) for the nuclear arms race, assume that $an - bm < 0$, so that the rest point lies in a quadrant other than the first one in the phase plane. Sketch the lines $dx/dt = 0$ and $dy/dt = 0$ in the phase plane and label them and their intercepts on the coordinate axes. Perform a graphical stability analysis to answer the following questions:
   a. Do any potential equilibrium levels for defense spending exist? List any such points and classify them as stable or unstable.
   b. Pick at least four starting points in the first quadrant and sketch their trajectories in the phase plane.
   c. What outcome for defense spending is predicted by your graphical analysis?
   d. From the point of view of Country 1, interpret qualitatively the outcome predicted by your graphical analysis in terms of the relative values of the parameters in the model (10.17).
2. Verify that the function (10.26) satisfies the differential equation (10.25).
3. a. Solve Equation (10.32) for y and substitute the results into the differential equation $dx/dt = -gxy$ from the model (10.30).

b. Separate the variables in the differential equation resulting from Part a and integrate using partial fractions to obtain the force level given by (10.34).

4. In the basic Lanchester model (10.29) assume the two forces are of equal effectiveness, so that $a = b$. The Y force initially has 50,000 soldiers. There are two geographically separate units of 40,000 and 30,000 soldiers that make up the X force. Use the basic model and the result (10.23) to show that if the commander of the Y force fights each of the X units separately, then he can bring about a draw.

5. Let X denote a guerilla force and Y denote a conventional force. The autonomous system:

$$\frac{dx}{dt} = -gxy$$

$$\frac{dy}{dt} = -bx$$

is a Lanchestrian model for conventional–guerilla combat, in which there are no operational loss rates and no reinforcements.

a. Discuss the assumptions and relationships necessary to justify the model. Does the model seem reasonable?

b. Solve the system and obtain the *parabolic law*:

$$gy^2 = 2bx + M$$

where $M = gy_0^2 - 2bx_0$.

c. What condition must be satisfied by the initial force levels x_0 and y_0 in order for the conventional Y force to win? If the Y force does win, how many survivors will there be?

6. a. Assuming that the single-weapon attrition rates a and b in (10.29) are constant over time, discuss the submodels

$$a = r_y p_y \quad \text{and} \quad b = r_x p_x$$

where r_y and r_x are the respective *firing rates* (shots/combatant/day) of the Y and the X force and p_y and p_x are the respective probabilities that a single shot kills an opponent.

b. How would you model the attrition rate coefficients g and h in the model (10.30)? It may be helpful to think of (10.30) as modeling guerilla–guerilla combat.

10.4 PROJECTS

Complete the requirements of the referenced UMAP module.

1. "The Richardson Arms Race Model," by Dina A. Zinnes, John V. Gillespie, and G. S. Tahim, UMAP 308. A model is constructed based upon the classical assumptions of Lewis Fry Richardson. Difference equations are introduced. Students gain experience in analyzing the equilibrium, stability, and sensitivity properties of an interactive model.

10.4 FURTHER READING

Callahan, L. G. "Do We Need a Science of War?" *Armed Forces Journal 106*, no. 36 (1969): 16–20.

Engel, J. H. "A Verification of Lanchester's Law." *Operations Research 2* (1954): 163–171.

Lanchester, F. W. "Mathematics in Warfare." In *The World of Mathematics*, vol. 4, edited by J. R. Newman, pp. 2138–2157. New York: Simon and Schuster, 1956.

McQuie, R. "Military History and Mathematical Analysis." *Military Review 50*, no. 5 (1970): 8–17.

Richardson, L. R. "Mathematics of War and Foreign Politics." In *The World of Mathematics*, vol. 4, edited by J. R. Newman, pp. 1240–1253. New York: Simon and Schuster, 1956.

U.S. General Accounting Office (GAO). "Models, Data, and War: A Critique of the Foundation Defense Analysis." PAD-80-21. Washington, D.C., March 1980.

INDEX